# SOIL
# PHOSPHORUS

# Advances in Soil Science

Series Editors: Rattan Lal and B. A. Stewart

## Published Titles

# Advances in Soil Science

# SOIL PHOSPHORUS

Edited by

## Rattan Lal
The Ohio State University
Columbus, USA

## B. A. Stewart
West Texas A&M University
Canyon, USA

CRC Press
Taylor & Francis Group
Boca Raton London New York

CRC Press is an imprint of the
Taylor & Francis Group, an **informa** business

CRC Press
Taylor & Francis Group
6000 Broken Sound Parkway NW, Suite 300
Boca Raton, FL 33487-2742

First issued in paperback 2021

© 2017 by Taylor & Francis Group, LLC
CRC Press is an imprint of Taylor & Francis Group, an Informa business

No claim to original U.S. Government works

Version Date: 20160216

ISBN 13: 978-1-03-209771-8 (pbk)
ISBN 13: 978-1-4822-5784-7 (hbk)

---

**Library of Congress Cataloging-in-Publication Data**

---

Names: Lal, R., editor, author. | Stewart, B. A. (Bobby Alton), 1932- editor, author.
Title: Soil phosphorus / editors: Rattan Lal and B.A. Stewart.
Other titles: Advances in soil science (Boca Raton, Fla.)
Description: Boca Raton, FL : CRC Press, Taylor & Francis Group, [2016] |
Series: Advances in soil science | Includes bibliographical references and index.
Identifiers: LCCN 2016006729 | ISBN 9781482257847
Subjects: LCSH: Phosphorus in agriculture. | Soils--Phosphorus content.
Classification: LCC S587.5.P56 S655 2016 | DDC 631.8/3--dc23
LC record available at http://lccn.loc.gov/2016006729
gbg

---

Visit the Taylor & Francis Web site at
http://www.taylorandfrancis.com

and the CRC Press Web site at
http://www.crcpress.com

# Contents

# Preface

Phosphorus (P) is an essential plant nutrient. The suboptimal availability and the nutrient imbalance in the root zone can adversely impact plant growth. While some have raised concerns about the limited supply of the proven reserves and the slogan "peak P," others (United States Geological Survey, Food and Agriculture Organization, International Plant Nutrition Institute, etc.) show that there are adequate phosphate rock sources to last for a few centuries. Furthermore, not only is the availability of P, even for a few centuries, a cause of serious concern, but so is the possibility that the cost of extraction may increase, thereby raising the price of fertilizer. An increase in cost can be a factor because the demand for P fertilizers is projected to increase with the increase in the world population from 7.2 billion now to 9.7 billion by 2050. Over and above the issue of supply and cost, there are also concerns regarding nonpoint source pollution and eutrophication of natural waters. It is important to realize that the major issues with P and water quality are attributed to the legacies of past management. For example, the rate of manure application on the basis of availability of N can lead to the misuse of P with the attendant impact on water quality. Furthermore, there exists a serious issue regarding the lack of P recycling strategies to minimize costs and address the need to enhance the P status on widely spread P-deficient soils. In addition to identifying strategies and developing techniques of recycling, it is important to assess the effect of P fertility on agronomic productivity. Cost-effective strategies are needed to enhance efficiency. Thus, there is a strong need to understand processes and practices to mitigate the role of agricultural P on eutrophication. The role of soil P needs to be critically deliberated and understood. Similarly, identifying appropriate systems of managing soil P and reducing the risks of eutrophication are needed to minimize the environmental risks, especially the transport of the scarce nutrient into oceans and other aquatic ecosystems. The formulation of P fertilizers and mode of their application are important to P dynamics in soil (e.g., solublization, fixation). Along with nitrogen (N), P is also needed in soil for the conversion of biomass-C into soil organic matter. P is essential to human health because of its role in several metabolic processes, and the lack of P is a contributor to malnutrition in the human population. In the human dimensions of the sustainable management of P, voluntary and regulatory mechanisms need to be objectively considered. This volume is specifically devoted to the availability and recycling of P, the regulatory/policy issues of sustainable use of P, and the management in agroecosystems in the context of maximizing the efficient use and minimizing the environmental risks of water quality.

This 13-chapter book discusses the global P cycle, the environmental consequences of anthropogenic disturbances in the P cycle, the impacts of agricultural ecosystems and P management on the eutrophication of surface waters, the importance of P in human health, the role of $^{32}P$ as a research tool in studying the fate of P in agroecosystems, the economics and policy issues of P management, and the research and development priorities. The book also provides a scientific basis of

P management to optimize agronomic production while minimizing eutrophication and other environmental hazards.

The editors thank all the authors for their outstanding contributions and for sharing their knowledge and experiences. Despite busy schedules and numerous commitments, the preparation of manuscripts in a timely manner by all authors is greatly appreciated. The editors also thank the editorial staff of Taylor and Francis for their help and support in publishing this book. The office staff of the Carbon Management and Sequestration Center provided support in the flow of the manuscripts between authors and editors and made valuable contributions; their help and support is greatly appreciated. In this context, special thanks are due to Ms. Laura Hughes who formatted the text and prepared the final submission. It is a challenge to thank by listing names of all those who contributed in one way or another to bringing this book to fruition. Thus, it is important to build upon the outstanding contributions of numerous soil scientists, agricultural engineers, and technologists whose research is cited throughout the book.

**Rattan Lal**
**B. A. Stewart**

# Editors

**Rattan Lal, PhD**, is a distinguished university professor of soil science, the director of the Carbon Management and Sequestration Center, The Ohio State University, and an adjunct professor of the University of Iceland. His current research focus is on climate-resilient agriculture, soil carbon sequestration, sustainable intensification, enhancing efficient use of agroecosystems, and sustainable management of soil resources of the tropics. He received the honorary degree of Doctor of Science from the Punjab Agricultural University, India (2001), the Norwegian University of Life Sciences, Norway (2005), the Alecu Russo Balti State University, Moldova (2010), and the Technical University of Dresden, Germany (2015). He was president of the World Association of the Soil and Water Conservation (1987–1990), the International Soil Tillage Research Organization (1988–1991), and the Soil Science Society of America (2005–2007), and at the time of this writing he is the president-elect of the International Union of Soil Science. Dr. Lal was a member of the Federal Advisory Committee on U.S. National Assessment of Climate Change (2010–2013), a member of the Strategic Environmental Research and Development Program Scientific Advisory Board of the U.S. Department of Defense (2011–), a senior science advisor to the Global Soil Forum of the Institute for Advanced Sustainability Studies, Potsdam, Germany (2010–), a member of the advisory board of the Joint Program Initiative of Agriculture, Food Security and Climate Change of the European Union (2013–2016), and the chair of the advisory board of the Institute for Integrated Management of Material Fluxes and Resources of the United Nations University, Dresden, Germany (2014–2017). Professor Lal was a lead author of the Intergovernmental Panel on Climate Change (1998–2000). He has mentored 104 graduate students and 54 postdoctoral researchers and hosted 145 visiting scholars. He has authored or coauthored 783 refereed journal articles, written 12 books, and edited or coedited an additional 61 books. In 2014 and 2016, Reuter Thomson listed Dr. Lal among the world's most influential scientific minds and as being among the top 1% of scientists in agricultural sciences cited in professional journals.

**B. A. Stewart** is the director of the Dryland Agriculture Institute and a distinguished professor of agriculture at West Texas A&M University, Canyon, Texas. He is a former director of the U.S. Department of Agriculture (USDA) Conservation and Production Laboratory at Bushland, Texas, a past president of the Soil Science Society of America, and a member of the 1990–1993 Committee on Long-Range Soil and Water Policy, National Research Council, National Academy of Sciences. He is a fellow of the Soil Science Society of America, American Society of Agronomy, and Soil and Water Conservation Society, a recipient of the USDA Superior Service Award, a recipient of the Hugh Hammond Bennett Award of the Soil and Water Conservation Society, an honorary member of the International Union of Soil

Sciences in 2008. In 2009, Dr. Stewart was inducted into the USDA Agriculture Research Service Science Hall of Fame. Dr. Stewart is very supportive of education and research on dryland agriculture. The B. A. and Jane Ann Stewart Dryland Agriculture Scholarship Fund was established in West Texas A&M University in 1994 to provide scholarships for undergraduate and graduate students with a demonstrated interest in dryland agriculture.

# Contributors

**Joseph Adu-Gyamfi**
International Atomic Energy Agency
Vienna, Austria

**Mahendra Bhandari**
Texas A&M University
Amarillo, Texas

**Otto C. Doering III**
Purdue University
West Lafayette, Indiana

**Gabriel M. Filippelli**
Indiana University—Purdue University
    Indianapolis (IUPUI)
Indianapolis, Indiana

**Daniel Sebastian Goll**
Laboratorie des Sciences du Climat et
    de l'Environment
Gif-sur-Yvette, France

**Sergei Katsev**
University of Minnesota Duluth
Duluth, Minnesota

**Rattan Lal**
Ohio State University
Columbus, Ohio

**Ruiqiang Liu**
Ohio State University
Columbus, Ohio

**Kaushik Majumdar**
International Plant Nutrition Institute
Gurgaon, India

**Andrew J. Margenot**
University of California, Davis
Davis, California

**Long Nguyen**
International Atomic Energy Agency
Vienna, Austria

**Pramod Pokhrel**
Texas A&M University
Canyon, Texas

**Rajendra Prasad**
Indian Agricultural Research Institute
New Delhi, India

**Samendra Prasad**
University of Virginia
Charlottesville, Virginia

**Idupulapati M. Rao**
International Center for Tropical
    Agriculture
Cali, Colombia

**Sasha C. Reed**
U.S. Geological Survey
Moab, Utah

**Yashbir Singh Shivay**
Indian Agricultural Research Institute
New Delhi, India

**Bal R. Singh**
Norwegian University of Life Sciences
Ås, Norway

**Rolf Sommer**
International Center for Tropical
    Agriculture
Nairobi, Kenya

**B. A. Stewart**
Texas A&M University
Canyon, Texas

**Tana E. Wood**
USDA Forest Service International
    Institute of Tropical Forestry
San Juan, Puerto Rico

**Felipe Zapata**
International Atomic Energy Agency
Vienna, Austria

# 1 The Global Phosphorus Cycle

*Gabriel M. Filippelli*

## CONTENTS

## 1.1 INTRODUCTION

Phosphorus (P) is a limiting nutrient for terrestrial productivity, and thus, it commonly plays a key role in the net carbon uptake in terrestrial ecosystems (Tiessen et al. 1984; Roberts et al. 1985; Lajtha and Schlesinger 1988). Unlike nitrogen (another limiting nutrient but one with an abundant atmospheric pool), the availability of new P in ecosystems is restricted by the rate of the release of this element during soil weathering. Because of the limitations of P availability, P is generally recycled to various extents in ecosystems depending on climate, soil type, and ecosystem level (e.g., Filippelli et al. 2006; Porder et al. 2007; Filippelli 2008). The release of P from apatite dissolution is a key control on ecosystem productivity (Cole et al. 1977; Tiessen et al. 1984; Roberts et al. 1985; Crews et al. 1995; Vitousek et al. 1997; Schlesinger et al. 1998), which in turn is critical to terrestrial carbon balances (e.g., Kump and Alley 1994; Adams 1995). Furthermore, the weathering of P from the terrestrial system and transport by rivers is the only appreciable source of P to the oceans. On longer timescales, this supply of P also limits the total amount of primary production in the ocean (Holland 1978; Broecker 1982; Smith 1984; Filippelli and Delaney 1994). Thus, understanding the controls on P weathering from land and

1

the transport to the ocean is important for models of global change. In this chapter, I will present the current state of knowledge of the natural (prehuman) and modern (synhuman) global P mass balances, including an in-depth examination of climatic and geologic controls on ecosystem dynamics and soil development.

## 1.2 GLOBAL PHOSPHORUS CYCLING

### 1.2.1 Natural (Prehuman) Phosphorus Cycle

The human impact on the global P cycle has been substantial over the last 150 years and will continue to dominate the natural cycle of P on the globe for the foreseeable future. Because this anthropogenic modification began well before the scientific efforts to quantify the cycle of P, we can only guess at the preanthropogenic mass balance of P. Several aspects of the preanthropogenic sources of sinks of P are relatively well constrained (Figure 1.1). The initial source of P to the global system is via the weathering of P during soil development, whereby P is released mainly from apatite minerals and is made soluble and bioavailable (this process will be discussed at considerable length later). In contrast to this process of chemical weathering, the physical weathering and the erosion of material from the continents results in P that is typically unavailable to biota. An exception to this, however, is the role that physical weathering plays in producing fine materials with extremely high surface area/mass ratios. Phosphorus and other components may be rapidly weathered if this fine material is deposited in continental environments (i.e., floodplains and delta systems) where it undergoes subsequent chemical weathering and/or soil development. Thus, the total amount of P weathered from continents may be very different from the amount of potentially bioavailable P.

Apatite minerals, the dominant source-weathering source of P, widely vary in chemistry and structure and can form in igneous, metamorphic, sedimentary, and biogenic conditions. All apatite minerals contain phosphate oxyanions linked by $Ca^{2+}$ cations to form a hexagonal framework, but they differ in elemental composition at

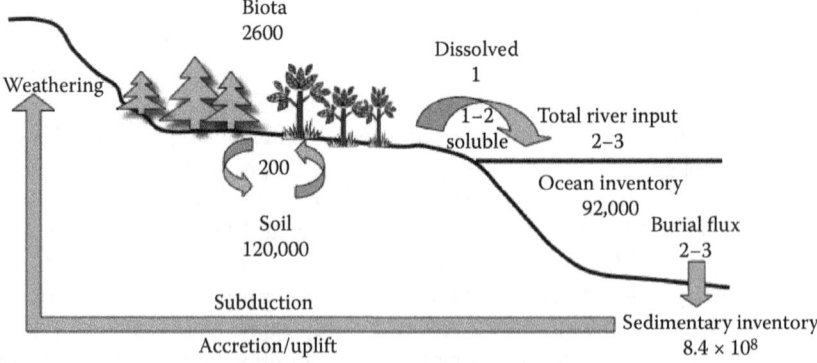

**FIGURE 1.1**   The natural (prehuman) phosphorus cycle, showing reservoirs (in teragram P) and fluxes (denoted by arrows, in teragram P per year) in the P mass balance.

**TABLE 1.1**

**Terminology and Crystal Chemistry of Apatites**

| Mineral Name | Chemical Composition |
|---|---|
| FAP | $Ca_5(PO_4)_3F$ |
| HAP | $Ca_5(PO_4)_3OH$ |
| Chlorapatite | $Ca_5(PO_4)_3Cl$ |
| Dahllite | $Ca_5(PO_4,CO_3OH)_3(OH)$ |
| CFAP | $Ca_5(PO_4,CO_3OH)_3(F)$ |

*Source:* Phillips, W. R., and D. T. Griffen, *Optical Mineralogy: The Nonopaque Minerals.* W. H. Freeman and Co., San Francisco, 1981. With permission.

the corners of the hexagonal cell (Table 1.1; McClellan and Lehr 1969). Fluorapatite (FAP) is by far the most abundant of the apatite minerals on land and is an early formed accessory mineral in igneous rocks, appearing as tiny euhedral crystals associated with ferromagnesian minerals (McConnell 1973).

### 1.2.1.1 Apatite Dissolution: Phosphate Rock and Synthetic Hydroxylapatite

The dissolution rate of apatite minerals has been the focus of a variety of relatively disparate studies. For example, the importance of P as a fertilizer and as an integral component of teeth and bone material has led to a wide range of P dissolution and retention studies. Most of the agricultural studies have focused on the dissolution of fertilizers, in which P has already been leached from a phosphate rock ore (usually a mixed lithology of marine sediments containing high percentages of francolite [CFAP], $Ca_5(PO_4,CO_3OH)_3(F)$; Filippelli 2011) and reprecipitated as highly soluble phosphoric salts. Some studies, however, have investigated the dissolution of phosphate rock in laboratory experiments (Smith et al. 1974, 1977; Olsen 1975; Chien et al. 1980; Onken and Matheson 1982). The medical studies have investigated the dissolution behavior of synthetic HAP (similar in composition to dahllite) in acidic solutions, especially to understand the processes of bone resorption and formation, as well as dental caries formation (Christoffersen et al. 1978; Fox et al. 1978; Nancollas et al. 1987; Constantz et al. 1995).

Several researchers have attempted to develop dissolution rate equations to model the apatite dissolution (Olsen 1975; Smith et al. 1977; Christoffersen et al. 1978; Fox et al. 1978; Lerman 1979; Chien 1980; Onken and Matheson 1982; Hull and Lerman 1985; Hull and Hull 1987; Chin and Nancollas 1991). The rate equations from these models are of forms that include zero order, first order, parabolic diffusion, mixed order, and others. One model (Hull and Hull 1987) focuses on surface dissolution geometry, which, the authors argue, fits the experimental results better than previous dissolution models. Clearly missing in these experiments and the dissolution rate equations derived from them are the experimental conditions that replicate the natural dissolution processes and agents in soils, as well as a reasonable range of apatite mineralogy likely to be exposed to natural weathering processes in soils.

### 1.2.1.2   Apatite Dissolution: Marine Sediments and CFAP

Perhaps the most comprehensive and geologically appropriate laboratory examination of apatite dissolution has been performed on CFAP. Researchers focused on the marine P cycle have performed several studies on CFAP dissolution (Lane and Mackenzie 1990, 1991; Tribble et al. 1995), as well as CFAP precipitation in laboratory settings (Jahnke 1984; Van Cappellen and Berner 1988; Van Cappellen 1991; Filippelli and Delaney 1993). These studies have focused on CFAP because of the importance of this authigenic marine mineral phase in terms of the oceanic P cycle (Ruttenberg 1993; Filippelli and Delaney 1996). Several general observations about apatite dissolution/precipitation have been made. For example, the presence of Mg has been determined to retard CFAP formation (Atlas and Pytkowicz 1977), but the presence of other trace elements (e.g., Fe and Mn) appears to have little effect (Filippelli and Delaney 1992, 1993).

In a study of the pH dependence of CFAP dissolution, Lane and Mackenzie (1990; 1991) used a fluidized bed reactor to determine the dissolution chemistry. As presented in the study by Tribble et al. (1995), a Ca- and F-depleted surface layer appears to form during early dissolution, followed by a later stage when stoichiometric dissolution is achieved. The incongruent initial dissolution is probably due to the removal of most or all of the F, and some Ca, in the depleted surface layer, and the formation of a hydrogen–calcium–phosphate phase, as has also been observed by other researchers (Atlas and Pytkowicz 1977; Smith et al. 1974; Driessens and Verbeeck 1981; Thirioux et al. 1990). This surface-controlled reaction eventually achieves a steady state, whereby the depleted surface layer does not change in depth, and the solid effectively dissolves congruently (Tribble et al. 1995). Ruttenberg (1990; 1992) performed extensive dissolution experiments on FAP, HAP, and CFAP, with a goal of developing an extraction scheme for the characterization of these mineral phases in marine sediments. Although geared toward extraction technique development, these studies showed increased dissolution rate with decreasing grain sizes and higher dissolution rates of HAP and CFAP than that of FAP with a sodium acetate–acetic acid solution at pH 4.0.

### 1.2.1.3   Phosphorus Cycling in Soils

The cycling of P in soils (see Figure 1.1) has received much attention, in terms of both fertilization and natural development of ecosystems. Of the approximately 122,600 Tg P within the soil/biota system on the continents, nearly 98% is held in soils in a variety of forms. The exchange of P between biota and soils is relatively rapid, with an average residence time of 13 years, whereas the average residence time of P in soils is 600 years (Figure 1.1). As noted earlier, the only significant weathering source for phosphorus in soils is apatite minerals. These minerals can be congruently weathered as a result of a reaction with dissolved carbon dioxide:

$$Ca_5(PO_4)_3OH + 4H_2CO_3 \rightarrow 5Ca^{2+} + 3HPO_4^{2-} + 4HCO_3^- + H_2O$$

In soils, P is released from mineral grains by several processes. First, the reduced pH produced from respiration-related $CO_2$ in the vicinity of both degrading organic matter and root hairs dissolves P-bearing minerals (mainly apatites) and releases P to the root pore spaces (e.g., Schlesinger 1997). Second, organic acids released by plant roots can also dissolve apatite minerals and release P to the soil pore spaces (Jurinak et al. 1986). Phosphorus is very immobile in most soils, and its slow rate of diffusion from the dissolved form in pore spaces strongly limits its supply to the rootlet surfaces (Robinson 1986). Furthermore, much of the available P in soils is in organic matter, which is not directly accessible for plant nutrition. Plants have developed two specific tactics to increase the supply of P to the roots. First, plants and soil microbes secrete phosphatase, an enzyme that can release bioavailable inorganic P from organic matter (McGill and Cole 1981; Malcolm 1983; Kroehler and Linkins 1988; Tarafdar and Claasen 1988). Second, the symbiotic fungi mycorrhizae can coat plant rootlets, excreting phosphatase and organic acids to release P and providing an active uptake site for the rapid diffusion of P from the soil pore spaces to the root surface (Antibus et al. 1981; Bolan et al. 1984; Dodd et al. 1987). In exchange, the plant provides carbohydrates to the mycorrhizal fungi (Schlesinger 1997).

The phosphorus in soils is present in a variety of forms, and the distribution of P between these forms dramatically changes with time and soil development. The forms of soil P can be grouped into refractory (not readily bioavailable) and labile (readily bioavailable). The refractory forms include P in apatite minerals and P coprecipitated with and/or adsorbed onto iron and manganese oxyhydroxides (termed *occluded* P). The reducible oxyhydroxides have large binding capacities for phosphate, due to their immense surface area and numerous delocalized positively charged sites (e.g., Smeck 1985; Froelich 1988). The labile forms include P in the soil pore spaces (as dissolved phosphate ion) and adsorbed onto the soil particle surfaces (these forms are termed *nonoccluded* P), as well as P incorporated in soil organic matter. On a newly exposed lithic surface, nearly all of the P is present as P in apatite. With time and soil development, however, P is increasingly released from this form and incorporated in the others (Figure 1.2). Over time, the total amount of P available in the soil profile decreases, as soil P is lost through the surface and the subsurface runoff. Eventually, the soil reaches a terminal steady state, when soil P is heavily recycled, and any P lost through the runoff is slowly replaced by new P weathered from apatites at the base of the soil column.

### 1.2.1.4 Riverine Transport of Particulate and Dissolved Phosphorus

The eventual erosion of soil material and transport by rivers delivers P to the oceans. Riverine P is found in two main forms: particulate and dissolved. Most of the P contained in the particulate load of rivers is held within mineral lattices and never participated in the active biogenic cycle of P. This will also be its unlucky fate once it is delivered to the oceans, because the dissolution rates in the high pH and heavily buffered waters of the sea are exceedingly low. Thus, much of the net P physically eroded from continents is delivered relatively unaltered to the oceans, where it is sedimented on the continental margins and in the deep sea, anxiously waiting for subduction or accretion for its second chance to participate in the P cycle. Some of the particulate P is adsorbed onto the soil surfaces, held within soil oxide, and incorporated into the particulate organic matter. This P likely interacted with the biotic P cycle on land, and its fate upon transfer to the ocean is poorly understood. For example, P adsorbed onto

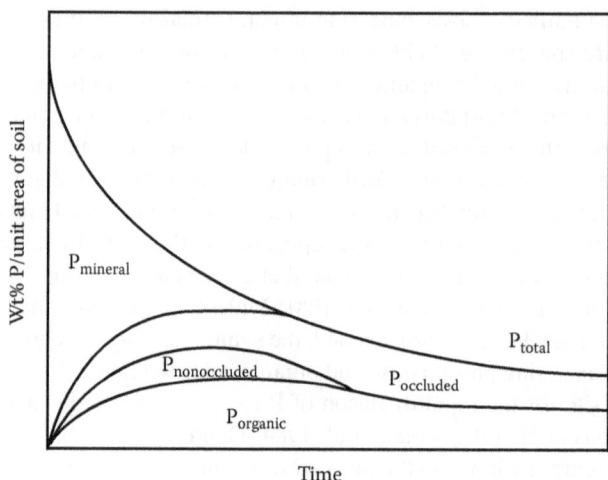

**FIGURE 1.2**  Modeled changes in soil phosphorus geochemistry over time (based on the study by Walker and Syers [1976]), showing the transformation of the mineral phosphorus into nonoccluded and organic forms before the eventual dominance of occluded (oxide-bound) and organic forms. The relative bioreactivity of phosphorus increases from mineral to occluded to organic forms of phosphorus. Note the continual loss of total phosphorus from system. (From Filippelli, G. M., and C. Souch, *Geology*, 27, 171–174, 1999. With permission.)

the soil surfaces may be effectively displaced by the high ionic strength of ocean water, providing an additional source of P into the ocean. Furthermore, a small amount of the P incorporated into terrestrial organic matter may be released in certain environments during bacterial oxidation after sediment burial. Finally, some sedimentary environments along the continental margins are suboxic or even anoxic, conditions favorable for oxide dissolution and release of the incorporated P. Several important studies have examined the transfer of P between terrestrial and marine environments (e.g., Froelich et al. 1982; Berner and Rao 1994; Ruttenberg and Goni 1997), but more work clearly needs to be done to quantify the interactions between the dissolved and particulate P forms and the aquatic/marine interface. The net prehuman flux of dissolved P to the oceans is ~1 Tg P per year, with an additional 1–2 Tg P per year of potentially soluble P, bringing the total to about 2–3 Tg P per year. Thus, the residence time of biologically available P on land is about 40–60 thousand years (kyr) with respect to export to the oceans. That this residence time is of a glacial timescale may be no coincidence—later I will discuss the control of climate on the terrestrial P cycle. But clearly, the interaction between biologically available P on land and loss of this P to the oceans is relatively dynamic and speaks to the relatively rapid cycling of P on land.

## 1.2.1.5  Marine Sedimentation

Once in the marine system, the dissolved P acts as the critical long-term nutrient limiting biological productivity. Phosphate concentrations are near zero in most surface waters, as this element is taken up by phytoplankton as a vital component of their photosystems (phosphate forms the base for adenosine triphosphate and adenosine diphosphate, required for photosynthetic energy transfer) and their cells (cell walls are composed of phospholipids). The dissolved P has a nutrient profile in the ocean, with a surface depletion and a deep enrichment. Furthermore, deep phosphate concentrations increase with the age of the deep water, and thus, the values in young deep waters of the Atlantic are typically ~1.5 μM, whereas those in the older Pacific are ~2.5 μM (Broecker and Peng 1982). Once incorporated into the plant material, P roughly follows the organic matter loop, undergoing active recycling in the water column and at the sediment/water interface.

Phosphorus input and output are driven to a steady-state mass balance in the ocean by biological productivity. As previously mentioned, one of the difficulties in accurately determining the preanthropogenic residence time of P in the ocean is that the input has nearly doubled due to anthropogenic activities, and thus, we must resort to the estimate of the burial output of P, a technique plagued by site–site variability in deposition rates and poor age control. Estimates of P burial rates have been performed by a variety of methods, including determination via P sedimentary sinks (e.g., Froelich et al. 1982; 1983) and riverine-suspended matter fluxes calibrated to the P geochemistry of those fluxes (Ruttenberg 1993). We have used an areal approach to quantifying the modern P mass balance (Table 1.2; Filippelli & Delaney 1996). Using this areal approach, we find that the output term for the P mass balance might be higher than previously estimated, and therefore, the P residence time might be shorter than previously thought (Table 1.2). Averaging across regions with varying P accumulation rates undoubtedly imparts a relative error of at least 50% to the burial rate value derived here. Also, some restricted environments exhibit quite high but spatially variable P burial rates (e.g., phosphatic continental margin environments [Froelich et al. 1982]; hydrothermal iron sediments [Froelich et al. 1977; Wheat et al. 1996]). The areal approach is a more direct route to quantifying the burial terms in the P mass balance and indicates that the reactive P burial in continental margin sediments accounts for about 60% of the oceanic output, with deep sea sediments nearly equivalent as a sink.

Although a promising and direct method for quantifying the modern P mass balance and potentially determining past variations in the balance, we are still greatly limited by the sparse data that presently exist for P geochemistry and accumulation rates, especially along the continental margins. Interestingly, the concentration of reactive P in sediments displays a relatively narrow range in both deep ocean and continental margin environments (Filippelli 1997). The major force driving the variations in the accumulation rates from site to site is the sedimentation rate, termed the *master variable* in P accumulation by Krajewski et al. (1994) (Figure 1.3). Thus, the P accumulation rate at any given site is most responsive to sedimentation rate—this also results in reactive P burial rates being approximately equally distributed between continental margin and deep ocean regions, even though continental margins represent only a fraction of the total ocean area (Filippelli & Delaney 1996; Compton et al. 2000).

## 1.2.2 MODERN PHOSPHORUS CYCLE

Human impacts have been manifested for decades at least in a variety of settings. Since the awareness that P is an important plant nutrient, the application of P directly

**TABLE 1.2**

**Phosphorus Accumulation, Burial, and Modern Oceanic Phosphorus Mass Balance**

| | Equat. Pacific[a] | All Mod./High Prod. Pelagic[b] | Low Prod. Pelagic Regions[c] | Cont. Margins[d] | Whole Ocean[e] | Previous Estimates Ruttenberg (1993) | Previous Estimates Froelich et al. (1982)[f] |
|---|---|---|---|---|---|---|---|
| P accumulation rate (μmol cm⁻² kyr⁻¹) | 19 | 28 | 10 | 500 | — | | |
| Area (10⁶ km²) | 30 | 276 | 60 | 25 | 361 | | |
| P burial rate[g] (10¹⁰ mol P/yr) | 0.6 | 7.7 | 0.6 | 12.5 | 21.0 | 8.0–18.5 | 5.0 |
| P residence time (kyr)[h] | | | | | 15 | 16–38 | 60 |

*Source:*  Reprinted from *Geochimica Cosmochimica Acta*, 60, Filippelli, G. M., and M. L. Delaney, Phosphorus geochemistry of equatorial Pacific sediments, 1479–1495, Copyright (1996), with permission from Elsevier.

a  Mean of all samples from 0 to 1 Ma in the study by Filippelli and Delaney (1996).

b  Defined as a region with a primary productivity greater than 35 g C m⁻² yr⁻¹ (Berger and Wefer 1991). The phosphorus accumulation rate is a mean of various moderate to high productivity pelagic regions (e.g., Southern Ocean, equatorial Atlantic, equatorial Pacific; compiled by Filippelli [1997]).

c  Defined as a region with a primary productivity less than 35 g C m⁻² yr⁻¹ (Berger and Wefer 1991). The phosphorus accumulation rate is a mean of various low productivity pelagic regions (e.g., pelagic red clay provinces; compiled by Filippelli 1997]).

d  Defined as the area above 200 m water depth (Menard and Smith 1966). The phosphorus accumulation rate is the low range value of various continental margin regions (e.g., upwelling and nonupwelling environments; compiled by Filippelli [1997]). The low range was used because many of the published phosphorus concentration values included nonreactive phosphorus components, which do not interact with the dissolved marine phosphorus cycle.

e  Filippelli and Delaney (1996). Whole ocean is the sum of all moderate to high productivity pelagic regions, low productivity pelagic regions, and continental margins.

f  As revised in the study by Froelich (1984).

g  Phosphorus accumulation rate multiplied by area.

h  Calculated as whole ocean phosphorus inventory (3.1 × 10¹⁵ mol P) divided by whole ocean phosphorus burial rate (21 × 10¹⁰ mol P/year).

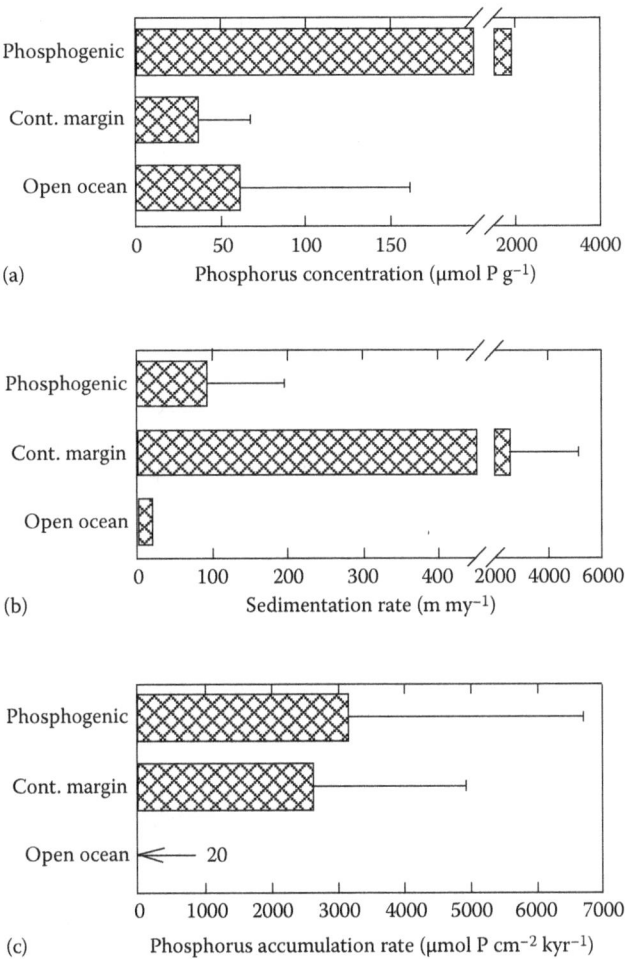

**FIGURE 1.3**  A plot of (a) phosphorus concentration, (b) sedimentation rate, and (c) phosphorus accumulation rate for open ocean, continental margin, and phosphogenic environments.

to fields to increase crop yields has increasingly dominated the P cycle. The early source of this fertilizer P was from rock phosphate, generally phosphate-rich sedimentary rocks (phosphorites) mined in the nineteenth century from a variety of settings in Europe and the United States. This P-rich rock was ground and applied to fields, where it significantly enhanced the crop yield after the initial application. Farmers soon found, however, that the crop production significantly decreased after several years of application of this rock fertilizer. Unbeknownst to the farmers, the P-rich rock they were applying to their fields was also rich in cadmium and uranium, both naturally found in the organic-rich marine sediments that are precursors to phosphorite rocks but also toxic to plants. Many previously productive fields became barren due to this practice, and the awareness of the heavy metal-enriched P fertilizers led to the development of a variety of leaching techniques to separate the beneficial P from the toxic heavy metals. The perfection of this technique and the coapplication of P with N, K, and other micronutrients in commercially available fertilizers boomed during the Green Revolution, the period after World War II that saw prosperity in some countries and exponential growth in global populations. The irony of the moniker *Green Revolution* has become apparent, as the production of enormous amounts of food fueled by fertilizers to feed a growing global population has caused a variety of detrimental environmental conditions, including eutrophication of surface water supplies, significant soil loss, and expanding coastal dead zones (regions of hypoxia and fish mortality, exemplified by the Gulf of Mexico near the outflow of the Mississippi River and likely caused by the fertilizer runoff from agricultural practices in the Great Plains and the Midwest of the United States).

The net increase in dissolved P release from land due to human activities also includes deforestation (+ concurrent soil loss) and sewage and waste sources (Figure 1.4). Deforestation, typically by burning after selective tree harvesting, converts the standing stock of P in plant matter to ash. This P is rapidly leached from the ash and transported as dissolved loads in rivers; this transfer can happen on timescales of a year or two (Schlesinger 1997). Furthermore, the lack of rooted stability on the landscape results in the loss of the relatively organic P-rich O and A soil horizons from many of these areas, from which some of the P is solubilized during transport. Sewage and waste are additional anthropogenic contributors to the terrestrial P cycle. Human waste, waste from processing of foodstuffs, and waste from industrial uses of detergents (which are now generally low P but in the past contained up to 7% P by weight) contribute roughly equally to deforestation and soil loss.

The synhuman terrestrial P cycle (Figure 1.4) is, therefore, substantially different from the prehuman cycle. This is evidenced by high loads of dissolved P in rivers (about two times the estimated natural values) and higher loads of P-bearing particulates. Together, the net input of dissolved P from land to the oceans is 4–6 Tg P per yr, representing a doubling of prehuman input fluxes. As previously noted, this increased flux of P does directly influence the coastal regions via eutrophication, but presumably, an impact will be seen on the whole ocean. One likely scenario is for

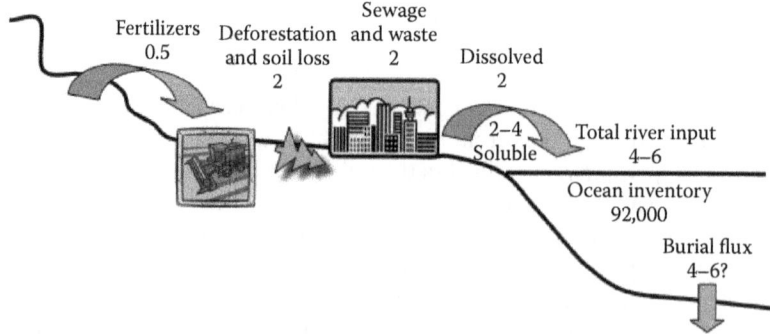

**FIGURE 1.4** The modern (synhuman) phosphorus cycle, showing reservoirs (in teragram P) and fluxes (denoted by arrows, in teragram P per year) associated with human activities, which have effectively doubled the natural dissolved P fluxes.

the ocean to achieve another steady state (e.g., Broecker and Peng 1982), in which both the P output flux and the oceanic P reservoir will increase in response to higher inputs. This effect will be minimal, however, given the limited sources of P available for exploitation (Filippelli 2008) and the relatively long residence time of P in the ocean.

## 1.3   ECOSYSTEM DYNAMICS AND SOIL DEVELOPMENT

The effect of climate and soil development on P availability has been a focus of several excellent papers (Walker and Syers 1976; Gardner 1990; Crews et al. 1995; Cross and Schlesinger 1995; Vitousek et al. 1997; Schlesinger et al. 1998; Chadwick et al. 1999). For the most part, these studies have used P extraction techniques to determine the biogeochemical forms of P within soils. These techniques differentiate P in similar fractions as displayed in Figure 1.2 (e.g., Tiessen and Moir 1993). The extraction techniques have been applied to depth and age profiles in soils, to assess the rate of soil P transformations, the role of climate on these processes, the bioavailability of P in these systems, and the limiting controls on plant productivity. As the current geochemical state of a given soil is an integration of all conditions acting since the soil development, most efforts have focused on settings in which the climate is likely to have been constant (i.e., tropical settings), and the beginning state of the system and its age are very well known (i.e., soil developing on lava flows). These studies have thus made the classic substitution of space for time, with all the inherent assumptions of constancy in climate and landscape history.

Another approach to assessing terrestrial P cycling is by examining the P geochemistry in lake sediments, using the same extraction techniques as the soil studies. This technique adds several dimensions to the soil work outlined earlier. First, lake sediment records allow us to examine an integrated record of watershed-scale processes associated with P cycling on the landscape. Second, it allows discrete temporal resolution at a given site, providing an actual record of local processes including landscape stability, soil development, and ecosystem development. Third, it extends our understanding of terrestrial P cycling to alpine and glaciated systems. The soil chronosequence approach is not likely to be successful here because of the climatic and slope variabilities between various sites (i.e., no substitution of space for time is possible). This third dimension is perhaps the most critical in terms of the P mass balance, as the greatest degree of variations in climate has occurred in these settings, and thus, they hold the key to understanding the terrestrial P cycle on glacial/interglacial timescales.

Several limitations also exist in this approach. First, the lake sediment records analyzed must be dominated by terrestrial input, with very little lake productivity (i.e., very oligotrophic) and/or diagenetic processes occurring in the lake sediments. This is critical because the clearest signal of the state of soils in the surrounding landscape will come from a lake receiving this sediment as the primary source. This situation holds for numerous small lakes in headwater catchments, where local surface sedimentation dominates, dissolved nutrient inputs are so low that in situ organic production is at a minimum, and low amounts of labile organic matter in the sediments limits the degree of diagenetic overprinting on the original sediment record. Lowland lakes that integrate several watersheds and streams have a complex sedimentation history and often have high rates of *in situ* productivity because of high nutrient inputs. Thus, these lakes are poor candidates for this approach. However, carefully chosen headwater lakes that characterize a relatively large range in climatic conditions, bedrock lithologies, and local landscape relief will provide important information on the weathering processes of P on land.

### 1.3.1 EXAMPLES OF THE LAKE HISTORY APPROACH TO TERRESTRIAL P CYCLING

We have applied the lake sediment approach described earlier to several settings and have found exciting results with important implications for the effects of climate and landscape development on the terrestrial P cycle. For example, Filippelli and Souch (1999) analyzed sediments from several small upland lakes in headwater catchments selected to represent contrasting climates and glacial/postglacial histories. The longest records (about 20,000 years) came from two different lakes in the western Appalachian Plateau of the midwestern United States: the Jackson Pond, in central Kentucky (289 m asl) and the Anderson Pond, in north central Tennessee (300 m asl). Climate has strongly affected the ecological development of both of these systems (Delcourt 1979; Mills and Delcourt 1991; Wilkins et al. 1991), although neither was directly glaciated. In contrast, a 11,500-year record from the Kokwaskey Lake, in British Columbia (1050 m asl), clearly reveals the deglacial history of the region and retains the effect of alpine glaciers throughout the record (Souch 1994).

The geochemical profile for the Jackson Pond reveals extreme changes through time (Figure 1.5). In the early part of the record, marked by full glacial conditions (Wilkins et al. 1991), mineralized P was the dominant form entering the lake, with occluded and organic forms of lesser importance. During this interval, the landscape was marked by thin soils, high surface runoff, and closed boreal forests (Wilkins et al. 1991). With landscape stabilization and onset of soil development (17–10 kiloanni [ka]), the dominant forms of P entering the lake are significantly changed, with the proportion of mineralized P declining and organic P increasing (occluded P varies but exhibits no clear trend with age during this interval). In this interval, the closed boreal forests give way to more open boreal woodland and a rapidly thickening soil cover (Wilkins et al. 1991). From the early mid-Holocene to the present, the concentration of mineral P shows little variation, while that of organic P and occluded P increases. Meanwhile, the ecosystem becomes dominated by deciduous hardwood forests and grasses, and a thick and stable soil exists (Wilkins et al. 1991). In terms of the percentage of total P reflected by each fraction, the early Holocene marks a stabilization of the system to one dominated by organic and occluded P. By comparison, the concentration of mineral P is higher throughout the Kokwaskey Lake record, with occluded P of secondary importance and organic P in relatively low concentrations (Figure 1.6). Mineral P exhibits a decrease during the interval of landscape stabilization and soil development (between about 9 and 6 ka), while the proportion of occluded P, and to a lesser extent organic P, increases during this interval. From the mid- to late Holocene (6–1 ka), characterized by cooler/wetter conditions (Pellatt and Mathewes 1997), each fraction varies slightly. The last 1000 years of this record,

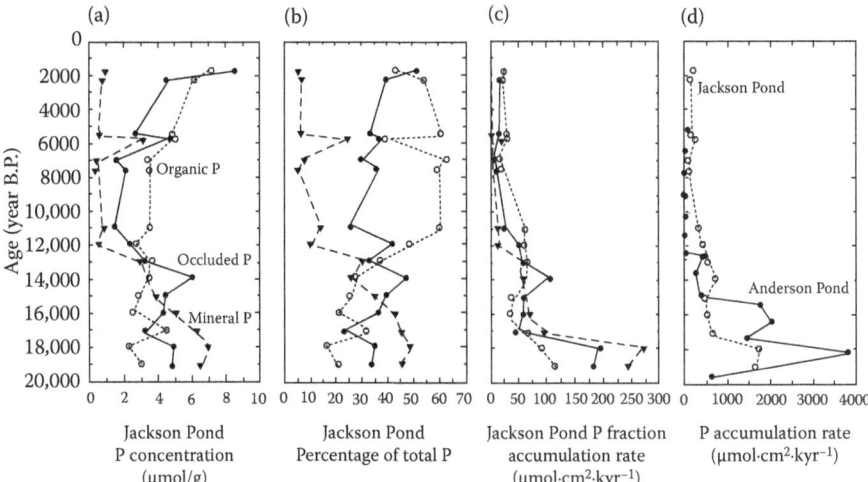

**FIGURE 1.5** (a) P concentration and (b) percentage of the total P for each P-bearing fraction in sediments from the Jackson Pond. (c) Accumulation rate of P fractions at the Jackson Pond. (d) Total P accumulation rate at the Anderson Pond and Jackson Pond. The nonoccluded fraction (Figure 1.2) is small and is combined with the occluded fraction in this analysis.

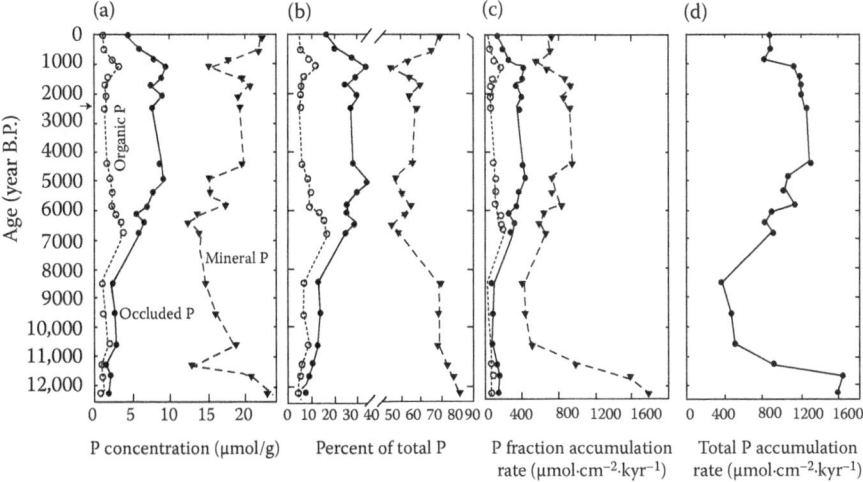

**FIGURE 1.6** (a) P concentration, (b) percentage of the total P for each P-bearing fraction, (c) accumulation rate of P fractions, and (d) total P accumulation rate in the Kokwaskey Lake sediments. Age-control points (Souch 1994) are shown as arrows on the age axis in a; lowermost point represents a $^{14}$C datum, and the other two points indicate tephra dates (Mazama and Bridge River).

marked by the most extensive Holocene neoglacial activity (Ryder and Thompson 1986), is characterized by a rapid return of P geochemistry to glacial/deglacial conditions. Throughout the record, the ecosystem in this site is dominated by boreal forest, and slopes are steep, relatively unstable, and have much thinner soils than in the western Appalachian example.

These two examples suggest several controlling factors on P geochemical cycles. First, the transformation of mineral P to other (more bioavailable) forms is more complete in a relatively wet, warm, and low relief system like the western Appalachian Plateau than in a wet but cold and high relief system like the Coast Mountains. Thus, mineralized P decreases from 50% to 5% of the total P at the Jackson Pond compared to 90–50% at the Kokwaskey Lake. This is consistent with the primary importance of physical weathering and erosion in the highly unstable environment in the Coast Mountains of British Columbia. Second, the relative starting points and stabilization points of these two systems are drastically different, driven by their contrasting landscape history and stability. The Kokwaskey Lake was completely covered in glacier at 12 ka, and thus, the initial starting point for this system is nearly completely mineral P from rock flour. Throughout the Holocene, the high relief in the watershed led to constant loss of surface sediments with poorly developed soils and relatively little organic P. In contrast, low relief and relatively rapid development of a stabilized landscape and soil profiles in the Jackson Pond watershed supports the development of a mature soil in terms of P geochemistry. And, because the Jackson Pond watershed started with a climate and an ecosystem similar to that of the Kokwaskey Lake during the interval from 6 to 1 ka, their comparative P geochemistries are similar in these intervals. The model presented earlier (Figure 1.2) is therefore useful for tracking soil P transformations from the lake cores presented here. For the western Appalachian setting, the soil system begins in a slightly preweathered stage and progresses to an end point dominated by the organic and occluded fractions (Figure 1.2). For the alpine setting, the soil system begins as rock flour and slightly shifts to the right then back again, as warming and more landscape stabilization occur in the mid-Holocene followed by extensive neoglaciation over the last 1000 years (Figure 1.6). Finally, the P geochemistry of these systems can be quickly reset. During the last 1000 years, the readvance of alpine glaciers during the Little Ice Age has been sufficient to return the P geochemical cycle in this region back to near glacial conditions (Figure 1.6).

The patterns in P weathering and release are similar to those observed in several tropical settings. For example, a chronosequence of Hawaiian soils reveals significant transitions in mineral P after about 1000 years of soil development (Crews et al. 1995; Vitousek et al. 1997), while significant transformations occur in P biogeochemistry during soil development on a Krakatau lava flow in just over 100 years (Schlesinger et al. 1998). In the temperate settings presented here, this decrease occurs over timescales of 3000–5000 years. One of the key differences is that the temperate settings are periodically reset due to glacial climate conditions. This resetting keeps these soil systems in a building phase. But like the tropical environments, the progressive loss of P during soil development leads to P limitation to terrestrial ecosystems.

Large changes in the accumulation rate of P over time are reflected in all the lake records. These accumulation rate changes are partly driven by changes in the bulk sedimentation rate, but several of the rapid shifts in P accumulation occur between age control points and are thus not just driven by rates of sedimentation. The two western Appalachian sites reveal a rapid pulse of P input during the glacial and initial deglacial intervals (Figure 1.5), a period of enhanced colluvial activity when soil development was just commencing. Upon landscape and soil stability (by 10 ka), P inputs had decreased 10- to 40-fold and remained low and constant to the top of the record (2 ka). The Kokwaskey alpine site also reveals significant changes in the P input (Figure 1.6), beginning with a rapid early deglacial pulse consisting mainly of mineralized P originating from glacial rock flour. From approximately 9–6 ka, landscape stabilization begins, although the landscape is still marked by high relief and rapid rates of erosion. Phosphorus input rates are high but slightly decrease from 4 to about 1 ka. In this high relief alpine setting, soil development is retarded by rapid denudation, and thus, the terrestrial P cycle is stuck in an initial development stage, with high mineral P and high P release rates from the landscape.

Although the temporal records differ between the sites, some generalizations may be made about terrestrial P cycling, climate, and landscape development as a function of glacial/interglacial cycles. First, the lake records reflect the input of solid-phase P released from watersheds. Some portion of this released P is bioreactive in terrestrial systems and upon transport, in the ocean. The P geochemical results indicate that the more bioreactive forms dominate in the later stages of soil development. However, this increase in the bioreactive nature of the P is offset by very low total P release at these times, as reflected by the total P accumulation rate records from the lakes. In the case of dissolved phosphate, the initial stage of soil development we relate to high solid-phase loss also likely leads to a relatively poor recycling of the dissolved phase (from a lack of oxide-bound occluded pools), a so-called leaky ecosystem. Second, the western Appalachian sites have reached a degree of landscape stabilization due to relatively stable climates. Although the elevation, the local relief, the mean temperature, the rainfall, and the ecosystem are quite different for these sites, the net result of soil stabilization is to reduce the rate of P loss from these systems. The alpine British Columbia site, on the other hand, has never achieved the relative stabilization of the other sites due to high relief and neoglacial activity. Thus, the trends in P release from this system lags, and P release remains relatively high and constant. We believe that the results presented here characterize mid and high latitude watersheds that have been directly glaciated or strongly affected by proximal glacial conditions. Tropical environments may experience less dramatic climate change through glacial/interglacial intervals, but recent terrestrial evidence reveals that even these systems experience notably cooler and drier conditions during glacial intervals (Thompson et al. 1997), which may lead to changes in the extent of soil development and P release in these systems as well.

The lake sediment records presented here indicate a terrestrial P mass balance that is not near steady state on glacial/interglacial timescales, with important implications for the functioning of terrestrial and oceanic systems. Coupling these records of solid-phase P changes over time with oceanic records and models of dissolved inputs from rivers (e.g., Tsandev et al. 2008) will eventually provide important constraints for the influence of climate on chemical weathering and the global P cycle.

## 1.4   MODERN PHOSPHORUS DEPOSIT RESOURCES AND THE IMPENDING P CRISIS

Phosphorus plays a very important but relatively unheralded role in global sustainability. The rates of exploitation of phosphate for fertilizer use has far outstripped the rate of formation of phosphate minerals, which poses a huge challenge for the capability of feeding a global population that is 7 billion and growing. Phosphate rocks, also referred to as phosphorites, are sedimentary deposits with high P concentrations. These rocks are one of the primary ore sources for P, which in turn is a critical and nonrenewable element for fertilizer production, upon which the global fertility depends. The startling limitation of phosphate rock reserves has led to a renewed intensity of research in mitigating P loss from landscapes, recycling P from waste streams, and exploring new P ore potentials. From a human perspective, this situation is one of taking an element from a high-concentration source and broadly distributing it—although the elemental P mass is conserved, it is now scattered across the globe and effectively diluted. The ecological impact of intense fertilization and loss from landscapes has been extensively documented in eutrophication literature (e.g., Bennett et al. 2001; Schindler et al. 2008), but the impacts from a resource conservation standpoint are just becoming apparent (Smil 2000, 2002; Cordell et al. 2009).

From a geologic standpoint, P is also a vital fertilizing agent for biological productivity on land and in the sea and indeed is considered the ultimate limiting nutrient in the ocean on timescales exceeding 1000 years (Tyrell 1999). As with all resources, geology had millions of years to act to form sedimentary deposits with high P concentrations, but humans have extracted these resources at rates so high that the current phosphate rock reserves might be largely depleted in this century (Cordell et al. 2009). Typical rocks have an average P concentration of about 0.1 wt%, whereas phosphate rocks have P concentrations 100 times that amount. Phosphate rocks are indeed unique from many perspectives, not least due to their high P content (e.g., Filippelli 2011).

## 1.5   CONCLUSIONS AND REMAINING QUESTIONS

Significant advances have been made in our understanding of the dynamics of the global P cycle. Interestingly, many of these advancements have derived from the widespread application of P geochemical techniques developed for soils and sediments over the last several decades, but only recently were they adopted by oceanographers. These techniques have allowed us to further elucidate the dynamics of P transformations during weathering, transport, and deposition; the role that P bioavailability plays on terrestrial and marine ecosystems; and the impacts of P availability on biotic systems and climate change in the past.

Despite these recent advances, several critical questions remain about terrestrial P cycling, including

- What is the role of P weathering and soil development on ecosystems?
- How has the terrestrial P cycle changed on glacial timescales?

- What was the terrestrial P cycle like during the establishment of land plants in the early Phanerozoic?
- What were the dynamics of the terrestrial P cycle during warmer intervals on earth?
- What will be the ultimate limitations on P resource exploitation, utilization, and global food production?

## REFERENCES

Adams, J. 1995. Weathering and glacial cycles. *Nature* 373:100.

Antibus, R. K., J. G. Croxdale, O. K. Miller, and A. E. Linkins. 1981. Ecotomycorrhizal fungi of *Salix rotundifolia*: III: Resynthesized mycorrhizal complexes and their surface phosphatase activities. *Canadian Journal of Botany* 59:2458–2465.

Atlas, E. L., and R. M. Pytkowicz. 1977. Solubility behavior of apatites. *Limnology & Oceanography* 22:290–300.

Bennett, E. M., S. R. Carpenter, and N. F. Caraco. 2001. Human impact on erodible phosphorus and eutrophication: A global perspective. *Bioscience* 51:227–234.

Berger, W. H., and G. Wefer. 1991. Productivity of the glacial ocean: Discussion of the iron hypothesis. *Limnology & Oceanography* 36:1899–1918.

Berner, R. A., and J.-L. Rao. 1994. Phosphorus in sediments of the Amazon River and estuary: Implications for the global flux of phosphorus to the sea. *Geochimica Cosmochimica Acta* 58:2333–2340.

Bolan, N. S., A. D. Robson, N. J. Barrow, and L. A. G. Aylmore. 1984. Specific activity of phosphorus in mycorrhizal and non-mycorrhizal plants in relation to the availability of phosphorus to plants. *Soil Biology and Biochemistry* 16:299–304.

Broecker, W. S. 1982. Ocean chemistry during glacial time. *Geochimica Cosmochimica Acta* 46:1689–1705.

Broecker, W. S., and T.-H. Peng. 1982. *Tracers in the Sea*. Eldigio Press, Palisades, New York.

Chadwick, O. A., L. A. Derry, P. M. Vitousek, B. J. Huebert, and L. O. Hedin. 1999. Changing sources of nutrients during four millions years of ecosystem development. *Nature* 397:491–497.

Chien, S. H., W. R. Clayton, and G. H. McClellan. 1980. Kinetics of dissolution of phosphate rocks in soils. *Soil Science Society of America Journal* 44:260–264.

Chin, K. O. A., and G. H. Nancollas. 1991. Dissolution of fluorapatite: A constant-composition kinetics study. *Langmuir: The ACS Journal of Surfaces and Colloids* 7:2175–2179.

Christoffersen, J., M. R. Christoffersen, and N. Kjaergaard. 1978. The kinetics of dissolution of calcium hydroxyapatite in water at constant pH. *Journal of Crystal Growth* 43:501–511.

Cole, C. V., G. S. Innis, and J. W. B. Stewart. 1977. Simulation of phosphorus cycling in semi-arid grasslands. *Ecology* 58:3–15.

Compton, J., D. Mallinson, C. Glenn, G. Filippelli, K. Föllmi, G. Shields, and Y. Zanin. 2000. Variations in the global phosphorus cycle. In *SEPM Special Publication 66, Marine Authigenesis: From Microbial to Global* (eds. Glenn, C., L. Prévôt-Lucas, and J. Lucas), pp. 21–34.

Constantz, B. R., I. C. Ison, and D. I. Rosenthal. 1995. Skeletal repair by in situ formation of the mineral phase of bone. *Science* 267:1796–1799.

Cordell, D., J.-O. Drangert, and S. White. 2009. The story of phosphorus: Global food security and food for thought. *Global Environmental Change* 19:292–305.

Crews, T. E., K. Kitayama, J. H. Fownes, R. H. Riley, D. A. Herbert, D. Mueller-Dombois, and P. M. Vitousek. 1995. Changes in soil phosphorus fractions and ecosystem dynamics across a long chronosequence in Hawaii. *Ecology* 76:1407–1424.

Cross, A. F., and W. H. Schlesinger. 1995. A literature review and evaluation of the Hedley fractionation: Applications to the biogeochemical cycle of soil phosphorus in natural ecosystems. *Geoderma* 64:197–214.

Delcourt, H. H. 1979. Late-Quaternary vegetation history of the eastern Highland Rim and adjacent Cumberland Plateau of Tennessee. *Ecological Monographs* 49:255–280.

Dodd, J. C., C. C. Burton, R. G. Burns, and P. Jeffries. 1987. Phosphatase activity associated with the roots and the rhizosphere of plants infected with vascular-arbuscular mycorrhizal fungi. *New Phytologist* 107:163–172.

Driessens, F. C. M., and R. M. H. Verbeeck. 1981. Metastable states in calcium phosphate-aqueous phase equilibria. *Journal of Crystal Growth* 53:55–62.

Filippelli, G. M. 1997. Controls on phosphorus concentration and accumulation in oceanic sediments. *Marine Geology* 139:231–240.

Filippelli, G. M. 2008. The global phosphorus cycle: Past, present and future. *Elements* 4(2):89–95.

Filippelli, G. M. 2011. Phosphate rock formation and marine phosphorus geochemistry: The deep time perspective. *Chemosphere* 84:759–766.

Filippelli, G. M., and M. L. Delaney. 1992. Quantifying cathodoluminescent intensity with an on-line camera and exposure meter. *Journal of Sedimentary Petrology* 62:724–725.

Filippelli, G. M., and M. L. Delaney. 1993. The effects of manganese (II) and iron (II) on the cathodoluminescence signal in synthetic apatite. *Journal of Sedimentary Petrology* 63:167–173.

Filippelli, G. M., and M. L. Delaney. 1994. The oceanic phosphorus cycle and continental weathering during the Neogene. *Paleoceanography* 9:643–652.

Filippelli, G. M., and M. L. Delaney. 1996. Phosphorus geochemistry of equatorial Pacific sediments. *Geochimica Cosmochimica Acta* 60:1479–1495.

Filippelli, G. M., and C. Souch. 1999. Effects of climate and landscape development on the terrestrial phosphorus cycle. *Geology* 27:171–174.

Filippelli, G. M., C. Souch, B. Menounos, S. Slater-Atwater, T. A. J. Jull, and O. Slaymaker. 2006. Alpine lake records reveal the impact of climate and rapid climate change on the biogeochemical cycling of soil nutrients. *Quaternary Research* 66:158–166.

Fox, J. L., W. I. Higuchi, M. B. Fawzi, and M. S. Wu. 1978. A new two-site model for hydroxyapatite dissolution in acidic media. *Journal of Colloid Interface Science* 67:312–330.

Froelich, P. N. 1988. Kinetic controls of dissolved phosphate in natural rivers and estuaries: A primer on the phosphate buffer mechanism. *Limnology & Oceanography* 33:649–668.

Froelich, P. N., M. L. Bender, and G. R. Heath. 1977. Phosphorus accumulation rates in metalliferous sediments on the East Pacific Rise. *Earth & Planetary Science Letters* 34:351–359.

Froelich, P. N., M. L. Bender, N. A. Luedtke, G. R. Heath, and T. DeVries. 1982. The marine phosphorus cycle. *American Journal of Science* 282:474–511.

Froelich, P. N., K.-H. Kim, R. Jahnke, W. C. Burnett, A. Soutar, and M. Deakin. 1983. Pore water flouride uptake in Peru continental margin sediments: Uptake from seawater. *Geochimica Cosmochimica Acta* 47:1605–1612.

Gardner, L. R. 1990. The role of rock weathering in the phosphorus budget of terrestrial watersheds. *Biogeochemistry* 11:97–110.

Holland, H. D. 1978. *The Chemistry of the Atmosphere and Oceans*. Wiley Interscience, New York.

Hull, A. B., and J. R. Hull. 1987. Geometric modeling of dissolution kinetics: Application to apatite. *Water Resources Research* 23:707–714.

Hull, A. B., and A. Lerman, A. 1985. The kinetics of apatite dissolution: Application to natural aqueous systems. *National Meeting of the American Chemical Society Division of Environmental Chemistry* 25:421–424.

Jahnke, R. A. 1984. The synthesis and solubility of carbonate fluorapatite. *American Journal of Science* 284:58–78.

Jurinak, J. J., L. M. Dudley, M. F. Allen, and W. G. Knight. 1986. The role of calcium oxalate in the availability of phosphorus in soils of semiarid regions: A thermodynamic study. *Soil Science* 142:255–261.

Krajewski, K. P., P. Van Cappellen, J. Trichet, O. Kuhn, J. Lucas, A. Martín-Algarra, L. Prévôt et al. 1994. Biological processes and apatite formation in sedimentary environments. *Eclogae Geologica Helvetica* 87:701–745.

Kroehler, C. J., and A. E. Linkins. 1988. The root surface phosphatases of *Eriophorum vaginatum*: Effects of temperature, pH, substrate concentration and inorganic phosphorus. *Plant and Soil* 105:3–10.

Kump, L. R., and R. B. Alley. 1994. Global chemical weathering on glacial time scales. In *Studies in Geophysics: Material Fluxes on the Surface of the Earth*. Board on Earth Sciences and Resources, National Research Council, National Academy of Sciences, Washington, DC, pp. 46–60.

Lajtha, K., and W. H. Schlesinger. 1988. The biogeochemistry of phosphorus cycling and phosphorus availability in a desert shrubland ecosystem. *Biogeochemistry* 2:29–37.

Lane, M., and F. T. Mackenzie. 1990. Mechanisms and rates of natural carbonate fluorapatite dissolution. *Bulletin of the Geological Society of America Abstract* A208.

Lane, M., and F. T. Mackenzie. 1991. Kinetics of carbonate fluorapatite dissolution: Application to natural systems. *Bulletin of the Geological Society of America Abstract* A151.

Lerman, A. 1979. *Geochemical Processes: Water and Sediment Environments*. John Wiley & Sons, New York, 481 pp.

Malcolm, R. E. 1983. Assessment of phosphatase activity in soils. *Soil Biology & Biochemistry* 15:403–408.

McClellan, G. H., and J. R. Lehr. 1969. Crystal chemical investigation of natural apatites. *American Mineralogist* 54:1372–1389.

McConnell, D. 1973. *Apatite: Its Crystal Chemistry, Mineralogy, Utilization, and Geologic and Biologic Occurrences*, Vol. 5. Springer-Verlag, New York.

McGill, W. B., and C. V. Cole. 1981. Comparative cycling of organic C, N, S, and P through soil organic matter. *Geoderma* 26:267–286.

Menard, H. W., and S. M. Smith. 1966. Hypsometry of ocean basin provinces. *Journal of Geophysical Research* 71:4305–4326.

Mills, H. H., and P. A. Delcourt. 1991. Quaternary geology of the Appalachian Highlands and interior low plateaus. In *Quaternary Non-Glacial Geology Conterminous U.S.* (ed. Morrison, R. B.) Geological Society of America, Boulder, CO, The Geology of North America, K-2, pp. 611–628.

Nancollas, G. H., Z. Amjad, and P. Koutsoukas. 1979. Calcium phosphates-speciation, solubility, and kinetic considerations. *American Chemical Society Symposium Series* 93:475–497.

Olsen, R. A. 1975. Rate of dissolution of phosphate from minerals and soils. *Soil Science Society of America Journal* 39:634–639.

Onken, A. B., and R. L. Matheson. 1982. Dissolution rate of EDTA-extractable phosphate from soils. *Soil Science Society of America Journal* 46:276–279.

Pellatt, M. G., and R. W. Mathewes. 1997. Holocene tree line and climate change on the Queen Charlotte Islands, Canada. *Quaternary Research* 48:88–99.

Phillips, W. R., and D. T. Griffen. 1981. *Optical Mineralogy: The Nonopaque Minerals*. W. H. Freeman and Co., San Francisco.

Porder, S., P. M. Vitousek, O. A. Chadwick, C. P. Chamberlain, and G. E. Hilley. 2007. Uplift, erosion, and phosphorus limitation in terrestrial ecosystems. *Ecosystems* 10:158–170.

Roberts, T. L., J. W. B. Stewart, and J. R. Bettany. 1985. The influence of topography on the distribution of organic and inorganic soil phosphorus across a narrow environmental gradient. *Canadian Journal of Soil Science* 65:651–665.

Robinson, D. 1986. Limits to nutrient inflow rates in roots and root systems. *Physiologia Plantarum* 68:551–559.

Ruttenberg, K. C. 1990. Diagenesis and burial of phosphorus in marine sediments: Implications for the marine phosphorus budget. PhD dissertation, Yale University, New Haven, CT.

Ruttenberg, K. C. 1992. Development of a sequential extraction method for different forms of phosphorus in marine sediments. *Limnology & Oceanography* 37:1460–1482.

Ruttenberg, K. C. 1993. Reassessment of the oceanic residence time of phosphorus. *Chemical Geology* 107:405–409.

Ruttenberg, K. C., and M. A. Goni. 1997. Phosphorus distribution, C:N:P ratios, and $\delta^{13}C_{oc}$ in arctic, temperate, and tropical coastal sediments: Tools for characterizing bulk sedimentary organic matter. *Marine Geology* 139:123–146.

Ryder, J. M., and B. Thompson. 1986. Neoglaciation in the southern Coast Mountains of British Columbia: Chronology prior to the late-Neoglacial maximum. *Canadian Journal Earth Sciences* 23:273–287.

Schindler, D. W., R. E. Hecky, D. L. Findlay, M. P. Stainton, B. R. Parker, M. J. Paterson, K. G. Beaty, M. Lyng, and S. E. M. Kasian. 2008. Eutrophication of lakes cannot be controlled by reducing nitrogen input: Results of a 37-year whole-ecosystem experiment. *Proceedings of the National Academy of Sciences USA* 105:11254–11258.

Schlesinger, W. H. 1997. *Biogeochemistry: An Analysis of Global Change.* Academic Press, San Diego, CA.

Schlesinger, W. H., L. A. Bruijnzeel, M. B. Bush, E. M. Klein, K. A. Mace, J. A. Raikes, and R. J. Whittaker. 1998. The biogeochemistry of phosphorus after the first century of soil development on Rakata Island, Krakatau, Indonesia. *Biogeochemistry* 40:37–55.

Smeck, N. E. 1985. Phosphorus dynamics in soils and landscapes. *Geoderma* 36:185–199.

Smil, V. 2000. Phosphorus in the environment: Natural flows and human interferences. *Annual Reviews of Energy and the Environment* 25:53–88.

Smil, V. 2002. Phosphorus: Global transfers. In *Encyclopedia of Global Environmental Change* (ed. Douglas, P. I.). John Wiley & Sons, Chichester, pp. 536–542.

Smith, S. V. 1984. Phosphorus versus nitrogen limitation in the marine environment. *Limnology & Oceanography* 29:1149–1160.

Smith, A. N., A. M. Posner, and J. P. Quirk. 1974. Incongruent dissolution of surface complexes of hydroxyapatite. *Journal of Colloid Interface Science* 48:442–449.

Smith, A. N., A. M. Posner, and J. P. Quirk. 1977. A model describing the kinetics of dissolution of hydroxyapatite. *Journal of Colloid Interface Science* 62:475–494.

Souch, C. 1994. A methodology to interpret downvalley lake sediments as records of neoglacial activity: Coast Mountains, British Columbia, Canada. *Geografiska Annaler* 76A:169–186.

Tarafdar, J. C., and N. Claasen. 1988. Organic phosphorus compounds as a phosphorus source for higher plants through the activity of phosphatase produced by plant roots and microorganisms. *Biology and Fertility of Soils* 5:308–312.

Thirioux, L., P. Baillif, J. C. Touray, and J. P. Ildefouse. 1990. Surface reactions during fluorapatite dissolution-recrystallization in acid media (hydrochloric and citric acids). *Geochimica Cosmochimica Acta* 54:1969–1977.

Thompson, L. G., T. Yao, M. E. Davis, K. A. Henderson, E. Mosley-Thompson, P.-N. Lin, H.-A. Synal, J. Cole-Dai, and J. F. Bolzan. 1997. Tropical climate instability: The last glacial cycle from a Qinghai-Tibetan ice core. *Science* 276:1821–1825.

Tiessen, H., and J. O. Moir. 1993. Characterization of available P by sequential extraction. In *Soil Sampling and Methods of Analysis* (ed. Carter, M.). Lewis, Boca Raton, FL, pp. 75–86.

Tiessen, H., J. W. B. Stewart, and C. V. Cole. 1984. Pathways of phosphorus transformations in soils of differing pedogenesis. *Soil Science Society of America Journal* 48:853–858.

Tribble, J. S., R. S. Arvidson, M. Lane, and F. T. Mackenzie. 1995. Crystal chemistry, and thermodynamic and kinetic properties of calcite, dolomite, apatite, and biogenic silica: Applications to petrologic problems. *Sedimentary Geology* 95:11–37.

Tsandev, I., C. P. Slomp, and P. Van Cappellen. 2008. Glacial-interglacial variations in marine phosphorus cycling: Implications for ocean productivity. *Global Biogeochemical Cycle* 22:GB4004.

Tyrell, T. 1999. The relative influence of nitrogen and phosphorus on oceanic primary productivity. *Nature* 400:525–531.

Van Cappellen, P. V. 1991. The formation of marine apatite: A kinetic study. PhD dissertation, Yale University, New Haven, CT.

Van Cappellen, P. V., and R. A. Berner. 1988. A mathematical model for the early diagenesis of phosphorus and fluorine in marine sediments: Apatite precipitation. *American Journal of Science* 288:289–333.

Vitousek, P. M., O. A. Chadwick, T. E. Crews, J. H. Fownes, D. M. Hendricks, and D. Herbert. 1997. Soil and ecosystem development across the Hawaiian Islands. *GSA Today* 7:1–8.

Walker, T. W., and J. K. Syers. 1976. The fate of phosphorus during pedogenesis. *Geoderma* 15:1–19.

Wheat, C. G., R. A. Feely, and M. J. Mottl. 1996. Phosphate removal by oceanic hydrothermal processes: An update of the phosphate budget in the oceans. *Geochimica Cosmochimica Acta* 60:3593–3608.

Wilkins, G. R., P. A. Delcourt, H. R. Delcourt, F. W. Harrison, and M. R. Turner. 1991. Paleoecology of central Kentucky since the last glacial maximum. *Quaternary Research* 36:224–239.

# 2 Positive and Negative Effects of Phosphorus Fertilizer on U.S. Agriculture and the Environment

*B. A. Stewart, Pramod Pokhrel,*
*and Mahendra Bhandari*

## CONTENTS

## 2.1   INTRODUCTION

Nitrogen and phosphorus are the two essential plant nutrients most widely added to fertilizers to increase crop production in the United States and the world. Worldwide, approximately 120 million Mg of N and 20 million Mg of P (1 unit P = 2.3 units $P_2O_5$) are added in chemical fertilizers (Food and Agriculture Organization Corporate Statistical Database [FAOSTAT], 2016) annually. Of these amounts, the United States uses about 11.3 million Mg N and 1.7 million Mg P (United States Department of Agriculture-Economic Research Service [USDA-ERS], 2013). The past supplies and sources of these nutrients, however, are very different. Prior to the end of World War II in 1945, N for crop production was supplied largely from decomposing soil organic matter, leguminous crops, and farmyard manure. Following World War II, many of the chemical plants that had used the Haber–Bosch process to produce anhydrous ammonia that could then be converted to ammonium nitrate for making bombs were converted to N fertilizer plants. The Haber–Bosch process synthesizes ammonia from hydrogen and nitrogen. Fritz Haber received the Nobel Prize for Chemistry in 1918 for inventing this method, and Carl Bosch translated the method into a large-scale process using high pressures in 1931 and shared a Nobel Prize jointly with Friedrich Bergius. Smil (1999) stated that the Haber–Bosch process was the most important invention of the twentieth century and suggested that without it, the world's population could not have grown from 1.6 billion in 1900 to 6 billion in 1999. Similarly, the late Nobel Laureate Norman Borlaug, the father of the Green Revolution, once stated that the world without chemical fertilizer would support no more than one-sixth of the world population (Roberts and Ryan, 2015). Stewart et al. (2005) attributed over 50% of crop yields to chemical fertilizer use, and Roberts and Tasistro (2012) estimated 40–60% of the world's cereal production was due to fertilizers. Figure 2.1 shows that the world production of cereals closely paralleled the use of synthetically produced N fertilizer. The world population is now 7.3 billion and is expected to reach 9 billion by 2050.

The rapidly increasing use of synthetic N fertilizer for cereal production caused a dramatic increase in the use of P fertilizer. Historically, manure was the primary source of P fertilizer, although guano and human excreta were also sometimes important supplies (Figure 2.2). These supplies became inadequate when large amounts of synthetic N fertilizer were added because P quickly became the limiting factor for grain production when synthetic N fertilizer became widely available and used. While manures supplied about 90% of the fertilizer P in 1950, it supplied only about 10% in 2010 as phosphate rock became the dominant source and the total use increased about fivefold (Figure 2.2).

While the uses of N and P fertilizers are closely linked, the dominant sources are vastly different. Since the Haber–Bosch process takes N from the atmosphere, and the atmosphere contains 78% N, there will never be a shortage of N fertilizer as long as there is energy available. In contrast, the dominant source of P is rock phosphate that is a finite source found in mines.

The Global Phosphorus Research Initiative estimated that the world's readily available phosphorus supplies will be inadequate to meet the agricultural demand within 30–40 years (*American Scientist*, 2010). Amundson et al. (2015) stated the

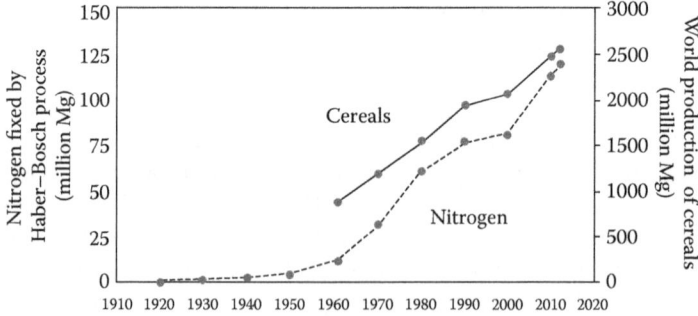

**FIGURE 2.1**  Nitrogen fixed by Haber–Bosch process (million Mg, left axis) and world production of cereals (million Mg, right axis). (From Smil, V., *Nature*, 36, 415, 1999; Stewart, B. A., X. Hou, and S. R. Yalla, *Principles of Sustainable Soil Management in Agroecosystems: Advances in Soil Science*, CRC Press, Boca Raton, Florida, 2013; FAOSTAT, Statistics Division, http://faostat3.fao.org/home/E, 2016. With permission.)

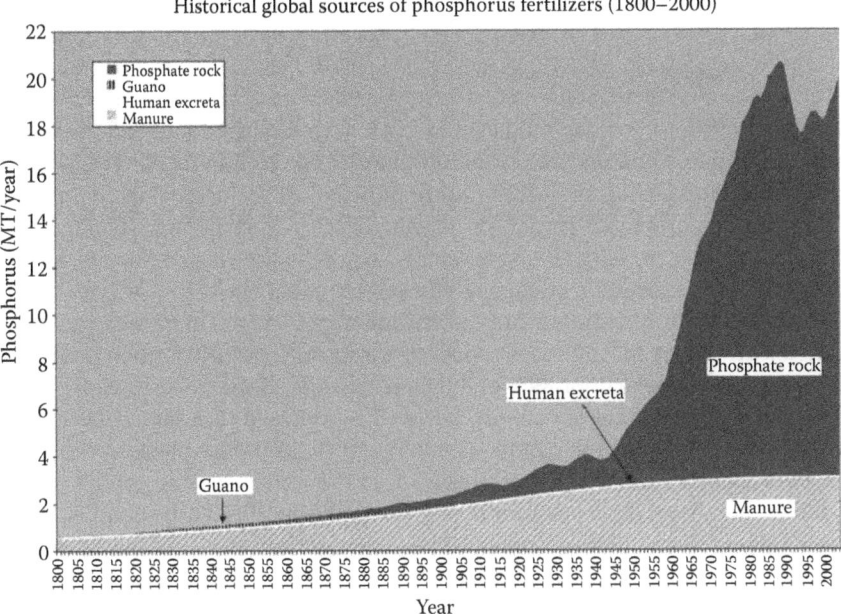

**FIGURE 2.2** Historical sources of phosphorus (million Mg year$^{-1}$) for use as fertilizers, including manure, human excreta, guano, and phosphate rock. (Reprinted from *Global Environmental Change*, 19, Cordell, D., J. Drangert, and S. White, The story of phosphorus: Global food security and food for thought, 292–305, Copyright (2009), with permission from Elsevier.)

United States has only about 2% of global P resources, and at current retrieval rates, the most productive mine will be depleted in 20 years, which will force the United States to become increasingly reliant on imports to sustain its agricultural and industrial sectors. Others, however, have estimated far greater supplies. For example, Scholz et al. (2013) stated that analyses by van Kauwenbergh (2010) and Jasinski (2009, 2012) suggest that the estimated world P reserves increased from 15 trillion Mg in 2008 to 71 trillion Mg in 2011. All seem to agree, however, that P fertilizer price increases recently seen on global markets will likely hit farmers hardest in the developing countries where essentially all the world population growth is occurring. Amundson et al. (2015) reported that the P fertilizer cost increased almost sixfold between 1961 and 2008, and while they have considerably fluctuated in recent years, the 2015 price is about 50% higher than in 2008. About 90% of the world's known reserves are controlled by five countries: Morocco, Jordan, South Africa, the United States, and China, with Morocco's share accounting for about 70%.

The production of cereal grains is of extreme importance and highly dependent on N and P fertilizers. Worldwatch Institute (2015), using the 2007 worldwide production of cereals, stated that on average humans get about 48% of their calories from grains, a share just slightly below the 50% average for the past four decades. Grain, particularly maize (*Zea mays*), in conjunction with soybeans (*Glycine max*), also forms the primary feedstock for industrial livestock production. In 2007, roughly

48% of the grain was directly consumed by people and about 35% by livestock, and about 17% was used to make ethanol and other fuels. More recent estimates (Nierenberg and Spoden, 2012) suggest that a smaller percentage of the grains produced are being directly consumed. They estimated that approximately 571 million Mg of the 2300 million Mg produced in 2011 were used for food, and the caloric intake from direct consumption of grain ranged from 23% in the United States to 60% in developing Asia, and 62% in North Africa.

Maize is the grain most rapidly increasing because in addition to its direct consumption, it is the primary component for meeting the growing worldwide demand for animal protein and for producing ethanol and other fuels. In 2013, the worldwide production of cereals totaled 2780 million Mg of which 716 million was wheat (*Triticum aestivum*), 741 million was rice (*Oryza sativa*), and 1018 million was maize. Together, they accounted for 89% of the total cereal production. Historically, wheat was produced in the largest quantity followed by rice and then maize. Nierenberg and Spoden (2012) stated that of the approximate 50,000 edible plants in the world, wheat, rice, and maize account for two-thirds of the world's food energy intake.

Figure 2.3 shows the relative increases from 1961 to 2013 for population, cereal production, synthetic N fertilizer, and P fertilizer derived from phosphate rock for China, India, United States, and the world. For the world, the population increased 2.3-fold; cereal production, 2.9-fold; P fertilizer use, 4.2-fold; and N fertilizer use, 10.3-fold. Assuming that the average grain N concentration was 1.6% (10% protein) and the P concentration was 0.3%, the amounts of N and P contained in the 2542 million Mg of grain produced in 2013 were 40.7 million Mg and 7.6 million Mg, respectively. Smil (2000) stated that nearly two-thirds of the supplied nutrients were applied to cereals. In 2013, the data shown in Figure 2.3 indicate that there were about 119 million Mg synthetic N used in 2013 which is 2.9 times the amount of N contained in cereal grains. In comparison, there were 20.2 million Mg of fertilizer P used that was 2.7 times the amount of P contained in the cereal grains. While these worldwide trends clearly show the relationship between fertilizer use and cereal production, the relationship between N fertilizer use and P fertilizer use is less clear. Cereal crops generally contain about one-fifth as much P as N. Although highly variable, plants utilize about 50% of the N applied as fertilizer. Phosphate fertilizers are usually less available than N fertilizers, and the uptake of added P during the year of application is often only about 20%. Therefore, even though plants contain only about one-fifth as much P as N, the ratio of fertilizer P added in relation to fertilizer N added was much lower. For the world, the N:P fertilizer ratio was 2.4:1 in 1961 and increased each decade to 5.8:1 in 2010 (Table 2.1). This suggests that perhaps P fertilizer was excessively applied in the early years of fertilization, but it also suggests that P fertilizer materials as well as methods of application have been improved to greatly increase P fertilizer use efficiency. The data for China, India, and the United States vary considerably from the world data and for different reasons (Figure 2.3). The widespread use of chemical fertilizers began much earlier in the United States, grew rapidly during the late 1950s and 1960s, but slowed sharply during the 1970s because of increasing costs and growing environmental concerns. The U.S. Environmental Protection Agency (EPA) was created in 1970, and there was increasing evidence that excessive fertilizer use was increasing the nitrate concentrations in groundwater

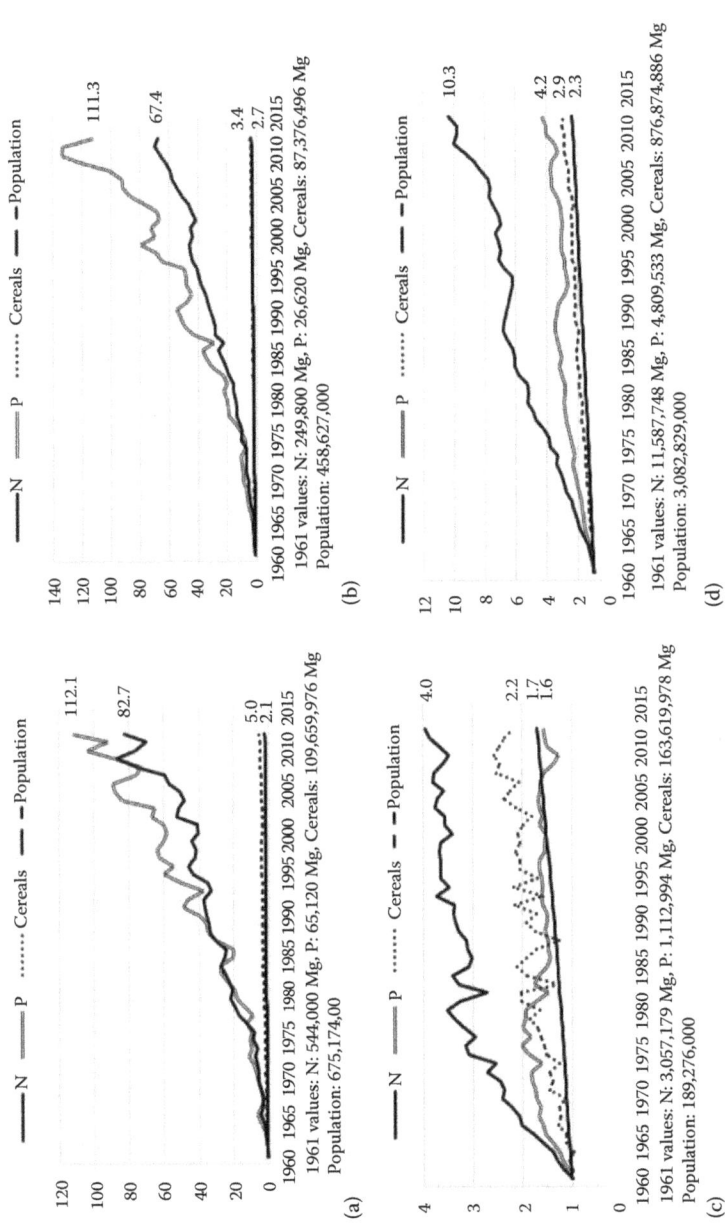

**FIGURE 2.3** Relative increases from 1961 to 2012 for population, cereal production, synthetic N fertilizer, and P fertilizer derived from phosphate rock for (a) China, (b) India, (c) the United States, and (d) the world (1961 values are shown at the bottom of each graph). (From FAOSTAT, Statistics Division, http://faostat3.fao.org/home/E, 2016. With permission.)

TABLE 2.1

Changes in Ratio of Amounts of N Fertilizer Used to P Fertilizer Used in the World and Selected Countries from 1961 to 2010 (Calculated Using FAOSTAT Data)

| | 1961 | 1970 | 1980 | 1990 | 2000 | 2010 |
|---|---|---|---|---|---|---|
| | Ratio of N Fertilizer Megagram Used to P Fertilizer Megagram Used | | | | | |
| United States | 2.7:1 | 3.9:1 | 4.9:1 | 6.0:1 | 6.2:1 | 6.8:1 |
| China | 8.4:1 | 8.4:1 | 10.1:1 | 7.6:1 | 5.8:1 | 6.1:1 |
| India | 9.4:1 | 6.2:1 | 6.8:1 | 5.5:1 | 5.9:1 | 4.6:1 |
| World | 2.4:1 | 3.4:1 | 4.3:1 | 4.9:1 | 5.7:1 | 5.8:1 |

and accelerating the eutrophication of surface waters. Phosphorus fertilizer use in the United States actually decreased by about 25% between 1970 and 2012 (Figure 2.3) In contrast, China used very little commercial fertilizer in 1961, and this was also the time of the Great Famine when cereal production was only 162 kg capita$^{-1}$. From 1961, even though the population increased 2.1 times, cereal production increased 5.0 times resulting in 386 kg capita$^{-1}$. However, to achieve this remarkable increase, fertilizer N use increased 82.7-fold and fertilizer P, 112.1-fold. India, on the other hand, has a high percentage of population who are vegetarians, and grain demand is considerably less. In 1961, cereal production was 191 kg capita$^{-1}$ and increased to 240 kg capita$^{-1}$ by 2012, even though the population increased 2.7 times. N fertilizer use increased 67.4 times, and P fertilizer use increased 111.3 times during the same time period. However, the cereal production in 2012 in India was only 297 million Mg compared to 548 million Mg in China with somewhat comparable populations. In sharp contrast to the United States where the annual growth rate of N and P fertilizer usage began to decline after about 30 years of usage, there is little evidence that the rates are significantly declining in China, India, or the world (Figure 2.3). Granted, the total cereal production increased only 2.2-fold from 1961 to 2012 for the United States compared to 2.9 for the world, 3.4 for India, and 5.0 for China.

It is also important to note in Figure 2.3 that while the world population increased 2.3 times between 1961 and 2012, the production of cereals increased 2.9 times. Thus, the per capita supply of cereals increased from 284 to 359 kg capita$^{-1}$ that not only reduced hunger and malnutrition for millions of people but also allowed millions of others to improve their diets with increasing amounts of animal protein. In all cases shown in Figure 2.3, P fertilizer usage closely followed with N usage, although at somewhat different ratios (Table 2.1). The United States in 1961 used only 2.7 kg N for every 1 kg P fertilizer compared to about 8 or 9 to 1 for China and India. This was likely due to the availability and the cost differences, but it suggests that P may have been excessively applied in the United States. In 2012, the ratios of N to P fertilizer usage for the world, the United States, China, and India were all close to 6:1, which seems reasonable based on plant requirements. Clearly, N and P fertilizer use has been a dominant factor in increasing cereal production, and China and India, the two most populous nations, were able to produce grain at a rate faster than the population growth rate. In 1961, these countries used less than 8% of the N fertilizer and less than 2% of the P. In 2012, they used 52% of the N fertilizer and 50% of the P fertilizer (Table 2.2). China alone used 38% of the N and 36% of the P fertilizers.

Phosphorus is an essential element of life and for growing crops. Smil (2000) states that P is absent in the N-rich amino acids that make up proteins of all living organisms, but neither proteins nor carbohydrate polymers can be made without phosphorus. He further stresses the importance of P in the formation of long chains of DNA and RNA, the nucleic acids that store and replicate all genetic information. The biological P cycle is driven by the essential role P plays in the energy transport in biological systems in the form of adenosine triphosphate and adenosine diphosphate (Smeck, 1985). Phosphorus is present in almost any food, and both a deficiency and an excess of phosphate in food may cause health issues (Elser, 2014; Scholz et al., 2014). Although P deficiency in humans is rare when they have sufficient food, there are 850 million people in the world that suffer from malnutrition (World Hunger Education Service, 2015). They state that the most relevant type of malnutrition is protein-energy malnutrition that leads to growth failure. It is basically a lack of calories and protein. Food is converted into energy by humans, and the energy contained in food is measured in calories. Protein is necessary for key body functions including the provision of essential amino acids and the development and maintenance of muscles. Protein-energy malnutrition is the more lethal form of malnutrition/hunger and is the type of malnutrition that is referred to when world hunger is discussed.

Although N is generally the first essential plant nutrient that limits crop yields, P deficiencies usually occur fairly soon after adequate N is supplied. Because N fertilizer can be synthetically produced by the Haber–Bosch process, the supply is limited only by price. In contrast, there is a finite supply of phosphorus resources needed for making fertilizer. Although the amount of this resource is not fully understood, its availability and ease of recovery is becoming more costly. The largest phosphate

---

## TABLE 2.2
### Percentages of World Population, Cereal Production, and Amounts of N and P Fertilizer Used in 1961 Compared to 2012 for the United States, China, and India (Calculated from FAOSTAT Data)

| | Population | N Fertilizer Use | P Fertilizer Use | Cereal Production |
|---|---|---|---|---|
| | | Percentage of World Values | | |
| United States | | | | |
| 1961 | 6.1 | 26.4 | 23.0 | 18.7 |
| 2012 | 4.5 | 10.0 | 8.8 | 14.0 |
| China | | | | |
| 1961 | 21.9 | 4.7 | 1.4 | 12.5 |
| 2012 | 20.2 | 38.0 | 36.0 | 21.5 |
| India | | | | |
| 1961 | 14.9 | 3.2 | 0.5 | 10.0 |
| 2012 | 17.4 | 14.0 | 14.0 | 11.6 |

mines in the United States are in Florida, and these reserves are dwindling fast. They are expected to run out within 25 years and the United States is already importing about 10% of its P fertilizer from Morocco (Philpott, 2013). As already described, most crops contain about 20% as much P as N, and there is no substitute for P. The common belief by most agronomists is that food and fiber needs for a growing and more prosperous world population can only be met by the use of synthetically produced N fertilizer, and this will depend on a corresponding supply of P. The largest uses of N and P fertilizers are for producing cereal grains, and the cereal production will need to increase at a faster rate than the population based on current trends. The Food and Agriculture Organization of the United Nations (2009) projected that overall food production will need to increase by 70% by 2050 to feed the expected 9.1 billion people. Food production, particularly cereals, needs to increase much faster than the population because improving economic conditions increase the demand for more animal protein that requires more grain. A paradox exists because while it appears future food supplies can only be met by the use of large amounts of N and P fertilizers, their use is resulting in serious environmental concerns. Higher than desirable levels of nitrate-N are occurring in some groundwater supplies and raising concerns. Soluble phosphorus compounds are increasingly becoming moved from agricultural fields to rivers, lakes, and other waters and causing accelerated eutrophication. Therefore, while a dwindling supply of P is considered by many as becoming a serious problem, an overabundance of P in water supplies is being looked on by others as an environmental crisis. The objective of this chapter is to briefly look at some of the P fluxes in U.S. agriculture with particular emphasis on the amount of P fertilizer, primarily manufactured from phosphate rock extracted from mines, that is applied each year to agricultural fields and where much of the P ultimately resides.

## 2.2 PHOSPHORUS INPUTS AND OUTPUTS IN THE UNITED STATES

### 2.2.1 SEWAGE

The primary role of agriculture is to provide food for people, and almost every food that people eat contains P. Phosphorus is essential for people just as it is for plants, but most of the P consumed by humans and animals is excreted in urine and feces. Of course, some P is retained in the bones and the flesh with growth, but this is a relatively small amount of the total consumed. Smil (2000) estimated that 98% of the P ingested by humans is excreted. He further estimated that the worldwide average ranged between 1.2 and 1.4 g P day$^{-1}$ capita$^{-1}$. Because the U.S. population consumes significantly more calories than most countries, we are using 1.5 g P day$^{-1}$ capita$^{-1}$ as an estimate for the United States. With a population of 320 million people, and each person excreting 1.5 g P day$^{-1}$ (0.55 kg year$^{-1}$), approximately 176,000 Mg of P was excreted meaning that about 180,000 Mg of P was contained in the food consumed. Therefore, the primary aim of U.S. agriculture is to produce enough food to supply 180,000 Mg of P in the food consumed. Of course, not all food produced is consumed so substantially more P is required. However, most of the excreted P ends up as sewage. In addition to the P excreted by humans, sewage contains P from industrial sources, detergents, dishwashing compounds, and other materials. Smil (2000) stated that it is unlikely that per capita industrial P discharges in affluent countries will fall below 2 g P day$^{-1}$. Based on this value, the U.S. yearly sewage contains 240,000 Mg of P in addition to the 180,000 Mg excreted from humans for a total of 420,000 Mg P. As will be discussed later, this is almost 25% as much P as added each year in the United States as fertilizer.

Historically, thousands of U.S. cities discharged treated wastewaters into rivers, lakes, and bays. These water bodies often became highly polluted and with increasing environmental concerns, regulations have been created that prevent such practices today. The Federal Water Pollution Control Act Amendments of 1972 placed restrictions on the discharge of wastewater to waterways and encouraged other disposal methods such as land application (Lu et al., 2012). Earlier, Logan and Chaney (1983) also reported that the public concerns associated with some municipal by-products were legislated by the Water Quality Act of 1972. The act mandated the development of technologies to treat, dispose, and recycle nutrients in wastewaters and biosolids in an environmentally sound manner. Even though many studies have shown that the land application of sewage biosolids can be safe and beneficial, many resist its use because of the presence of heavy metals and restrictions. For example, certified organic foods cannot be produced on land where sewage has been applied within the past 3 years.

Hue (1995) stated that 25% of U.S. sludge was land applied, 25% landfilled, and 14% incinerated. However, landfilling and incineration methods were considered disposal methods and were being faced with more stringent regulations. A 1990 survey on sludge management in the United States showed that New Jersey had completely banned the landfilling of sludge except in an emergency; North Carolina no longer allowed disposing of sludge on active landfills, and Missouri, Ohio, Oregon, and Washington had <10% sludge land-filled (Hue, 1995). Even when sludge was applied to land, the emphasis was often more on the disposal than on the efficient recycling of nutrients. Hue (1995) reported that sludge applications to land ranged from 2–70 Mg ha$^{-1}$, with 15 Mg ha$^{-1}$ year$^{-1}$ being typical. These application rates often resulted in excess amounts of P because a review of literature showed that the dry weight percentages of N and P in sewage sludge were in the order of 3.5 and 2%, respectively, and that sludge P was mainly present in inorganic (mostly bioavailable) form (Hue, 1995). Most plants require only about 15% as much P as N for optimum growth.

Basta (2000) reported that the land application of the total biosolids produced in the United States increased from 20% in 1972 (600,000 Mg dry) to 54% in 1995 (3,100,000 Mg dry). He further reported that typical N and P concentrations were 4.3 and 2.3%, similar to, but somewhat higher than, those reported earlier by Hue (1995) and pointed out that most research studies on sewage sludge focused on the benefits from plant nutrients or on the associated environmental impacts to soil, water, and crop qualities. The evidence is clear that the land application of wastewater, biosolids, and municipal solid wastes can benefit crop production and improve soil quality in a sustainable manner, but the applications require careful scrutiny and management. Lu et al. (2012) stated that the most recent national biosolids survey conducted in 2004 indicated that about 6,000,000 Mg dry biosolids were produced in the United

States, and approximately 60% was applied to soil. Even though this sounds positive, the U.S. EPA estimates that biosolids are applied only to about 0.1% of the available agricultural land in the United States on an annual basis (Committee on Toxicants and Pathogens in Biosolids and National Research Council, 2002). Therefore, only a small percentage of P in food consumed by people and later excreted as urine and feces is efficiently recycled. Much of it continues to be incinerated or placed in landfills (Lu et al., 2012).

## 2.2.2 FERTILIZER

In 2012, the phosphorus fertilizer consumption in the United States was 1,750,000 Mg P. While this is 40% more than used in 1961, it is 20% less than used during the 1975-to-1980 period. As already discussed, from 1961 to 2012, the U.S. population increased 1.7-fold; the cereal production, about 2.2-fold; N fertilizer, use fourfold; and P fertilizer use, 1.6-fold (Figure 2.3). The data in Figure 2.3 for the United States clearly suggest that N and P fertilizers were applied in excess of plant needs during the 1960s and the early 1970s. During this period, the fertilizer costs were relatively low and the detrimental environmental effects of using excess fertilizer had not been clearly documented. As these effects became better understood, and with the creation of the EPA in 1970 coupled with the formation of the Organization of Petroleum Exporting Countries in 1973 that significantly increased the energy prices making N and P fertilizers much more costly, there was a concerted effort to increase fertilizer use efficiency. Between 1980 and 2012, U.S. cereal production increased to about 40%, while N use increased to only about 15%, and P fertilizer use decreased by 20% (Figure 2.3). These trends will likely change in the future, however, because the applied ratio of N to P was 6.9 in 2012 compared to 2.7 in 1961. This suggests that excessive P was applied in the early years resulting in a buildup of P in the soil that is being used later. As stated earlier, most crops contain about 20% as much P as N, so if N fertilizer use continues at the same rate or increases, P application rates will likely tend to increase because it is unlikely that the P uptake efficiency will become greater than the N uptake efficiency.

Maize is the U.S. crop most widely fertilized with P. Using 2010 data (USDA-ERS, 2013) as an example, 78% percent of maize cropland was fertilized at an average rate of 67 kg P ha$^{-1}$ for a total usage of about 780,000 Mg. Although only 23% of soybean cropland received P fertilizer at an average rate of 52 kg ha$^{-1}$, there was approximately 155,000 Mg P used. Therefore, almost 55% of all P fertilizer applied to cropland in 2010 was on land growing maize and soybeans. Another 10% was applied to wheat where 62% of the land was fertilized at an average rate of 15 kg ha$^{-1}$. Maize, soybean, and wheat cropland received 1,110,000 Mg of the total 1,750,000 Mg P from commercial fertilizer that was either produced from phosphate rock extracted from mines in the United States or imported from other countries, primarily Morocco. In addition, some, if not most, of the estimated 240,000 Mg of P in sewage each year from industrial uses is extracted from mines as well. Therefore, between 1,750,000 and 2,000,000 Mg P is added to the environment in the United States each year.

Assuming an average P content in the harvested maize, soybean, and wheat of 0.30, 0.58, and 0.36%, respectively, and harvested amounts of 318 million Mg for maize, 86 million Mg for soybean, and 55 million Mg for wheat, the amounts of P removed from the fields were approximately 954,000 Mg with maize, 498,000 Mg with soybean, and 196,000 Mg with wheat for a total of 1,648,000 Mg. Since 1,110,000 Mg P were added to these three crops as fertilizer, there was 67% as much P added as there was removed with the harvested grain and soybeans. Of course, much of the P added as fertilizer is not taken up by the crop in the year it is applied, and, depending on the soil characteristics, some of it will become chemically or physically fixed in the soil so that it is never available for plant uptake. Nevertheless, a significant amount of P contained in the harvested maize, soybean, and wheat comes from applied fertilizer P.

Although variable from year to year, approximately 20% of the maize, 35% of the soybean, and 50% of the wheat produced in the United States are exported. Therefore, about 460,000 Mg of P is exported from the U.S. environment annually from these three crops, which is about 25% as much P as is introduced into the landscape each year from fertilizer P.

### 2.2.3 ANIMAL WASTE

Historically, essentially all livestock and poultry were produced at the farm level, and the wastes were recycled on the cropland. The P exported from the farm was mostly limited to grain and animal products that were marketed. Legumes were often grown to supply additional N, but P demands were mostly met by soil P and manure. However, when synthetic N and P from phosphate rock became available at economic prices, soil fertility was no longer dependent on manure and legumes. This resulted in more available land for growing crops, particularly maize and soybeans, and the yields also increased from the applied N and P fertilizers. These changes soon resulted in large supplies of grain and soybeans, so concentrated livestock and poultry operations became common. There are single facilities for feeding as many as 100,000 beef animals, milking 10,000 dairy cows, and growing thousands of hogs and poultry animals. Enormous amounts of manure are produced, but the manure is generally not recycled to the land as it was in historical times.

Grace Communications Foundation (2015) stated that 60% of the maize and 47% of the soybean produced in the United States are consumed by livestock. These products are the bases for animal feed because they are protein rich, thus causing animals to gain weight fast and making it more profitable than providing other feedstuffs that do not produce as rapid a weight gain. Poultry and swine raised in concentrated feeding operations are there from birth to harvest, whereas beef animals are generally placed into concentrated feeding facilities only for a fattening period generally ranging between 120 and 200 days. Unfortunately from the standpoint of recycling nutrients, concentrated animal feeding operations are increasingly being located greater distances from where the feedstuffs, particularly maize and soybean, are produced. Figure 2.4 shows where maize and soybean are produced in the United States, and where the various classes of livestock are grown. The cattle data in Figure 2.4 are only for the cattle sold, which are largely those that were fattened for harvest and those sold at the farm and ranch levels to move into the feeding facilities. Breeding cows and calves on ranches and farms are not included. The data for milk cows include

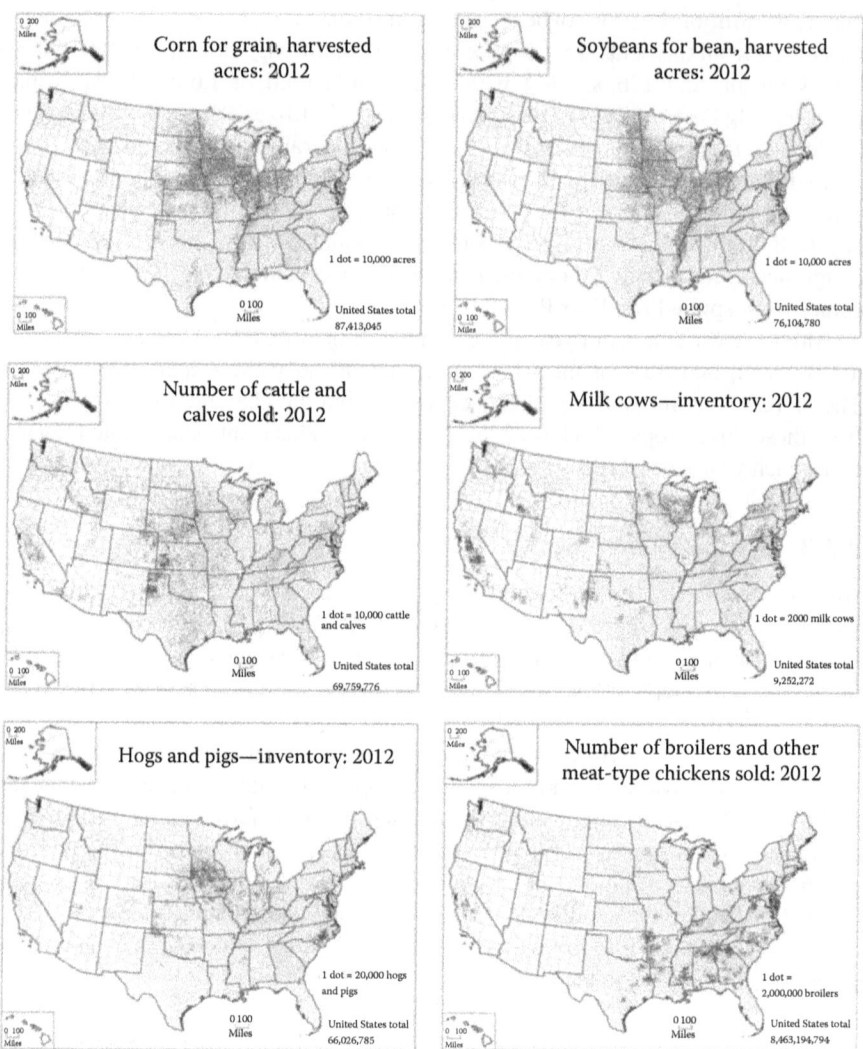

**FIGURE 2.4** Major areas in the United States where maize and soybeans are grown and animals are reared. (From USDA, Census of Agriculture, U.S. Department of Agriculture, Washington, DC, 2012. With permission.)

only those being milked, so replacement heifers and others are not included. This information is important because it affects how much of the manure produced from these various classes of livestock can potentially be recovered. The US Department of Agriculture-Natural Resources Conservation Service (USDA-NRCS) (1995) estimated that 95% of the manure from poultry broilers and layers and 75% from turkeys could be recovered. For dairy cows, the estimate was 80%, 75% for hogs and pigs, and 80% for beef animals in feedlots. For grazing animals, less than 10% is considered recoverable, but the manure from grazing animals is largely recycled naturally.

Earlier, we estimated that the P content of maize grain produced in the United States was 954,000 Mg, and that the P content of soybeans was 498,000 Mg. Using the estimates that 60% of the maize and 47% of the soybeans are fed to animals (Grace Communications Foundation, 2015), this would result in animals consuming 806,000 Mg of P from these two crops. Most of this P is excreted in urine and feces. Withers and Sharpley (1995) reported that 70–80% of the P fed to dairy cows, sheep, and feeder pigs is excreted, and 87% of the P fed to poultry is excreted. Similarly, Kissinger et al. (2006) reported that 83% of P fed to beef cattle is excreted. Assuming an 80% value, about 645,000 Mg of P from maize and soybeans mostly fed in concentrated animal feeding operations would be excreted as urine and feces. It was estimated earlier that 935,000 Mg of the total 1,750,000 Mg of fertilizer P used in 2012 was applied for production of maize and soybeans. Therefore, the amount of P excreted in manure from just maize and soybeans fed in concentrated animal feeding operations was about 70% of the amount applied to land producing those crops and about 37% of the total fertilizer P used.

A more complete estimate of the amount of P excreted by livestock was calculated using methods and conversion factors developed by USDA-NRCS (1995) along with numbers of animal units for various classes of cattle, swine, and poultry animals from the U.S. Department of Agriculture (USDA) Census of Agriculture (2012). Animal units were used as a method of aggregating across animal types and life stages. The USDA Census animal unit numbers were gathered for cattle, swine, chickens (broilers and layers), and turkeys. For cattle, three categories were used: beef cows, milk cows, and others (beef animals being fattened in feedlots for slaughter, and young females and males for replacements of breeding stock and feedlots). Swine animal units were separated into market hogs and breeder hogs. Poultry animal units were summarized as broiler chickens, layer chickens, turkeys. Using the American Society of Agricultural Engineers (ASAE) (2003) standard values for fresh manure produced per animal unit for each class of animals and the P content of the fresh manure for each animal class, the amounts of fresh manure and P contents were estimated for the United States. These calculated estimates are shown in Table 2.3. Cattle produced about 738.5 million Mg of fresh manure and contained an estimated 1,092,650 Mg P. Swine produced 120 million Mg fresh manure containing 252,000 Mg P, and poultry estimates were 72.9 million Mg fresh manure that contained 289,050 Mg P. The total estimated P excreted from livestock in 2012 was 1,633,700 Mg (Table 2.3), which is almost equal to the 1,700,000 Mg of fertilizer P applied (USDA-ERS, 2013).

Data for animal wastes on a worldwide basis are not as available as data for the United States, but Sheffield (2000) estimated that the world's farm livestock and poultry includes about 1 billion cattle, 0.8 billion pigs, 0.9 billion sheep and goats and, on average, 8 billion chickens (broilers and layers). He stated a rough estimate was 1.7 billion Mg of dry manure annually. Similarly, Oliver (2008) reported that according to the United Nations' Food and Agriculture Organization, there were 1.3 billion cattle worldwide (one for every 5 people), slightly more than 1 billion sheep, around 1 billion pigs, 800 million goats, and 17 billion chickens. In total, about 13 billion Mg of manure was produced annually. Assuming the manure averaged about 85% water, about 2 billion Mg of dry manure remains, which is close to

## TABLE 2.3
## Estimated Fresh Manure Produced in 2012 and Amounts of P Contained from Various Classes of Animals in the United States

| Animal Type | Animals (AU$^{-1}$)[a] | AUs[b] (M) | Fresh Manure[c] (million Mg) | P[c] (%) | P (Mg) |
|---|---|---|---|---|---|
| Beef cows | 1.00 | 28.90 | 277.4 | 0.16 | 443,840 |
| Dairy cows | 0.74 | 12.50 | 177.9 | 0.11 | 195,690 |
| Other cattle | 1.69 | 29.50 | 283.2 | 0.1 | 453,120 |
| Breeder hogs | 2.67 | 2.09 | 29.1 | 0.21 | 61,110 |
| Market hogs | 9.09 | 6.54 | 90.9 | 0.21 | 190,890 |
| Chicken layers | 250 | 1.37 | 14.6 | 0.47 | 68,620 |
| Chicken broilers | 455 | 3.31 | 46.6 | 0.35 | 163,100 |
| Turkeys | 67 | 1.50 | 11.7 | 0.49 | 57,330 |
| Total | | 85.71 | 931.4 | | 1,633,700 |

[a]  An animal unit (AU) is defined as an animal equivalent of 455 kg live weight and equates to one beef animal (U.S. EPA, 2013).
[b]  United States Department of Agriculture (2012).
[c]  ASAE (2003).

the 1.7 billion Mg that Sheffield estimated several years earlier. Assuming 2 billion Mg dry manure and further assuming that the average P in dry manure is 1%, then about 20 million Mg of P would be annually excreted in animal waste worldwide. This amount is similar, although somewhat less, than estimated by MacDonald et al. (2011). They estimated that 9.6 million Mg of manure P were annually applied to cropland worldwide, and that this was only 40% of the total P excreted by livestock, which would indicate approximately 24 million Mg of P was excreted by livestock.

It was estimated earlier that about 7.6 million Mg of P would be removed with the 2542 million Mg of cereals produced worldwide, and that 20 million Mg of chemical fertilizer P was used and with the 8–10 million Mg manure P applied to cropland, it is clear that P applied to cropland far exceeds crop removal. Even so, P deficits (amount of P removed less amount added as fertilizer or manure) covered almost 30% of the world cropland in year 2000 (MacDonald et al., 2011; Scholz et al., 2014). Of course, much of the 20 million Mg of P excreted in animal waste cannot be recycled, but animal waste is a major source of P for cropland, and it is the only source that many small-holder farmers in developing countries have because P fertilizers are not available or too expensive. Worldwide, MacDonald et al. (2011) estimated that 40% of the manure produced is applied to cropland. In the United States, manure is annually applied to only about 5% of the cropland, and more than half is applied to maize land (MacDonald et al., 2009).

Historically, much of the animal waste in the United States was recycled on the land that produced the feed that was given to the animals that generated the manure. Unfortunately, this is no longer the case because much of the maize and the soybeans are moved from areas where they are grown to areas where large concentrated feeding operations are located (Figure 2.4). As previously stated, 55% of P fertilizer is applied to maize and soybeans, but these products are moved in huge quantities to other areas to feed livestock. Cattle feedlots are concentrated in the central and the southern Great Plains, while poultry production, particularly broilers, is largely in the southeastern United States (Figure 2.4). This results in large amounts of livestock manure being produced in areas where there is limited need for enhancing soil fertility. Therefore, the manure is often not applied to cropland where the P can be efficiently recycled. In some cases, manure is stockpiled and, in other cases, applied to land other than cropland simply for disposal. Based on 2012 U.S. Agriculture Census data and ASAE (2003) manure standards, approximately 51% of the P in poultry manure was produced in Arkansas, Mississippi, North Carolina, South Carolina, Alabama, and Georgia. However, much of the feed fed to animals in these states is grown in the Corn Belt states and then transported to these states. While most of the meat and the eggs produced from these large operations is shipped out of the area, most of the P remains where the animals are fed because they excrete about 80–90% of the P that they ingest (Withers and Sharpley, 1995). Therefore, a large portion of the P extracted each year to produce P fertilizer is eventually moved in feed to areas where it ends up in manure that is not needed in the quantities produced. Then, the process is repeated because the P removed with the maize and the soybeans from the land where they were produced must be replaced to maintain production. The overall effect is that huge amounts of P are being added each year to the environment rather than recycling more of the P already present in the environment. Even in areas where large amounts of manure are produced, most producers use chemical fertilizer because it is more convenient and more easily applied in the proper amount and place. In many instances, it is also more economical because of the high cost of transporting and spreading manure. Of the 158 million ha of cropland in the United States, more than 100 million ha are annually treated with chemical fertilizer, lime, or soil amendments compared to less than 9 million ha that have manure applied (Figure 2.5). Therefore, while there is almost as much P produced each year in manure as applied as chemical fertilizer, only a small percentage is effectively recycled. Based on the studies by Gilbertson et al. (1979) and the USDA-NRCS (1995), it is reasonable to assume that about 80% of poultry, dairy cows, beef animals in feedlots, and swine manure is harvestable. Therefore, of the more than 1.6 million Mg P annually produced in livestock, it is likely that at least 50% of this could potentially be recycled.

Data are not available regarding the amount of manure applied to the nearly 9 million ha of cropland (Figure 2.5) or to what crops. However, a major concern with managing P in manure is that many producers base manure rates to supply adequate N for crop production, and the N:P ratios in manure do not generally match the N:P needs of the crop. For example, the P contents in a stockpiled feedlot manure or a poultry litter are about one-half of the N contents (Zhang et al., 2012). The crop requirements for N are about 5–6 times those of P. When producers add manure to supply adequate N for crop production, P is added in excess, and it can lead to environmental problems. Therefore, even when manure is applied to cropland, it may not be efficiently recycled.

The consequences of the additional P in the environment are not fully understood. Perhaps even more important, there are no easy solutions. The importance of P in accelerating water eutrophication also became widely emphasized during

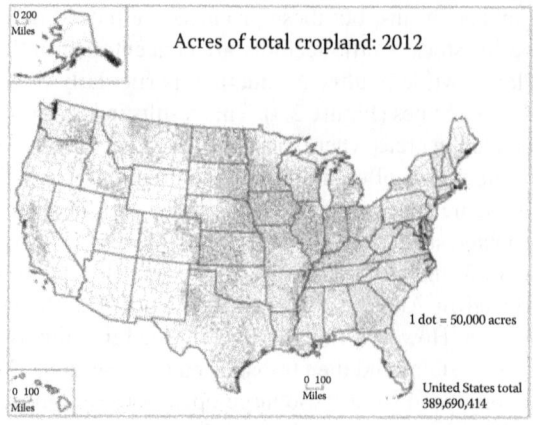

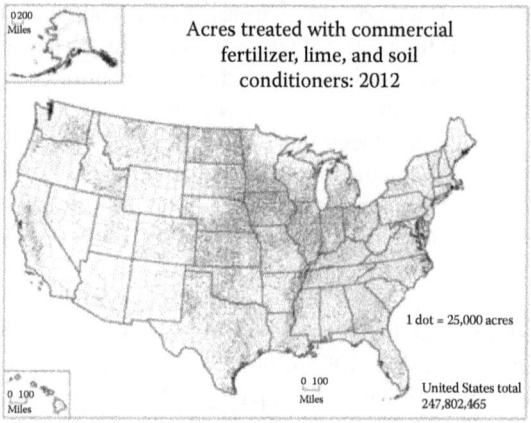

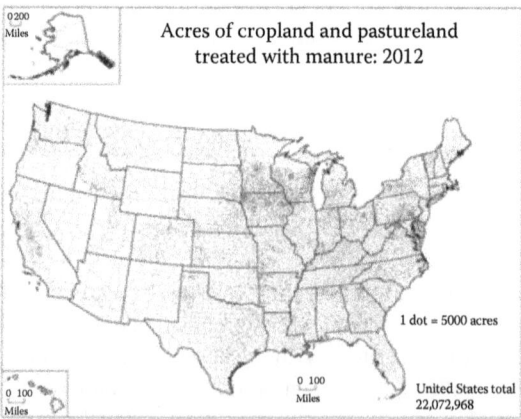

**FIGURE 2.5** Areas of U.S. cropland (157,769,000 ha), cropland treated with commercial fertilizer, lime, and soil conditioners (100,325,000 ha), and cropland treated with manure (8,936,400 ha) in 2012. (From USDA, Census of Agriculture, U.S. Department of Agriculture, Washington, DC, 2012. With permission.)

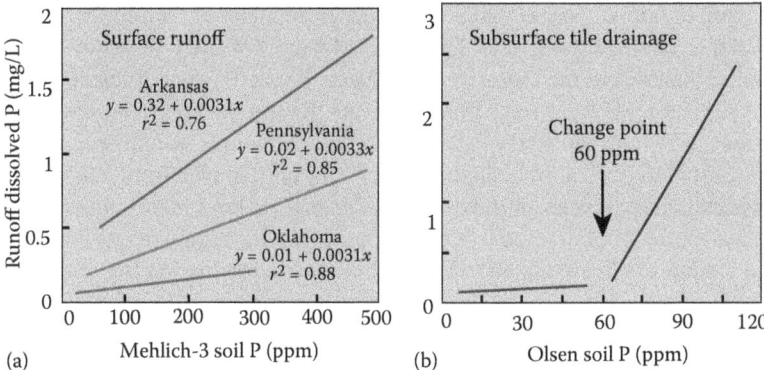

**FIGURE 2.6** Effect of soil P on the dissolved P concentration of (a) surface runoff from pasture watersheds and (b) subsurface tile drainage from Broadbalk fields. (From Sharpley, A. N., T. Daniel, T. Sims, J. Lemunyon, R. Stevens, and R. Perry, *Agricultural Phosphorus and Eutrophication*, USDA Agricultural Research Service ARS-149, Washington, DC, 2003. With permission.)

the growing awareness of environmental problems that led to the establishment of the U.S. EPA in 1970. In recent years, P from fertilizers and animal wastes has been associated with water quality problems in the Gulf of Mexico and Chesapeake Bay. It has long been recognized that P is adsorbed to soil particles, and when soil is eroded by wind and water, P moves with the soil. However, the common view was that the P movement through the soil profile and eventually into streams, rivers, and lakes was not a serious problem. Recent studies and viewpoints, however, suggest that these problems may be a serious concern because even though the concentration of P in water moving through soil may be low, the impact may be great. While the recent studies are documenting the importance and the environmental consequences of the P movement through the soil profile, as a grad student, Andrew Sharpley documented P moving through tile drains in New Zealand more than 30 years ago (Fisher, 2015). Studies have shown that additions of P to soils increase the amounts of P measured by methods that estimate plant-available P. Sharpley et al. (2003) showed that the concentrations of dissolved P in runoff water or drainage water linearly increased with increasing estimates of plant-available P (Figure 2.6). Therefore, the accumulations of P in soil are considered as environmental risks particularly in areas where runoff water or drainage water can enter water bodies that can be damaged by accelerated eutrophication. Notable examples where agricultural practices are considered as having negative effects on water quality include the Gulf of Mexico, Chesapeake Bay, and Lake Erie.

## 2.3 CONCLUSION

The use of P in the United States has both positive and negative effects. Crop yields have dramatically increased by the addition of P fertilizer made from phosphate rock extracted from mines. Much of the P fertilizer is used to produce maize and soybeans that are extensively used to produce food, particularly animal-based products. The

P contained in human wastes and animal wastes annually exceed the 1,750,000 Mg of P added as fertilizer each year. Unfortunately, most of the P in wastes ends up in areas other than where the crops were produced, so the P is not efficiently recycled. Historically, animals were raised on the farms that produced the animal feed, and the manure was returned to the land. Today, large amounts of maize and soybeans are transported to large animal feeding operations far removed from where the crops are produced, and even when the manure is spread on land, the P is not efficiently used. Then, more P is extracted from the mines to produce more maize and soybeans so huge amounts of P are added to the environment each succeeding year. This has resulted in unintended consequences to the environment, particularly accelerated eutrophication of surface waters. Therefore, a paradox has resulted in that while there is concern that rock phosphate reserves in the United States are quite limited, there is even more concern that P concentrations in surface waters are becoming unmanageable. This is a complex problem that was not caused by any single group and cannot be solved by any single group. The fact is that crop production areas and animal production areas have become more and more separated as efficiencies have become greater. As this developed, the negative effects on the environment were not fully understood, and even today, the long-term effects are not well known and, perhaps more importantly, not accepted by all the groups. In the meantime, additional studies are underway to develop practices and strategies for addressing the problem. The problem can and will only be addressed by society as a whole if and when environmental problems are perceived as unacceptable. Voluntary, policy, or regulatory actions will then be implemented that can perhaps slowly focus on increasing the recycling of P rather than just adding more P to the environment each year from mines.

## REFERENCES

American Scientist. 2010. Does peak phosphorus loom? *American Scientist* 98 (4): 91.

American Society of Agricultural Engineers (ASAE). 2003. D384.1: Manure production and characteristics. ASAE, St. Joseph, MI.

Amundson, R., A. S. Berhe, J. W. Hopmans, C. Olson, A. E. Sztein, and D. L. Sparks. 2015. Soil and human security in the 21st century. *Science* 348 (6235). doi:10.1126 /science.1261071.

Basta, N. T. 2000. Examples and case studies of beneficial reuse of municipal by-products. In Power J. F., and W. A. Dick (eds.) *Land Application of Agricultural, Industrial, and Municipal By-Products*. SSSA Book Series No. 6. Soil Science Society America, Madison, WI, pp. 481–504.

Committee on Toxicants and Pathogens in Biosolids Applied to Land and National Research Council (eds.). 2002. *Biosolids Applied to Land: Advancing Standards and Practices*. National Academies Press, Washington, DC.

Cordell, D., J. Drangert, and S. White. 2009. The story of phosphorus: Global food security and food for thought. *Global Environmental Change* 19: 292–305.

Elser, J. 2014. Health dimensions of phosphorus. In Scholz, R. W., A. H. Roy, F. S. Brand, D. T. Hellums, and A. E. Ulrich (eds.) *Sustainable Phosphorus Management: A Global Transdisciplinary Roadmap*. Springer, Berlin, pp. 229–231.

Fisher, M. 2015. Subsoil phosphorus loss: A complex problem with no easy solutions. *CSA News* 60(2): 4–10. American Society Agronomy, Madison, WI.

Food and Agriculture Organization. 2009. *High Level Expert Forum—How to Feed the World in 2050. 12–13 October, 2009*. Food and Agriculture Organization, Rome.

Food and Agriculture Organization. 2011. Current world fertilizer trends and outlook to 2015. Food and Agriculture Organization, Rome. Available at ftp://ftp.fao.org/ag/agp/docs/cwfto15.pdf.

Food and Agriculture Organization Corporate Statistical Database (FAOSTAT). 2016. Statistics Division. Food and Agriculture Organization of the United Nations, Rome. Available at http://faostat3.fao.org/home/E.

Gilbertson, C. B., F. A. Nordstadt, A. C. Mathers, R. F. Holt, A. P. Barnett, T. M. McCalla, C. A. Onstad, and R. A. Young. 1979. Animal Waste Utilization on Cropland and Pastureland: A Manual for Evaluating Agronomic and Environmental Effects. U.S. Department of Agriculture, Science and Education Administration, Hyattsville, MD, USDA Report No. URR-6.

Grace Communications Foundation. 2015. Animal feed. Available at http://www.sustainable table.org/260/animal-feed.

Hue, N. V. 1995. Sewage sludge. In Rechcigl, J. E. (ed.) *Soil Amendments and Environmental Quality*. CRC Press, Boca Raton, FL. pp. 199–248.

Jasinski, S. M. 2009. Phosphate rock. In *Mineral Commodity Summaries*. U.S. Geological Survey, U.S. Department of the Interior, Washington, DC, pp. 120–121.

Jasinski, S. M. 2012. Phosphate rock. In *Mineral Commodity Summaries*. U.S. Geological Survey, U.S. Department of the Interior, Washington, DC, pp. 118–119.

Kissinger, W. F., G. E. Erickson, T. J. Klopfenstein, and R. K. Koelsch. 2006. Managing phosphorus in beef feedlot operations. Nebraska Beef Cattle Report, Report 1-1-2006. University of Nebraska, Lincoln. Available at http://digitalcommons.unl.edu/animalscinbcr/136/.

Logan, T. J., and R. L. Chaney. 1983. Utilization of municipal wastewater and sludge on land-metals. In Page, A. L., T. L. Gleason III, J. Smith, J. E. Iskandar, I. K. Iskandar, and L. E. Sommers (eds.) *Utilization of Municipal Wastewater and Sludge on Land*. University of California, Riverside, CA, pp. 235–295.

Lu, Q., Z. L. He, and P. J. Stoffella. 2012. Land application of biosolids in the USA: A review. *Applied and Environmental Soil Science* 2012 (201462). doi:10.1155/2012/201462.

MacDonald, G. K., E. M. Bennett, P. A. Potter, and N. Ramakutty. 2011. Agronomic phosphorus imbalances across the world's croplands. *Proceedings of the National Academy of Sciences* 108 (7): 1–9.

MacDonald, J. M., M. O. Ribaudo, M. J. Livingston, J. Beckman, and W. Huang. 2009. Manure use for fertilizer and for energy. Economic Research Service, USDA, Washington, DC, Report to Congress.

Nierenberg, D., and K. Spoden. 2012. Global grain production at record high despite extreme climatic events. *Vital Signs*. Available at http://link.springer.com/chapter /10.5822/978-1-61091-457-4_11#page-1.

Oliver, R. 2008. Animal waste: Future energy, or just hot air. Available at http://www.cnn .com/2008/WORLD/asiapcf/01/07/eco.about.manure/index.html?iref=allsearch.

Philpott, T. 2013. You need phosphorus to live and we're running out. *Mother Jones*. Available at http://www.cnn.com/2008/WORLD/asiapcf/01/07/eco.about.manure/index.html?iref =allsearch.

Roberts, T. L., and J. Ryan. 2015. Soil and food security. *Better Crops with Plant Food* 99 (1): 4–6.

Roberts, T. L., and A. S. Tasitro. 2012. The role of plant nutrition in supporting food security. In Bruulsema, T., P. Heffer, R. M. Welch, I. Cakmak, and K. Moran (eds.) *Fertilizing Crops to Improve Human Health*. International Fertilizer Industry Association (IFA), Paris; and International Plant Nutrition Institute (IPNI), Norcross, GA.

Scholz, R. W., A. H. Roy, and D. T. Hellums. 2014. Sustainable phosphorus management: A transdisciplinary challenge. In Scholz, R. W., A. H. Roy, F. S. Brand, D. T. Hellums, and A. E. Ulrich (eds.) *Sustainable Phosphorus Management: A Global Transdisciplinary Roadmap*. Springer, Berlin, pp. 1–113.

Scholz, R. W., A. E. Ulrich, M. Eilitta, and A. Roy. 2013. Sustainable use of phosphorus: A finite resource. *Science of the Total Environment* 461–462: 799–803.

Sharpley, A. N., T. Daniel, T. Sims, J. Lemunyon, R. Stevens, and R. Perry. 2003. *Agricultural Phosphorus and Eutrophication*. Second edition. USDA Agricultural Research Service ARS-149, Washington, DC.

Sheffield, J. 2000. Farm animal manure is an important sustainable renewable energy resource. Available at http://web.ornl.gov/~webworks/cpr/pres/107931_.pdf.

Smeck, N. E. 1985. Phosphorus dynamics in soils and landscapes. *Geoderma* 36: 185–199.

Smil, V. 1999. Detonator of the population explosion. *Nature* 400: 415.

Smil, V. 2000. Phosphorus in the environment: Natural flows and human interferences. *Annual Review of Energy and the Environment* 25: 53–88.

Stewart, B. A., X. Hou, and S. R. Yalla. 2013. Facts and myths of feeding the world with organic farming methods. pp. 87–108. In Lal, R., and B. A. Stewart (eds.) *Principles of Sustainable Soil Management in Agroecosystems: Advances in Soil Science*. CRC Press, Boca Raton, FL.

Stewart, W., D. W. Dibb, A. E. Johnston, and T. J. Smyth. 2005. The contribution of commercial fertilizer nutrients to food production. *Agronomy Journal* 97: 1–6.

United States Department of Agriculture (USDA). 2012. Census of agriculture. U.S. Department of Agriculture, Washington, DC. Available at http://www.agcensus.usda.gov/Publications/2012/index.php#full_report.

United States Department of Agriculture—Economic Research Service (USDA-ERS). 2013. Fertilizer use and price. U.S. Department of Agriculture, Washington, DC. Available at http://www.ers.usda.gov/data-products/fertilizer-use-and-price.aspx.

United States Department of Agriculture—Natural Resources Conservation Service (USDA-NRCS). 1995. Animal manure management. U.S. Department of Agriculture, Washington, DC. Available at http://www.nrcs.usda.gov/wps/portal/nrcs/detail//?cid=nrcs143_014211.

United States Environmental Protection Agency (U.S. EPA). 2013. Literature review of contaminants in livestock and poultry manure and implications for water quality. U.S. EPA, Washington, DC, EPA 820-R-13-002.

van Kauwenbergh, S. J. 2010. World phosphate rock reserves and resources. Technical Bulletin IFDC-T-75. International Fertilizer Development Center, Muscle Shoals, AL, 48 pp.

Withers, P. J., and A. N. Sharpley. 1995. Phosphorus fertilizers. In Rechcigl, J. E. (ed.) *Soil Amendments and Environmental Quality*. CRC Press Inc., Boca Raton, FL, pp. 65–107.

World Hunger Education Service. 2015. 2015 World hunger and poverty facts and statistics. Available at http://www.worldhunger.org/articles/Learn/world%20hunger%20facts%20 2002.htm.

Worldwatch Institute. 2015. Grain harvest sets record, but supplies still tight. Available at http://www.worldwatch.org/node/5539.

Zhang, H., G. Johnson, and M. Fram. 2012. Managing phosphorus from animal manure. Oklahoma Cooperative Extension Service F-2249, Division of Agricultural Sciences and Natural Resources, Oklahoma State University, Stillwater, OK. Available at http://pods.dasnr.okstate.edu/docushare/dsweb/Get/Rendition-5084/unknown.

# 3 Coupled Cycling of Carbon, Nitrogen, and Phosphorus

*Daniel Sebastian Goll*

## CONTENTS

## 3.1 INTRODUCTION

### 3.1.1 OBJECTIVES

The foundation to our understanding of the coupling between the cycles of carbon (C), nitrogen (N), and phosphorus (P) was laid by the work of Justus von Liebig in the early nineteenth century. Based on extensive analyzes of the chemical composition (stoichiometry) of plants, animals, and soils, Liebig was first to recognize that a plant tissue is assimilated from atmospheric C, in addition to a limited set of other chemical elements from organic and inorganic sources in soils (Liebig 1862).

About 100 years later, the understanding of biogeochemical coupling was greatly facilitated by the work of oceanographer Alfred C. Redfield. In 1958, he published evidence of what have become two powerful principles in biogeochemistry: that marine plankton consists of C, N, and P in a characteristic molar ration and that the abundance of C, N, and P is regulated by the interactions between the marine organisms and the ocean environment.

Earlier Redfield discovered that on average, the ratio of N and P in planktonic biomass was similar to the ratio of N and P in marine water, which seemed unlikely to him to be "a mere coincidence" (Redfield 1934). He proposed two mutually non-exclusive mechanisms, which could lead to such a relation. The first states that the composition of the plankton community is a result of the composition of the sur-rounding, as individuals with different nutrient requirements compete for a limited supply of resources. The second hypothesis states that the composition of the marine water is the result of the nutrient requirement of marine organisms. He suggested a thermostat-like scenario, in which the biota exerts a strong control on the ocean N budget by which the N:P ratio of seawater is balanced (Redfield 1958). The latter explanation since then has been widely accepted and supported by modeling studies (Tyrrell 1999; Lenton and Watson 2000; Klausmeier et al. 2004; Weber and Deutsch 2012). These findings make up the core of our understanding of the coupling between cycles of C, N, and P via biological processes.

Since the work by Redfield, scientists have searched for similar patterns and rela-tionships in terrestrial and freshwater ecosystems, and stoichiometric concepts have been widely applied in ecology. Over the last 20 years, experimental evidence for relationships between soils, water, and organismic stoichiometry with essential eco-logical and physiological traits has increased, and these relationships now play a central role in ecological research (Elser et al. 1996, 2000; Sterner and Elser 2002; Raubenheimer and Simpson 2004). New theories combine physical principles, such as the conservation of mass and the dissipation of energy, with biological principles, such as natural selection, importance of trade-offs in energy metabolism, and growth at the biochemical and individual levels. These theories link stoichiometry to bio-logical processes from molecular to global levels and have been introduced to global models of the earth system in which they serve as an important constraint on future climate change (Thornton et al. 2007; Goll et al. 2012; Buendía et al. 2014; Prentice et al. 2014). The focus of this chapter is on the cycles of C, N, and P as these are usu-ally regarded as major constraints on the functioning of the biosphere. Nonetheless, other elements such as sulfur, potassium, iron, or boron could be of comparable significance, but the current understanding is limited (Astolfi 2009; Sardans and Peñuelas 2015a; Steidinger 2015).

In this chapter, the current understanding of the biological coupling of biogeo-chemical cycles of C, N, and P is reviewed. The processes among the scales of bio-logical organization are linked to the global patterns of nutrient availabilities, the functioning of the earth system, and its response to the human alteration of the cycles.

The first part gives an overview of the processes governing the coupling of the cycles among the scales of biological organization. It focuses on the fundamental constraints on the elemental composition of organisms and ecosystems and links the processes among the scales of organization.

The second part discusses the extent to which the biota is able to influence its environment in respect to the availability of C, N, and P. The focus lies on mecha-nisms that enhance and balance the availabilities on ecosystem level.

The third part deals with the alteration of the biogeochemical cycles by human activities leading to an imbalance between the cycles of C, N, and P. It further dis-cusses the implications of the imbalance for the functioning of the earth system.

## 3.2 THE COUPLING AMONG SCALES OF ORGANIZATION: FROM MOLECULES TO THE EARTH SYSTEMS

The biological coupling of the biogeochemical cycles is the sum of multiple processes operating on different levels of biological organization (Sterner and Elser 2002). The analysis of the elemental composition (stoichiometry) of the units of organization is key to understanding the biological coupling. For example, on an ecosystem scale, the C:N:P ratio of planktonic biomass (Redfield ratio) is well constrained, whereas at a species level, there can be substantial deviations from the Redfield ratio of 106:16:1 (Klausmeier et al. 2004). The differences in the stoichiometric flexibility among the scales of organization can be linked to the differences in the variability of environmental conditions (Klausmeier et al. 2004). The stoichiometric theory is also used to link the processes among the scales of biological organization. Loladze and Elser (2011) found that the optimal ratio between RNA and protein synthesis for photosynthesis on a cellular level corresponds to a molar N:P ratio of 16:1, which is similar to the ecosystem scale ratio.

In this section, the major processes underlying the coupling of the cycles of C, N, and P are discussed from the lowest level of organization, the biological polymers, to the highest level, the ecosystem.

### 3.2.1 BIOLOGICAL POLYMERS

On the lowest level of biological organization, the level of molecules, four major chemical classes of biological polymers are distinguished: carbohydrates, lipids, proteins, and nucleic acids. Each of these classes serves a different functional role in organisms.

Carbohydrates are the main substrate of the energy metabolism and serve a structural purpose in the form of cellulose, long chains of carbohydrates. Cellulose is the single most abundant organic polymer on earth (Klemm et al. 2005) illustrating the central role of C in the biosphere. Lipids mainly serve as energy storage compounds, but a special type of P-rich lipids (phospholipids) is the main constituent of cell membranes. Proteins are the most complex biological compounds, comprising sequences of amino acids. Proteins serve multiple purposes, such as structure, energy storage, metabolism, and signaling. The most prominent member is Rubisco, a metabolic protein that catalyzes the fixation of C from the atmosphere during photosynthesis. Rubisco is at the core of the coupling between cycles of C and N and energy conversion. It is the most abundant protein in the leaves of plants and believed to be the most abundant protein on earth (Ellis 1979; Losh et al. 2013; Raven 2013). The fourth class is nucleic acids, which are the building blocks of DNA and RNA, the central compounds in the storage, transport, and translation of genetic information. In the form of adenosine triphosphate (ATP), nucleic acids serve as energy transporters and in signaling on a cellular level. In general, nucleic acids play an important role in the synthesis and the activation of proteins.

The four classes of biological polymers have distinctive but constrained chemical compositions, which allow to relate stoichiometry to physiological functions (Sterner and Elser 2002). Carbohydrates and lipids are rich in C but low in N and P, in contrast to proteins and nucleic acids. While nucleic acids are rich in both N and P, proteins are rich in N and poor in P.

There is a strong correlation between the amount of N in photosynthetically active (autotrophic) organisms and their photosynthetic capacity (Field and Mooney 1983; Losh et al. 2013), as a large fraction of N in autotrophs directly functions in capturing energy in photosynthesis in the form of Rubisco (Evans 1989; Evans and Seemann 1989). The regeneration of Rubisco depends on ATP and RNA, thereby linking P availability to photosynthesis (Reich et al. 2009; Walker et al. 2014). The ratio between protein content and protein turnover in general links stoichiometry to growth strategies. Fast-growing organisms have a lower N:P ratio than slow-growing organism, as their faster protein turnover needs more nucleic acids relative to proteins (Sterner and Elser 2002; Reef et al. 2010; Sardans et al. 2012).

## 3.2.2 CELLULAR LEVEL

There is a considerable variation in the relative abundance of elements on the level of biological polymers, which is comparable to the variation in the inorganic world. At the cellular level, the variation narrows down (Sterner and Elser 2002). As a functional cell needs a full set of life-sustaining functions, its composition is more restricted than the compositions of its single constituents. Nonetheless, there is considerable stoichiometric variation among taxa and in between taxa as a result of differing growing conditions and growth strategies (Marschner 1998; Rien and Chapin 2000). The major differences among taxa are due to the different modes of nutrition, autotrophy, and heterotrophy (Berman-Frank and Dubinsky 1999). Autotrophs show large variations in stoichiometry, in contrast to heterotrophs (Table 3.1). In autotrophs, the acquisition of energy from the atmosphere is physiologically decoupled from the acquisition of nutrients from soils. Heterotrophic organisms, in contrast, rely on a single source (organic matter) for energy and nutrients.

This fundamental difference is reflected in the composition of cells (Sterner and Elser 2002). The cells of autotrophs differ from those of heterotrophs, in that they have cell organelles, vacuoles, which allow the storage of compounds without

## TABLE 3.1
## Variation in Stoichiometry (Molar Ratios) on Ecosystem Scale Given by the Range of Ecosystem Averages

|                                         | C:N       | C:P          | N:P       | Reference            |
| --------------------------------------- | --------- | ------------ | --------- | -------------------- |
| Marine plankton and organic matter      | 5.8–14    | 67.0–226.9   | 13.3–37.6 | Martiny et al. 2013  |
| Photosynthetic tissues of land plants   | 35.0–59.5 | 923.0–2457.6 | 22.1–44.2 | McGroddy et al. 2004 |
| Soil                                    | 10.5–31.5 | 7.7–1353.9   | 2.2–61.9  | Xu et al. 2013       |
| Soil microbes                           | 5.8–9.3   | 30.9–131.5   | 4.4–8.8   | Xu et al. 2013       |

*Note:* In case of marine plankton and organic matter, the variation is given by the range of averages over 10 latitudinal bands between 60 N and 60 S.

affecting the cell metabolism. The vacuoles allow autotrophs to buffer temporal and spatial disbalances between nutrient demand and supply, thereby enabling autotrophs to cope with two uncoupled sources of nutrition. In addition, autotrophs have C-rich cell walls, in contrast to heterotrophs. The imbalance between the stoichiometry of autotrophs and heterotrophs, which feed on the former, is a major factor shaping trophic webs (Sterner and Elser 2002), which play a fundamental role in the recycling of nutrients on an ecosystem level (Cherif and Loreau 2009; Fanin et al. 2013).

### 3.2.3 MULTICELLULAR ORGANISMS

Multicellular organisms consist of different kinds of tissue, functional groups of specialized cells, which have contrasting stoichiometries. Therefore, multicellular organisms more strongly differ in their stoichiometry among each other than single-celled organisms. A particular tissue can differ in its stoichiometry among taxa; for example, herbs have on average a slightly lower N:P ratio in photosynthetic tissue than do woody plants (Reich and Oleksyn 2004; Sardans et al. 2012). However, the largest variation in stoichiometry is found between different tissues (Table 3.2).

Each tissue serves a different functional purpose; thus, the proportions of the four classes of biological polymers vary among tissues. In higher plants, most of the variations in the stoichiometry among taxa and species can be explained by the differences in the distribution of plant biomass among plant organs (Lawlor et al. 1981). While stems and branches consist of structural tissue, which has hardly any N and P, leaves are rich in N and P. The allocation of biomass to the different organs is highly

---

**TABLE 3.2**

**The Stoichiometric Composition of Components of Marine, Freshwater, and Terrestrial Ecosystems**

|  | C:N | C:P | N:P | Reference |
|---|---|---|---|---|
| Marine plankton and organic matter | 6.64 | 105.81 | 15.97 | Redfield 1934 |
| Freshwater plankton | 8.61 | 154.73 | 17.98 | Kahlert et al. 1998 |
| Photosynthetic tissue of land plants | 43.62 ± 3.50 | 1334.00 ± 137.45 | 27.86 ± 1.33 | McGroddy et al. 2004 |
| Structural tissue of land plants wood | 172.59 ± 1.17 | 3937.80 ± 67.05 | 24.77 ± 0.44 | Sardans and Peñuelas 2015b |
|  |  |  | 35.38 | Kattge et al. 2011 |
| Soil | 16.44 | 288.82 | 17.47 | Xu et al. 2013 |
|  | 14.34 ± 0.47 | 186.96 ± 12.89 | 13.05 ± 0.88 | Cleveland and Liptzin 2007 |
| Soil microbes | 7.58 | 42.55 | 5.53 | Xu et al. 2013 |
|  | 8.63 ± 0.35 | 59.57 ± 3.61 | 6.90 ± 0.44 | Cleveland et al. 1999 |

*Note:* Shown are globally averaged molar ratios (± standard error) of ecosystem properties.

flexible among taxa and can change during different life stages, leading to considerable stoichiometric variability in time and space.

The need for structure in higher organisms results in a substantial amount of structural tissue (wood, bones, collagen), which is reflected in their stoichiometry. A structural tissue is either high in P (bones), N (collagen), or C (wood), leading to deviations from the stoichiometric relationships derived from metabolic constraints on a cellular level. The amount of structural tissue strongly varies among taxa. Stems and branches can make up 75% of the total biomass of a tree (Ovington 1957) but are very low in herbs and grasses. The relatively high proportions of structural tissue in autotrophs, which compete with each other for light, drives the differences in the C: nutrient ratio between autotrophs and heterotrophs.

### 3.2.4 ECOSYSTEM

On the ecosystem level, the variation in stoichiometry narrows down (McGroddy et al. 2004; Elser et al. 2007). On this scale, the thermostat-like mechanism proposed by Redfield acts. Depending on the N balance, organisms which either make atmospheric N available (N fixers) or organisms which release N to atmosphere are favored, resulting in ecosystem stoichiometries which are close to the Redfield ratio. There is a trend of increasing stoichiometric variation from marine, to river, to terrestrial systems (Elser et al. 2007; Sardans et al. 2012) (Table 3.1), which can be attributed to different physical–chemical properties of water compared to soils and atmosphere.

- **Marine ecosystems:** Marine ecosystems are relatively homogeneous in their stoichiometry (Table 3.1). On average, the N:P ratio in open areas of all major ocean is remarkably similar to the N:P ratio of plankton. The majority of studies support the Redfield ratio being a result of the trade-offs between competitive equilibrium and the buffer effect of N fixers and denitrification (Sardans et al. 2012). The processes of biological N fixation (BNF) and denitrification occur at very different intensities in different oceanic areas (Gruber and Sarmiento 1997), but large oceanic currents mix waters from different areas, fostering similar N:P ratios throughout the world (Weber and Deutsch 2012). The maintenance of the Redfield ratio is favored by the long duration of residence of these two elements in the ocean ($10^4$ yr) relative to the ocean's overturning time ($10^3$) (Falkowski and Davis 2004).

  Going from open oceans to coastal zone to lakes, the variation increases due to variations in external nutrient inputs (Sterner and Elser 2002). The variations in coastal waters are mainly linked to freshwater inputs of N and P from human activity (Slomp and Van Cappellen 2004; Yin and Harrison 2007) or linked to local sediments (Fox et al. 1986). The causes that determine C:N:P ratios in freshwater are different, at least to some extent, from the causes in oceans. The lower volume of water in lakes compared to that of oceans means that they are more likely to be influenced by the particular traits of the surrounding environment such as rock type or human impact.

Therefore, the C:P and N:P ratios of particulate matter are more variable (and generally higher) in lake particles than in marine particles (Hecky et al. 1993).

- **Terrestrial ecosystems:** In comparison to aquatic systems, terrestrial systems are characterized by larger variations in stoichiometry, which are related to differences in soils and climate among terrestrial ecosystems (Table 3.1). Whereas in oceans, regional differences in N and P availability are alleviated by ocean circulation, in terrestrial systems such a mechanism is lacking. Although the transport of nutrients in animals as flesh and dung can counteract stoichiometric heterogeneity, this mechanism is spatially restricted and less efficient than ocean circulation (de Mazancourt and Schwartz 2010; Rückamp et al. 2010; Doughty et al. 2013; Wolf et al. 2013).

  The age of soils is a key determinant for nutrient availability, in particular, in case of P. The single source of P to the biosphere derives from the weathering of P-bearing minerals. As soils age, P becomes increasingly unavailable due to mineralogical transformation and depletion of P in minerals, usually on a timescale of $10^3$–$10^5$ years (Schlesinger et al. 1998; Wardle et al. 2004; Crews et al. 2010; Buendía et al. 2014). Therefore, young soils tend to be rich in P and old highly weathered soils to be poor in P.

N availability is predominantly controlled by succession stage and climate (Cleveland et al. 1999; Vitousek and Howarth 2007). While inputs of N from N-rich sedimentary rocks can make a significant contribution to the N budget of specific sites (Holloway 2002; Morford et al. 2011), the major inputs are from the biological fixation of atmospheric N, which is in ample supply (Vitousek et al. 2013). Therefore, the availability of N increases with soil age, as N from biological fixation progressively accumulates in ecosystems. Because of the large variations in soil age and succession stages of ecosystems, the N and P availability is very heterogeneous among terrestrial ecosystems.

An additional factor, although less certain, that contributes to higher stoichiometric variations in terrestrial systems compared to aquatic systems is the limited availability of water on land (Sardans et al. 2012). Among terrestrial systems, the availability of water substantially differs, implying different physiological responses and adaptations between plants of different ecosystems, which likely affect the allocation of C, N, and P in plants. In drier ecosystems, allocation strategies should theoretically aim at enhancing the efficiency of water use, whereas in humid ecosystems, allocation should be more directly related to growth rate and competitive efficiency.

Besides the larger stoichiometric flexibility of terrestrial systems, the stoichiometry of terrestrial systems also deviates on average from the marine Redfield ratio. The higher demand for structural tissue in terrestrial plants compared to aquatic organisms results in a higher C-to-nutrient ratio (Sterner and Elser 2002; Cleveland and Liptzin 2007) (Table 3.2). Also the N:P ratio of the photosynthetical tissue of plants is elevated compared to an ocean ratio of 16 (Hedin 2004). This trend could be due to the biophysical C concentration mechanism of marine phytoplankton which enhances the efficiency of Rubisco, thereby minimizing the N need of photosynthesis, whereas terrestrial autotrophs lack or rely on biochemical concentration mechanisms, which are energetically less efficient. The finding that the cellular concentration of Rubisco is lower in marine organisms than that in terrestrial plants by Losh et al. (2013) supports this view.

## 3.3   THE BIOLOGICAL CONTROL ON THE CYCLES OF NITROGEN AND PHOSPHORUS

Inspired by Redfield's work, James Lovelock proposed that over the long run of geological time, life controls the planetary environment to generate and sustain "homeostasis by and for the biosphere" (Gaia hypothesis) (Lovelock and Margulis 1974). Since then, the original formulation of the hypothesis has been contradicted by observations and theory (Kirchner 2002). Nonetheless, weak forms of the hypothesis in which the biota "influences atmospheric composition, temperature and nutrient availability" are widely supported (Kirchner 2002; Elser et al. 2007; Ciais et al. 2013).

The extent to which the biota can balance the availability of N and P to their own needs primarily depends on its control (1) on external sources and (2) on losses of nutrients, as well as (3) on the efficiency with which nutrients are internally recycled. The relative importance of the different processes driving nutrient availability varies as an ecosystem matures to a steady state (Menge et al. 2009) and differs between N and P. Figure 3.1 shows a mature tropical forest on a highly weathered soil, as an example for an ecosystem with suboptimal nutrient conditions. In the following, the biological controls on N and P cycles and their major constraints are discussed.

FIGURE 3.1   (See color insert.) A mature lowland rainforest limited by nutrient availability located in northern French Guiana. The ecosystem is characterized by a high diversity of species, low P concentration in tree leaves, a highly weathered soil depleted in nutrients and a tight coupling between cycles of C, N, and P (Barantal et al. 2012). (Courtesy of Oriol Grau.)

### 3.3.1 Nitrogen

The N cycle, compared to the P cycle, is a rather leaky cycle with large fluxes connecting biospheric reservoirs with abiotic reservoirs. Of the annual N requirement for global net primary productivity, approximately 16% is met by N from the ample supply of atmospheric dinitrogen via BNF (Cleveland et al. 2013). In contrast, only 1–2% of the net primary productivity is supported by P from external reservoirs (Goll et al. 2012; Cleveland et al. 2013).

Several factors are known that prevent ecosystems from achieving sufficiently high BNF rates to avoid N limitation (Vitousek et al. 2013). The major cause of insufficient rates of BNF in a terrestrial system is commonly attributed to low temperatures (Cleveland et al. 1999). The temperature dependence of the enzyme system underlying BNF is unusual in that it has strongly biphasic kinetics, with a low activation energy of 0.65 eV above 22°C and a very high activation energy of 2.18 eV below 22°C (Ceuterick et al. 1978). Accordingly, there is a strong increase in the energetic cost of BNF rates when the temperature drops under 22°C. Thus, the N fixers become competitively inferior to nonfixers at low temperatures, leading to suboptimal N conditions in many high altitude systems (Reich and Oleksyn 2004; Zaehle and Dalmonech 2011). However, in marine systems, the energetic constrains on BNF cannot be ruled out but seem to be rarely responsible for low rates of BNF (Vitousek and Howarth 2007).

Additionally, other factors that affect the competitive strength of N-fixing species can cause N deficiency in aquatic and terrestrial systems. Among these factors are shade intolerance of N fixers, preferential grazing on N fixers, and higher non-N nutrient demand of N fixers than nonfixers (Vitousek and Howarth 2007; Vitousek et al. 2013). Here, a low supply of elements needed for the enzyme system underlying BNF, like P, molybdenum, or iron, was shown to constrain the BNF in a range of terrestrial and aquatic systems (Howarth and Cole 1985; Vitousek and Field 1999; Barron et al. 2009). The dependence of BNF on P availability (Figure 3.2) provides a mechanism potentially explaining the observations of widespread N and P colimitation in aquatic and terrestrial systems (Elser et al. 2007). Recently, an awareness arose that the conditions of low N availability might be stabilized by plant–microbial relationships due to the volatility of individual partners, trapping an ecosystem in N limitation (Franklin et al. 2014; van der Heijden et al. 2015). This suggests that symbiotic relationships between trees and their mycorrhizae may not be as mutually beneficial as commonly assumed.

Nitrogen is easily lost from terrestrial systems due to its high mobility in soils and makes it susceptible to losses by leaching or denitrification. Phosphorus in contrast is relatively immobile in soil, and losses are marginal. Therefore, natural disturbances, like fire, as well as anthropogenic land use changes can trigger N losses, which outweigh P losses (Davidson et al. 2007). Periodical changes in climate, like the El Niño/southern oscillation, can lead to temporary losses of N, which cannot be counterbalanced due to the inertia of the BNF leading to a sustained disequilibrium between N and P availability in terrestrial and aquatic systems (Altabet et al. 1995; Hedin et al. 2009).

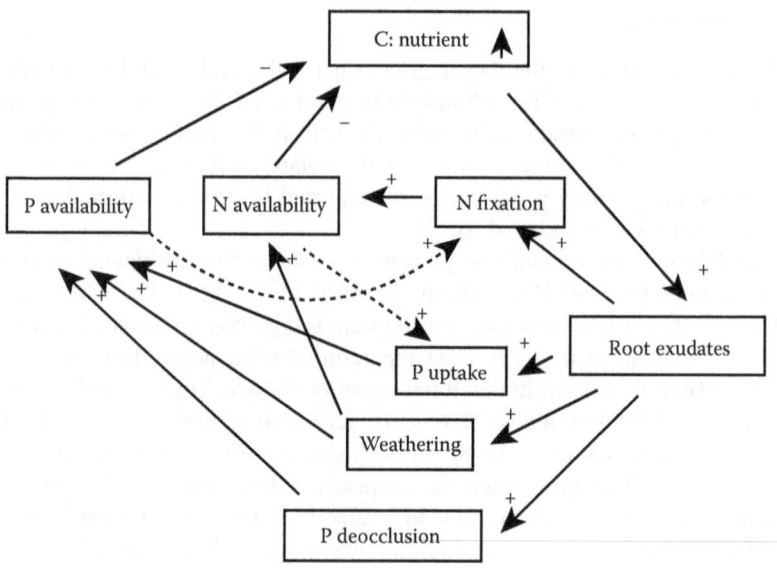

**FIGURE 3.2** Conceptual diagram showing the feedback processes by vegetation which can affect the availability of N and P—balancing the availability of C, N, and P. In case the nutrients are limiting (elevated C: nutrient ratio), plants invest the surplus C in the exudation of energy-rich carbohydrates to enhance microbial turnover and root activity. Increasing soil respiration potentially leads to (1) enhanced rock weathering releasing P, (2) P deocclusion from secondary minerals, (3) more efficient P uptake by means of symbiosis with root mycorrhiza and phosphatase production, and (4) stimulation of BNF. The interlinkages between acquisition processes and the availability of the respective other nutrients balance the availability of N and P (dotted lines). In case the nutrient availability cannot be enhanced sufficiently, photosynthesis, and therefore C availability, is down-regulated restoring the balance between C, N, and P (not shown). (Inspired from Buendía, C., Arens, S., Hickler, T., Higgins, S. I., Porada, P., and Kleidon, A., *Biogeosciences*, 11, 3661–3683, 2014. With permission.)

### 3.3.2 PHOSPHORUS

The P cycle has a very slow geochemical cycle mainly driven by tectonic processes, which makes P the major constraint on biosphere productivity on a geological timescale (Walker and Syers 1976). The single source of P for the biosphere is the weathering of minerals, which are replenished by a tectonic uplift on a timescale of millions of years. Once released from its geological source, P is quickly converted into less available forms or, to a lesser extent, transported from the land to the ocean by erosion and leaching (Walker and Syers 1976). As minerals get depleted of P over time, inputs decline, causing terrestrial ecosystems to run into P limitation on an average timescale of 100,000 years in models (Buendía et al. 2014) and observations (Wardle et al. 2004; Crews et al. 2010). Nonetheless, P can become limiting in soils as young as 100 years (Schlesinger et al. 1998). The actual timescale on which P becomes limiting primarily depends on weathering intensity, tectonic uplift of fresh material, and losses (Buendía et al. 2010, 2014).

There are several mechanisms by which the biota can influence weathering and associated P release (Schwartzman and Volk 1991; Buendía et al. 2014). The formation and stabilization of organic soils generates an environment, which maintains aqueous contact with a high surface area of mineral grains, thus maximizing weathering (Schwartzman and Volk 1991). Vegetation enhances the hydrological cycle, increasing precipitation, thereby enhancing weathering (White and Blum 1995). Additional mechanisms, operating on shorter timescales, are associated with the allocation of energy-rich carbohydrates to the roots and the associated microbes (Figure 3.2). Root respiration and heterotrophic respiration raise the acid level in soils enhancing chemical weathering (Schwartzman and Volk 1991). Plants and their symbionts directly acidify the soil environment by the exudation of organic acids and chelating agents, like citrate and oxalate, promoting weathering (Schwartzman and Volk 1991) and make P occluded in secondary minerals available.

The symbioses of plants and fungi in the form of mycorrhiza are crucial to keep P biologically bound and thereby protected from losses by leaching or wind (Buendía et al. 2014; van der Heijden et al. 2015). The minimization of losses is a major factor delaying the onset of P limitation due to P depletion of minerals (Buendía et al. 2014). The hypha of the mycorrhiza extends the volume of soil from which P can be acquired by the exudation of acids and chelating agents; thereby, the availability of P is kept low in soils, minimizing the losses. Symbioses are widespread throughout the plant kingdom. About 86% of plant species have mycorrhizal associations (Wang and Qiu 2006; Lambers et al. 2008; van der Heijden et al. 2015) and 4 to 50% of the C gained from photosynthesis can be allocated to symbiotic partners (Wright et al. 1998; Allen et al. 2003; van der Heijden et al. 2015). In return, the mycorrhizal fungi provide up to 90% of plant P (van der Heijden et al. 2015).

The rate by which P is released from rocks is small compared to the demand by the biota, indicated by a high recycling ratio of P of 50, which is the highest among any element (Volk 2003). The exudation of phosphatases, enzymes that specifically cleave out P from organic matter, by plants and microbes amplifies the turnover and the release of P from organic matter (Walker and Syers 1976; Houlton et al. 2008; Goll et al. 2012). The extent to which the turnover can be enhanced is assumed to be constrained by N availability, as phosphatases are rich in N (Olander and Vitousek 2000; Houlton et al. 2008; Marklein and Houlton 2012). The tight coupling between the N availability and the amplification of the P turnover (Figure 3.2) provides a mechanism for the observations of widespread N and P colimitation (Elser et al. 2007). It could also explain the high N-fixing capacities of tropical ecosystems despite the ample supply of mineral N, due to the benefits in P acquisition of a high N investment in phosphatases (Houlton et al. 2008).

Marine systems rely on the P lost from the terrestrial system and the upwelling of deep waters. Both sources of P are out of their control. Therefore, marine systems rely on the amplification of P turnover and the minimization of losses to sustain P availability. The losses of P from surface waters are minimized by several mechanisms that keep P biologically bound. Comparable to terrestrial systems, the exudation of enzymes increases the affinity and the rate of P uptake and therefore enhances the turnover of P in surface waters (de Mazancourt and Schwartz 2010). The turnover of inorganic P surface waters is of the order of hours (Van Mooy et al. 2009), allowing aquatic systems to sustain productivity with a minimum amount of P. The losses to deep waters are minimized by the preferential synthesis of polyphosphate, which is more readily recycled in surface ocean than other P-containing biochemicals, in case of low marine P concentration (Martin et al. 2014).

In general, the widespread occurrence of colimitation by N and P among aquatic and terrestrial systems (Elser et al. 2007; Paytan and McLaughlin 2007) illustrates that the extent to which biota can increase nutrient availability is limited. Moore et al. (2013) illustrate that the extent to which the biota can increase nutrient availability is limited. Nonetheless, it underlines the ability of the biota to balance the availability of N and P. Recent advances in the theoretical understanding of the interactions between the nutrient cycles provide mechanisms that are able to balance their availability (Houlton et al. 2008; Morford et al. 2011; Buendía et al. 2014; Franklin et al. 2014). However, large gaps in process understanding remain (Elser et al. 2007; Vitousek et al. 2013), and the quantification of nutrient fluxes on regional to global scales has large uncertainties. This severely hampers the ability to predict the consequences of the anthropogenic disturbances of the cycles, which are discussed in the next section.

## 3.4 ANTHROPOGENIC ACCELERATION OF THE BIOGEOCHEMICAL CYCLES

Human activities have significantly accelerated the cycling of C, N, and P, in particular, by the combustion of fossil fuels, the mining of P, and the industrial fixation of N from the atmosphere. The high rate of exploitation compared to the low rate at which these reservoirs have accumulated might potentially push the earth system into a new state with unforeseeable consequences (Steffen et al. 2015).

Currently we experience atmospheric $CO_2$ concentrations, which are unprecedented during the last 2 million years (Lüthi et al. 2008; Hönisch et al. 2009) and projected to increase at a rate unseen in the geological past (Ciais et al. 2013). As C becomes more abundant, the productivity of ecosystems becomes increasingly limited by the availability of nutrients in theory (Hungate et al. 2003; Luo et al. 2004; Thornton et al. 2007; Goll et al. 2012) and observations (Dukes et al. 2005; Norby et al. 2010; Reich et al. 2014). The enhanced growth of long-lived plant biomass under elevated $CO_2$ and the increasing C storage in soils can cause nutrients to be sequestered in organic matter, progressively decreasing soil nutrient availability (Luo et al. 2004; Finzi et al. 2006; Norby et al. 2010). The resulting imbalance between the C cycle and the nutrient cycles has profound effects on the functioning of ecosystems (Peñuelas et al. 2013).

The invention of industrial N fixation in the twentieth century provided an inexhaustible supply of reactive N for agricultural, industrial, and military uses (Galloway et al. 2013). Due to industrial N fixation, extensive fossil fuel combustion, and cultivation of legumes, by the beginning of the twenty-first century, anthropogenic sources of N were two to three times those of the natural terrestrial sources of 40–100 Tg N year$^{-1}$ (Galloway et al. 2013; Vitousek et al. 2013). As much as 120 Tg N year$^{-1}$ is industrially fixed and applied as fertilizer to agricultural land (Peñuelas et al. 2012; Galloway et al. 2013), of which a significant fraction reaches natural ecosystems (Galloway et al. 2004) due to the high mobility of N in soils. Losses of superfluous N to natural systems are enhanced by the preferential use of urea in developing countries rather than expensive composite fertilizers, which include a whole set of essential nutrients (Glibert et al. 2006; MacDonald et al. 2011). The cultivation of legumes capable of BNF is an additional source of 32–60 Tg N year$^{-1}$ to agricultural land (Galloway et al. 2013). The combustion of fossil fuels emits 2533 Tg N year$^{-1}$ of reactive N, which reaches a wide range of ecosystems (Peñuelas et al. 2012).

Phosphorus is mined at a global rate, which is nearly an order of magnitude larger (17–20 Tg P year$^{-1}$) than the natural release of P from rock weathering of 1–2 Tg P year$^{-1}$ (Goll et al. 2014). Currently, the combustion of fossil fuels and biomass release 12 Tg P year$^{-1}$ to the atmosphere, which is deposited over a wide range of ecosystems (Mahowald et al. 2008; Wang et al. 2014). Thereby, approximately 30% of current P deposition is from anthropogenic source (Wang et al. 2014).

The consequences of increasing P availability for the biosphere, although uncertain, are expected to be long lasting due to the long residence time of P in the biosphere (Goll et al. 2012). On a shorter timescale, the growing imbalance between the C and N cycles and the P cycle is likely to affect the functioning of the biosphere as well as the agricultural sector. Since 1975, the global average N:P ratio of fertilizer application has increased by 51% (Peñuelas et al. 2013) due to higher costs of P-containing compound fertilizer compared to P-free urea resulting in low application rates of P fertilizer of 22–26 Tg P year$^{-1}$ globally (MacDonald et al. 2011; Peñuelas et al. 2012). Due to high cost, phosphorus deficits occur across about 30% of the agricultural land in low-income (and food-deficient) countries, whereas P fertilizer is often used in excess in North America, western Europe, and eastern and southern Asia, including China and India (Obersteiner et al. 2013). Reductions in P wastage could free up this resource for low-income, food-deficient countries. However, current business practice is favoring the short-term interests of those able to pay for high-priced fertilizers over the concerns of the poor countries, the future generations, and the environment (Obersteiner et al. 2013). Climatic changes, increasing $CO_2$ concentration and limited supply of mineable P reserves are likely widening the imbalance between the P and the C and N cycles until the end of the twenty-first century (Goll et al. 2012; Obersteiner et al. 2013).

Due to the important role on organism and ecosystem functioning and structure, C cycle, climate, and agriculture, the understanding of the imbalance between the cycles are and will be vital. Nevertheless, there is little knowledge of where, how, and to what degree the imbalance in P and N additions to the ecosystems will affect the biota and how this will impact the climate system (Vitousek 1997; Peñuelas et al. 2013).

## 3.5 CONCLUSION

The coupling of the cycles of C, N, and P is linked to processes among the scales of biological organization. Functional constraints on the elemental composition of cells and organism result in stoichiometric variability of organisms, which is reduced compared to the stoichiometry of the abiotic environment (Table 3.2). The mode of nutrition (autotrophy or heterotrophy) is a major factor explaining the differences in stoichiometry among taxa. Additionally, fundamental differences in the abiotic environment are reflected in the stoichiometry. For example, the larger heterogeneity in abiotic conditions on land compared to those in marine and limnic systems is connected to a higher stoichiometric flexibility of terrestrial ecosystems (Table 3.1). In general, the stoichiometric needs of organisms and their interactions with the biotic and abiotic environments are considered to be primary drivers of the biogeochemical cycles on the ecosystem scale.

On the ecosystem level, the availabilities of N and P are rather balanced by the biota on the long term (Figure 3.2). Hereby, autotrophs, alone and in symbiosis with microbes, play a fundamental role in balancing the availability of the elemental cycles. Nonetheless, there are major constraints on the ability of the biota to influence the biogeochemical cycles, which are related to the limited control of nutrient influxes, recycling, and losses: for example, global patterns in the stoichiometry of terrestrial ecosystems can be linked to climatic constraints on BNF and the weathering status of soils. The occurrence of wide spread nutrient limitation indicates that the extent to which the biota enhances the availability of N and P is limited.

Human activities, in particular the combustion of fossil fuels, the mining of P, and the industrial fixation of N from the atmosphere, are making C, N, and P increasingly more available to the biota. However, the increases are not balanced leading to a growing imbalance between the cycles of C and N and the P cycle. In particular, the high costs of P fertilizer and the limited and unevenly distributed reserves of mineable P are playing a central role for the imbalances between P and N. The human alteration of the cycles has a potentially profound impact on the functioning of the whole earth system with unforeseeable consequences for humankind.

## ACKNOWLEDGMENTS

The author is funded by the IMBALANCEP project of the European Research Council (ERC2013SyG610028). The author thanks Dr. Jordi Sardans for providing a compilation of estimates of ecosystem stoichiometries from literature and Dr. Oriol Grau for contributing the picture of a lowland rainforest in northern French Guiana.

## REFERENCES

Allen, M. F., Swenson, W., Querejeta, J. I., Egerton-Warburton, L. M., and Treseder, K. K. 2003. Ecology of mycorrhizae: A conceptual framework for complex interactions among plants and fungi. *Annual Review of Phytopathology*, 41, 271–303.

Altabet, M. A., Francois, R., Murray, D. W., and Prell, W. L. 1995. Climate-related variations in denitrification in the Arabian Sea from sediment 15N/14N ratios. *Nature*, 373, 506–509.

Astolfi, S. 2009. *Sulfur in Plants: An Ecological Perspective*. Springer, Dordrecht.

Barantal, S., Schimann, H., Fromin, N., and Hättenschwiler, S. 2012. Nutrient and carbon limitation on decomposition in an Amazonian moist forest. *Ecosystems*, 15, 1039–1052.

Barron, A. R., Wurzburger, N., Bellenger, J. P., Wright, S. J., Kraepiel, A. M. L., and Hedin, L. O. 2009. Molybdenum limitation of asymbiotic nitrogen fixation in tropical forest soils. *Nature Geoscience*, 2, 42–45. Available at http://dx.doi.org/10.1038/ngeo366.

Berman-Frank, I., and Dubinsky, Z. 1999. Balanced growth in aquatic plants: Myth or reality? *BioScience*, 49, 29.

Buendía, C., Arens, S., Hickler, T., Higgins, S. I., Porada, P., and Kleidon, A. 2014. On the potential vegetation feedbacks that enhance phosphorus availability: Insights from a process-based model linking geological and ecological timescales. *Biogeosciences*, 11, 3661–3683. Available at http://www.biogeosciences.net/11/3661/2014/.

Buendía, C., Kleidon, A., and Porporato, A. 2010. The role of tectonic uplift, climate, and vegetation in the long-term terrestrial phosphorous cycle. *Biogeosciences*, 7, 2025–2038.

Ceuterick, F., Peeters, J., Heremans, K., De Smedt, H., and Olbrechts, H. 1978. Effect of high pressure, detergents and phospholipase on the break in the Arrhenius plot of Azotobacter nitrogenase. *European Journal of Biochemistry/FEBS*, 87, 401–407.

Cherif, M., and Loreau, M. 2009. When microbes and consumers determine the limiting nutrient of autotrophs: A theoretical analysis. *Proceedings of the Royal Society B: Biological Sciences*, 276, 487–497.

Ciais, P., Sabine, C., Bala, G., Bopp, L., Brovkin, V., Canadell, J., Chhabra, A. et al. 2013. Carbon and other biogeochemical cycles. In *Climate Change 2013: The Physical Science Basis. Contribution of Working Group I to the Fifth Assessment Report of the Intergovernmental Panel on Climate Change*, edited by Stocker, T., Qin, D., Plattner, G. K., Tignor, M., Allen, S. K., Boschug, J., Nauels, A., Xia, Y., Bex, V., and Midgley, P. M. Cambridge University Press, Cambridge and New York, pp. 465–570.

Cleveland, C. C., and Liptzin, D. 2007. C:N:P stoichiometry in soil: Is there a "Redfield ratio" for the microbial biomass? *Biogeochemistry*, 85, 235–252.

Cleveland, C. C., Houlton, B. Z., Smith, W. K., Marklein, A. R., Reed, S. C., Parton, W., Del Grosso, S. J., and Running, S. W. 2013. Patterns of new versus recycled primary production in the terrestrial biosphere. *Proceedings of the National Academy of Sciences*, 110, 12733–12737.

Cleveland, C. C., Townsend, A. R., Schimel, D. S., Fisher, H., Hedin, L. O., Perakis, S., Latty, E. F., Fischer, C. V., Elseroad, A., and Wasson, M. F. 1999. Global patterns of terrestrial biological nitrogen (Nz) fixation in natural ecosystems. *Global Biochemical Cycles*, 13, 623–645.

Crews, T. E., Kitayama, K., Fownes, J. H., Riley, R. H., Darrell, A., Muellerdombois, D., and Vitousek, P. M. 2010. Changes in soil phosphorus fractions and ecosystem dynamics across a long chronosequence in Hawaii. *Ecology*, 76, 1407–1424.

Davidson, E. A., de Carvalho, C. J. R., Figueira, A. M., Ishida, F. Y., Ometto, J. P. H. B., Nardoto, G. B., Sabá, R. T. et al. 2007. Recuperation of nitrogen cycling in Amazonian forests following agricultural abandonment. *Nature*, 447, 995–998.

de Mazancourt, C., and Schwartz, M. W. 2010. A resource ratio theory of cooperation. *Ecology Letters*, 13, 349–359. Available at http://www.ncbi.nlm.nih.gov/pubmed/20455920.

Doughty, C. E., Wolf, A., and Malhi, Y. 2013. The legacy of the Pleistocene megafauna extinctions on nutrient availability in Amazonia. *Nature Geoscience*, 6, 761–764.

Dukes, J. S., Chiariello, N. R., Cleland, E. E., Moore, L. A., Rebecca Shaw, M., Thayer, S., Tobeck, T., Mooney, H. A., and Field, C. B. 2005. Responses of grassland production to single and multiple global environmental changes. *PLoS Biology*, 3, 1829–1837.

Ellis, R. 1979. The most abundant protein in the world. *Trends in Biochemical Sciences*, 4, 241–244.

Elser, J. J., Bracken, M. E. S., Cleland, E. E., Gruner, D. S., Harpole, W. S., Hillebrand, H., Ngai, J. T., Seabloom, E. W., Shurin, J. B., and Smith, J. E. 2007. Global analysis of nitrogen and phosphorus limitation of primary producers in freshwater, marine and terrestrial ecosystems. *Ecology Letters*, 10, 1135–1142. Available at http://www.ncbi.nlm.nih.gov/pubmed/17922835.

Elser, J., Dobberfuhl, D., MacKay, N., and Schampel, J. 1996. Organism size, life history, and N: P stoichiometry: Towards a unified view of cellular and ecosystem processes. *Bioscience*, 46, 674–684.

Elser, J. J., Sterner, R. W., Gorokhova, E., Fagan, W. F., Markow, T. A., Cotner, J. B., Harrison, J. F., Hobbie, S. E., Odell, G. M., and Weider, L. J. 2000. Biological stoichiometry from genes to ecosystems. *Ecology Letters*, 3, 540–550.

Evans, J. R. 1989. Photosynthesis and nitrogen relationship in leaves of C3 plants. *Oecologia*, 78, 9–19.

Evans, J., and Seemann, J. 1989. The allocation of nitrogen in the photosynthetic apparatus: Costs, consequences and control. In *Photosynthesis*, edited by Briggs, W. Alan R. Liss, New York, pp. 183–205.

Falkowski, P. G., and Davis, C. S. 2004. Natural proportions. *Nature*, 431, 131.

Fanin, N., Fromin, N., Buatois, B., and Hättenschwiler, S. 2013. An experimental test of the hypothesis of non-homeostatic consumer stoichiometry in a plant litter–microbe system. *Ecology Letters*, 16, 764–772.

Field, C., and Mooney, H. A. 1983. The photosynthesis–nitrogen relationship in wild plants. In *On the Economy of Plant Form and Function*, edited by Givnish, T. Cambridge University Press, Cambridge; New York, pp. 25–55.

Finzi, A. C., Moore, D. J. P., DeLucia, E. H., Lichter, J., Hofmockel, K. S., Jackson, R. B., Kim, H. S. et al. 2006. Progressive nitrogen limitation of ecosystem processes under elevated $CO_2$ in a warm temperate forest. *Ecology*, 87, 15–25.

Fox, L. E., Wofsy, S. C., and Sager, S. L. 1986. The chemical control of soluble phosphorus in the Amazon estuary. *Geochimica et Cosmochimica Acta*, 50, 183–794.

Franklin, O., Näsholm, T., Högberg, P., and Högberg, M. N. 2014. Forests trapped in nitrogen limitation an ecological market perspective on ectomycorrhizal symbiosis. *New Phytologist*, 203, 657–666. Available at http://www.ncbi.nlm.nih.gov/pubmed/24824576.

Galloway, J. N., Dentener, F. J., Capone, D. G., Boyer, E. W., Howarth, R. W., Seitzinger, S. P., Asner, G. P. et al. 2004. Nitrogen cycles: Past, present, and future. *Biogeochemistry*, 70, 153–226.

Galloway, J. N., Leach, A. M., Bleeker, A., and Erisman, J. W. 2013. A chronology of human understanding of the nitrogen cycle. *Philosophical Transactions of the Royal Society of London: Series B: Biological Sciences*, 368 (20130), 1–11. Available at http://www.pubmedcentral.nih.gov/articlerender.fcgi?artid=3682740&t.

Glibert, P. M., Harrison, J., Heil, C., and Seitzinger, S. 2006. Escalating worldwide use of urea: A global change contributing to coastal eutrophication. *Biogeochemistry*, 77, 441–463. Available at http://link.springer.com/10.1007/s1053300530705.

Goll, D. S., Brovkin, V., Parida, B. R., Reick, C. H., Kattge, J., Reich, P. B., Van Bodegom, P. M., and Niinemets, U. 2012. Nutrient limitation reduces land carbon uptake in simulations with a model of combined carbon, nitrogen and phosphorus cycling. *Biogeosciences*, 9, 3547–3569. Available at http://www.biogeosciences.net/9/3547/2012/.

Goll, D. S., Moosdorf, N., Hartmann, J., and Brovkin, V. 2014. Climate-driven changes in chemical weathering and associated phosphorus release since 1850: Implications for the land carbon balance. *Geophysical Research Letters*, 41, 3553–3558.

Gruber, N., and Sarmiento, J. L. 1997. Global patterns of marine nitrogen fixation and denitrification. *Global Biogeochemical Cycles*, 11, 235–266.

Hecky, R. E., Campbell, P., and Hendzel, L. L. 1993. The stoichiometry of carbon, nitrogen, and phosphorus in particulate matter of lakes and oceans. *Limnology and Oceanography*, 38, 709–724. Available at http://www.aslo.org/lo/toc/vol_38/issue_4/0709.html.

Hedin, L. O. 2004. Global organization of terrestrial plant-nutrient interactions. *Proceedings of the National Academy of Sciences of the United States of America*, 101, 10849–10850. Available at http://www.pnas.org/content/101/30/10849.short.

Hedin, L. O., Brookshire, E. J., Menge, D. N., and Barron, A. R. 2009. The nitrogen paradox in tropical forest ecosystems. *Annual Review of Ecology, Evolution, and Systematics*, 40, 613–635. Available at http://www.annualreviews.org/doi/abs/10.1146/annurev.ecolsys.37.091305.110246.

Holloway, J. M. 2002. Nitrogen in rock: Occurrences and biogeochemical implications. *Global Biogeochemical Cycles*, 16, 65-1–65-17. Available at http://doi.wiley.com/10 .1029/2002GB001862.

Hönisch, B., Hemming, N. G., Archer, D., Siddall, M., and McManus, J. F. 2009. Atmospheric carbon dioxide concentration across the mid-Pleistocene transition. *Science*, 324, 1551–1554.

Houlton, B. Z., Wang, Y. P., Vitousek, P. M., and Field, C. B. 2008. A unifying framework for dinitrogen fixation in the terrestrial biosphere. *Nature*, 454, 327–U34. Available at http://www.ncbi.nlm.nih.gov/pubmed/18563086.

Howarth, R. W., and Cole, J. J. 1985. Molybdenum availability, nitrogen limitation, and phytoplankton growth in natural waters. *Science*, 229, 653–655.

Hungate, B. A., Dukes, J. S., Shaw, M. R., Luo, Y., and Field, C. B. 2003. Atmospheric science: Nitrogen and climate change. *Science*, 302, 1512–1513. Available at http://www.sciencemag .org/content/302/5650/1512.short.

Kahlert, M., and Kahlert, M. 1998. C:N:P ratios of freshwater benthic algae. In *Hydrobiology Special Issues–Advances in Limnology*, edited by Forsberg, C., and Pettersson, K. Schweizerbart Science Publishers, Stuttgart, vol. 51, pp. 105–114.

Kattge, J., Ogle, K., Bönisch, G., Díaz, S., Lavorel, S., Madin, J., Nadrowski, K., Nöllert, S., Sartor, K., and Wirth, C. 2011. A generic structure for plant trait databases. *Methods in Ecology and Evolution*, 2, 202–213.

Kirchner, J. W. 2002. The Gaia hypothesis: Fact, theory, and wishful thinking. *Climatic Change*, 52, 391–408.

Klausmeier, C. A., Litchman, E., Daufresne, T., and Levin, S. A. 2004. Optimal nitrogen-to-phosphorus stoichiometry of phytoplankton. *Nature*, 429, 171–174.

Klemm, D., Heublein, B., Fink, H. P., and Bohn, A. 2005. Cellulose: Fascinating biopolymer and sustainable raw material. *Angewandte Chemie International Edition*, 44, 3358–3393. Available at http://www.ncbi.nlm.nih.gov/pubmed/15861454.

Lambers, H., Raven, J. A., Shaver, G. R., and Smith, S. E. 2008. Plant nutrient acquisition strategies change with soil age. *Trends in Ecology and Evolution*, 23, 95–103. Available at http://www.ncbi.nlm.nih.gov/pubmed/18191280.

Lawlor, D. W., Day, W., Johnston, A. E., Legg, B. J., and Parkinson, K. J. 1981. Growth of spring barley under drought: Crop development, photosynthesis, dry matter accumulation and nutrient content. *The Journal of Agricultural Science*, 96, 167.

Lenton, T. M., and Watson, A. J. 2000. Redfield revisited: 1: Regulation of nitrate, phosphate, and oxygen in the ocean. *Global Biogeochemical Cycles*, 14, 225. Available at http://doi .wiley.com/10.1029/1999GB900065.

Liebig, J. V. 1862. *Die Chemie in ihrer Anwendung auf Agrikultur und Physiologie*. Verlag von Friedrich Vieweg und Sohn, Braunschweig.

Loladze, I., and Elser, J. J. 2011. The origins of the Redfield nitrogen-to-phosphorus ratio are in a homoeostatic protein-to-rRNA ratio. *Ecology Letters*, 14, 244–250. Available at http://www.ncbi.nlm.nih.gov/pubmed/21244593.

Losh, J. L., Young, J. N., and Morel, F. M. M. 2013. Rubisco is a small fraction of total protein in marine phytoplankton. *New Phytologist*, 198, 52–58. Available at http://www.ncbi .nlm.nih.gov/pubmed/23343368.

Lovelock, J. E., and Margulis, L. 1974. Atmospheric homeostasis by and for the biosphere: The gaia hypothesis. *Tellus A*, 26, 1–2.

Luo, Y., Su, B., Currie, W. S., Dukes, J. S., Finzi, A., Hartwig, U., Hungate, B. et al. 2004. Progressive nitrogen limitation of ecosystem responses to rising atmospheric carbon dioxide. *BioScience*, 54, 731.

Lüthi, D., Le Floch, M., Bereiter, B., Blunier, T., Barnola, J. M., Siegenthaler, U., Raynaud, D. et al. 2008. High resolution carbon dioxide concentration record 650,000800,000 years before present. *Nature*, 453, 379–382.

MacDonald, G. K., Bennett, E. M., Potter, P. A., and Ramankutty, N. 2011. Agronomic phosphorus imbalances across the world's croplands. *Proceedings of the National Academy of Sciences of the United States of America*, 108, 3086–3091.

Mahowald, N., Jickells, T. D., Baker, A. R., Artaxo, P., Benitez-Nelson, C. R., Bergametti, G., Bond, T. C. et al. 2008. Global distribution of atmospheric phosphorus sources, concentrations and deposition rates, and anthropogenic impacts. *Global Biogeochemical Cycles*, 22, 1–19.

Marklein, A. R., and Houlton, B. Z. 2012. Nitrogen inputs accelerate phosphorus cycling rates across a wide variety of terrestrial ecosystems. *New Phytologist*, 193, 696–704. Available at http://doi.wiley.com/10.1111/j.14698137.2011.03967.x.

Marschner, H. 1998. *Mineral Nutrition of Higher Plants*. Academic, London.

Martin, P., Dyhrman, S. T., Lomas, M. W., Poulton, N. J., and Van Mooy, B. A. S. 2014. Accumulation and enhanced cycling of polyphosphate by Sargasso Sea plankton in response to low phosphorus. *Proceedings of the National Academy of Sciences of the United States of America*, 111, 8089–94.

Martiny, A. C., Pham, C. T. A., Primeau, F. W., Vrugt, J. A., Moore, J. K., Levin, S. A., and Lomas, M. W. 2013. Strong latitudinal patterns in the elemental ratios of marine plankton and organic matter. *Nature Geoscience*, 6, 279–283. Available at http://dx.doi.org/10.1038/ngeo1757nnpapers2://publication/doi/10.1038/2013.

McGroddy, M. E., Daufresne, T., and Hedin, L. O. 2004. Scaling of C:N:P stoichiometry in forest worldwide: Implications of terrestrial redfield-like ratios. *Ecology*, 85, 2390–2401.

Menge, D. N. L., Pacala, S. W., and Hedin, L. O. 2009. Emergence and maintenance of nutrient limitation over multiple timescales in terrestrial ecosystems. *The American Naturalist*, 173, 164–175.

Moore, C. M., Mills, M. M., Arrigo, K. R., Berman-Frank, I., Bopp, L., Boyd, P. W., Galbraith, E. D. et al. 2013. Processes and patterns of oceanic nutrient limitation. *Nature Publishing Group*, 6, 701–710. Available at http://dx.doi.org/10.1038/ngeo1765nnpapers://dee23d a0e34b4588b62 2013.

Morford, S. L., Houlton, B. Z., and Dahlgren, R. A. 2011. Increased forest ecosystem carbon and nitrogen storage from nitrogen rich bedrock. *Nature*, 477, 78–81. Available at http://www.nature.com/nature/journal/v477/n7362/full/nature10415.html.

Norby, R. J., Warren, J. M., Iversen, C. M., Medlyn, B. E., and McMurtrie, R. E. 2010. $CO_2$ enhancement of forest productivity constrained by limited nitrogen availability. *Proceedings of the National Academy of Sciences of the United States of America*, 107, 19 368–19 373.

Obersteiner, M., Peñuelas, J., Ciais, P., van der Velde, M., and Janssens, I. A. 2013. The phosphorus trilemma. *Nature Geoscience*, 6, 897–898. Available at http://www.nature.com/doifinder/10.1038/ngeo1990.

Olander, L. P., and Vitousek, P. M. 2000. Regulation of soil phosphatase and chitinase activity by N and P availability. *Biogeochemistry*, 49, 175–190.

Ovington, J. D. 1957. Dry matter production by *Pinus sylvestris* L. Annals of Botany, XXI, 288–314.

Paytan, A., and McLaughlin, K. 2007. The oceanic phosphorus cycle. *Chemical Reviews*, 107, 563–576. Available at http://www.ncbi.nlm.nih.gov/pubmed/17256993.

Peñuelas, J., Poulter, B., Sardans, J., Ciais, P., van der Velde, M., Bopp, L., and Boucher, O. 2013. Human-induced nitrogen phosphorus imbalances alter natural and managed ecosystems across the globe. *Nature Communications*, 4, 2934. Available at http://www.ncbi.nlm.nih.gov/pubmed/24343268.

Peñuelas, J., Sardans, J., Rivasubach, A., and Janssens, I. A. 2012. The humaninduced imbalance between C, N and P in Earth's life system. *Global Change Biology*, 18, 3–6. Available at http://doi.wiley.com/10.1111/j.13652486.2011.02568.x.

Prentice, I. C., Liang, X., Medlyn, B. E., and Wang, Y.-P. 2014. Reliable, robust and realistic: The three R's of next-generation land surface modeling. *Atmospheric Chemistry and Physics Discussions*, 14, 24811–24861. Available at http://www.atmos-chem-phys-dis cuss.net/14/24811/2014/.

Raubenheimer, D., and Simpson, S. J. 2004. Organismal stoichiometry: Quantifying nonindependence among food components. *Ecology*, 85, 1203–1216.

Raven, J. A. 2013. Rubisco: Still the most abundant protein of Earth? *New Phytologist*, 198, 1–3.

Redfield, A. C. 1934. On the proportions of organic derivatives in sea water and their relation to the composition of plankton. In *University Press of Liverpool, James Johnstone Memorial Volume*, edited by Daniel, R. University Press of Liverpool, Liverpool, pp. 177–192.

Redfield, A. C. 1958. The biological control of chemical factors in the environment. *American Scientist*, 46, 205–221.

Reef, R., Ball, M. C., Feller, I. C., and Lovelock, C. E. 2010. Relationships among RNA DNA ratio, growth and elemental stoichiometry in mangrove trees. *Functional Ecology*, 24, 1064–1072. Available at http://doi.wiley.com/10.1111/j.13652435.2010.01722.x.

Reich, P. B., and Oleksyn, J. 2004. Global patterns of plant leaf N and P in relation to temperature and latitude. *Proceedings of the National Academy of Sciences of the United States of America*, 101, 11 001–11 006.

Reich, P. B., Hobbie, S. E., and Lee, T. D. 2014. Plant growth enhancement by elevated $CO_2$ eliminated by joint water and nitrogen limitation. *Nature Geoscience*, 7, 1–5.

Reich, P. B., Oleksyn, J., and Wright, I. J. 2009. Leaf phosphorus influences the photosynthesis–nitrogen relation: A cross-biome analysis of 314 species. *Oecologia*, 160, 207–212.

Rien, A., and Chapin, F. 2000. The mineral nutrition of wild plants revisited: A reevaluation of processes and patterns. *Advances in Ecological Research*, 30, 1–67.

Rückamp, D., Amelung, W., Theisz, N., Bandeira, A. G., and Martius, C. 2010. Phosphorus forms in Brazilian termite nests and soils: Relevance of feeding guild and ecosystems. *Geoderma*, 155, 269–279.

Sardans, J., and Peñuelas, J. 2015a. Potassium: A neglected nutrient in global change. *Global Ecology and Biogeography*. Available at http://doi.wiley.com/10.1111/geb.12259.

Sardans, J., and Peñuelas, J. 2015b. Trees increase their P:N ratio with size. *Global Ecology and Biogeography*, 24, 147–156. Available at http://doi.wiley.com/10.1111/geb.12231.

Sardans, J., Rivas-Ubach, A., and Peñuelas, J. 2012. The elemental stoichiometry of aquatic and terrestrial ecosystems and its relationships with organismic lifestyle and ecosystem structure and function: A review and perspectives. *Biogeochemistry*, 111, 1–39.

Schlesinger, W. H., Bruijnzeel, L. A., Bush, M. B., Klein, E. M., Mace, K. A., Raikes, J. A., and Whittaker, R. J. 1998. The biogeochemistry of phosphorus after the first century of soil development on Rakata Island, Krakatau, Indonesia. *Biogeochemistry*, 40, 37–55.

Schwartzman, D. W., and Volk, T. 1991. Biotic enhancement of weathering and surface temperatures on earth since the origin of life. *Palaeogeography, Palaeoclimatology, Palaeoecology*, 90, 357–371.

Slomp, C. P., and Van Cappellen, P. 2004. Nutrient inputs to the coastal ocean through submarine groundwater discharge: Controls and potential impact. *Journal of Hydrology*, 295, 64–86.

Steffen, W., Richardson, K., Rockström, J., Cornell, S., Fetzer, I., Bennett, E., Biggs, R. et al. 2015. Planetary boundaries: Guiding human development on a changing planet. In press. *Science*, 347(6223), 1259855. doi: 10.1126/science.1259855.

Steidinger, B. 2015. Qualitative differences in tree species distributions along soil chemical gradients give clues to the mechanisms of specialization: Why boron may be the most important soil nutrient at Barro Colorado Island. *New Phytologist*, 277(5325), 1–5.

Sterner, R. W., and Elser, J. J. 2002. *Ecological Stoichiometry*. Princeton University Press, Princeton, NJ.

Thornton, P. E., Lamarque, J. F., Rosenbloom, N. A., and Mahowald, N. M. 2007. Influence of carbon-nitrogen cycle coupling on land model response to $CO_2$ fertilization and climate variability. *Global Biogeochemical Cycles*, 21, 1–15.

Tyrrell, T. 1999. The relative influences of nitrogen and phosphorus on oceanic primary production. *Nature*, 400, 525–531.

van der Heijden, M. G. A., Martin, F. M., and Sanders, I. R. 2015. Tansley review: Mycorrhizal ecology and evolution: The past, the present, and the future. *New Phytologist,* 205(4), 1406–1423.

Van Mooy, B. A. S., Fredricks, H. F., Pedler, B. E., Dyhrman, S. T., Karl, D. M., Koblízek, M., Lomas, M. W. et al. 2009. Phytoplankton in the ocean use non-phosphorus lipids in response to phosphorus scarcity. *Nature*, 458, 69–72.

Vitousek, P. M., Mooney, H. A., Lubchenco, J., and Melillo, J. M. 1997. Human domination of earth's ecosystems. *Science*, 277(5325), 494–499.

Vitousek, P. M., and Field, C. B. 1999. Ecosystem constraints to symbiotic nitrogen fixers: A simple model and its implications. *Biogeochemistry*, 46, 179–202.

Vitousek, P. M., and Howarth, R. W. 2007. Nitrogen limitation on land and in the sea: How can it occur? *Biogeochemistry*, 13, 87–115.

Vitousek, P. M., Menge, D. N. L., Reed, S. C., and Cleveland, C. C. 2013. Biological nitrogen fixation: Rates, patterns and ecological controls in terrestrial ecosystems. *Philosophical Transactions of the Royal Society of London. Series B: Biological Sciences*, 368, 20130119. Available at http://www.ncbi.nlm.nih.gov/pubmed/23713117, 2013.

Volk, T. 2003. *Gaia's Body: Toward a Physiology of Earth*. Copernicus. New York.

Walker, T., and Syers, J. 1976. The fate of phosphorus during pedogenesis. *Geoderma*, 15, 1–19.

Walker, A. P., Beckerman, A. P., Gu, L., Kattge, J., Cernusak, L. A., Domingues, T. F., Scales, J. C. et al. 2014. The relationship of leaf photosynthetic traits V cmax and J max to leaf nitrogen, leaf phosphorus, and specific leaf area: A metaanalysis and modeling study. *Ecology and Evolution*, 4(16), 3218–3235. Available at http://doi.wiley.com/10.1002/ece3.1173.

Wang, B., and Qiu, Y. L. 2006. Phylogenetic distribution and evolution of mycorrhizas in land plants. *Mycorrhiza*, 16, 299–363.

Wang, R., Balkanski, Y., Boucher, O., Ciais, P., Peñuelas, J., and Tao, S. 2014. Significant contribution of combustionrelated emissions to the atmospheric phosphorus budget. *Nature Geoscience*, 8, 48–54. Available at http://www.nature.com/doifinder/10.1038/ngeo2324.

Wardle, D. A., Walker, L. R., and Bardgett, R. D. 2004. Ecosystem properties and forest decline in contrasting longterm chronosequences. *Science (New York, N.Y.)*, 305, 509–513. Available at http://www.ncbi.nlm.nih.gov/pubmed/15205475.

Weber, T., and Deutsch, C. 2012. Oceanic nitrogen reservoir regulated by plankton diversity and ocean circulation. *Nature*, 489, 419–422. Available at http://www.ncbi.nlm.nih.gov/pubmed/22996557.

White, A. F., and Blum, A. E. 1995. Effects of climate on chemical weathering in watersheds. *Geochimica et Cosmochimica Acta*, 59, 1729–1747.

Wolf, A., Doughty, C. E., and Malhi, Y. 2013. Lateral diffusion of nutrients by mammalian herbivores in terrestrial ecosystems. *PLoS ONE*, 8, e71 352.

Wright, D. P., Read, D. J., and Scholes, J. D. 1998. Mycorrhizal sink strength influences whole plant carbon balance of Trifolium repens L. *Plant, Cell and Environment*, 21, 881–891.

Xu, X., Thornton, P. E., and Post, W. M. 2013. A global analysis of soil microbial biomass carbon, nitrogen and phosphorus in terrestrial ecosystems. *Global Ecology and Biogeography*, 22, 737–749.

Yin, K., and Harrison, P. J. 2007. Influence of the Pearl River estuary and vertical mixing in Victoria Harbor on water quality in relation to eutrophication impacts in Hong Kong waters. *Marine Pollution Bulletin*, 54, 646–656.

Zaehle, S., and Dalmonech, D. 2011. Carbon–nitrogen interactions on land at global scales: Current understanding in modelling climate biosphere feedbacks. *Current Opinion in Environmental Sustainability*, 3, 311–320.

# 4 Phosphorus in Soil and Plants in Relation to Human Nutrition and Health

*Rajendra Prasad, Samendra Prasad, and Rattan Lal*

## CONTENTS

## 4.1 INTRODUCTION

There are 14 essential mineral nutrients for plant health. Of these, 6 are macronutrients and include nitrogen (N), phosphorus (P), potassium (K), calcium (Ca), magnesium (Mg), and sulfur (S). Essential micronutrients required for vigorous plant growth include iron (Fe), manganese (Mn), boron (B), copper (Cu), zinc (Zn), nickel (Ni), molybdenum (Mo), and chlorine (Cl) (Marschner 1995; Mengel et al. 2001; Fageria 2009; White and Brown 2010). Nutrients considered beneficial for plant growth include sodium (Na), selenium (Se), cobalt (Co), silicon (Si), and aluminum (Al) (Epstein 1999; White and Brown 2010). Additional mineral nutrients associated with animal and human nutrition and health include iodine (I), vanadium (V), fluorine (F), lithium (Li), lead (Pb), arsenic (As), cadmium (Cd), and mercury (Hg); the last six could be toxic when present in more than tolerable mounts (Graham et al. 2007; Lal 2009; Stein 2010). It may be added that sodium (Na) and chlorine (Cl) are major mineral nutrients for humans, while Cl is a micromineral and Na is only a beneficial nutrient for plants.

The deficiencies of minerals in diet, which cause hidden hunger or malnutrition, reported so far include those of Fe, Zn, I, Se, Ca, Mg, and Cu (Stein 2010), and soil degradation is the principal cause of malnutrition (Lal 2009). In general, Fe, Zn, and Cu are deficient in calcareous soils, Mg in sandy and acidic soils, Se in soils derived from igneous rocks (White and Brown 2010), and most mineral nutrients are deficient in eroded and depleted soils, especially those of the tropics (Lal 2009). Agronomic techniques of soil and nutrient management are used to enhance the uptake of essential elements (Graham et al. 2007; White and Brown 2010). Rather than the total concentration of these elements in soil, it is the low phytoavailability and the uptake by plant roots that is often a major constraint.

In addition to deficiency caused by low phytoavailability, there are also adverse effects caused by toxic levels of some heavy metals. For example, phosphate rocks (used to manufacture P fertilizers) contain toxic concentrations of Cd, Pb, Hg, U, Cr, and As. Superphosphates can contain high levels of these elements (Dissanayake and Chandrajith 2009). The high concentration of Cd in P fertilizers is linked with a disease called *Itai-Itai*. High concentration of U in P fertilizers is also related to several adverse effects leading to gastrointestinal, pulmonary, and kidney ailments (Kuttio et al. 2002; Bleise et al. 2003; Mendes et al. 2006; Oyedele et al. 2006; Takeda et al. 2006; Dissanayake and Chandrajith 2009; Chandrajith et al. 2010).

High dietary intake of P has also several adverse effects. Calvo and Uribarri (2013) reported that increasing the P content in the American diet (partly related to the growing use of processed food containing preservatives) is responsible for declining renal, cardiovascular, and bone health of the general population.

There has been considerable research on the amount, the nature, and the plant availability of native soil phosphorus and P applied as fertilizer to soil all over the world, and volumes of literature are available on the subject (Khasawneh et al. 1986; Harrison 1987; Prasad and Power 1997). In recent years, due attention has also been given to the pollution aspects, both the eutrophication of surface waters and the cadmium enrichment of soil (Chien et al. 2011). However, we have overlooked the fact that it is soil P that contributes along with soil calcium to the formation of bones and teeth in humans and other mammals through consumption of grains, vegetables, milk, poultry, and meat produced from the crops raised on farm fields and pastures. The entire skeleton is formed mostly by the mineral hydroxyapatite (HA) $(Ca_{10}(PO_4)_6(OH)_2)$, which is one of the forms in which it is found in natural phosphate rock deposits; of course, the major component in these deposits is fluorapatite FA $(Ca_{10}(PO_4)_6.F_2)$. The skeletons from animals and humans when buried in soil complete the P cycle (soil–plant–animal/humans–soil) in nature (Ashley et al. 2011).

Since its discovery by the German alchemist Henning Brandt, the importance of P to human health and environment quality has been widely recognized. This review briefly focuses on the journey of P from soil to plants and from plants to humans, with specific objectives of describing the impact of P in a diet on human health and well-being.

## 4.2 PHOSPHORUS IN SOIL

Phosphorus is the tenth most abundant element and constitutes about 0.12% of the earth's crust (Van Wazer 1958). The total P content in the surface soil and the subsoil may vary from a few milligrams per kilogram to over 1 g kg$^{-1}$ (Prasad and Power 1997), and the two major forms are organic-P and inorganic-P. Organic-P may vary from 20 to 65% or even more of the total P; the higher values are for organic soils (histosols) (Table 4.1). Further organic-P in surface soils is a smaller fraction of the total soil P in the warm regions than it is in the cool regions of the world; about 35.2% as organic-P in equatorial regions (countries between 40° parallels) and 48.6% in colder regions (Harrison 1987). Inorganic-P in soil is generally Al-, Fe-, and Ca- bound (Chang and Jackson 1957; Levy and Schlesinger 1999; Jalali and Tabar 2011). This is because soil P in its water-soluble form $\left(H_2PO_4^-\right)$ is highly reactive and forms insoluble precipitates of Ca, Fe, and Al (Williams 1950; Sample et al. 1986; Tomar 2000). The concentration of P in a soil solution is therefore generally very low (about 10 μM). Also because of its high reactivity, P is not absorbed by plant roots by mass flow as nitrates but by diffusion, which is again very slow ($10^{-12}$– $10^{-15}$ m$^2$ s$^{-1}$) (Schachtman et al. 1998). Only a very small fraction of the total soil P becomes available to growing plants, and a number of soil tests have been proposed to estimate the available P in soils (Kamprath and Watson 1980; Cox 2001; Prasad 2013). Most of these tests, however, focus on determining the release of Ca-, Al-, and Fe-bound P, although some organic-P is also extracted. Ron-Vaz et al. (1993) and Shand et al. (1994) opine that organic forms of P are predominant in a soil solution. These organic forms of P are derived from the turnover of soil microorganisms, have relatively greater mobility than orthophosphates (Hellal and Dessler 1989; Seeling and Zasoski 1993), and therefore are of critical importance to soil P dynamics and

## TABLE 4.1
## Organic P Content of Surface Mineral/Organic Soils in Relation to Soil Texture

| Soil Texture | Number of Samples | Organic P (mg kg$^{-1}$ Soil) | Percentage of Total P |
|---|---|---|---|
| **Mineral Soils** | | | |
| Sands | 194 | 121 | 34.1 |
| Loams | 663 | 250 | 39.9 |
| Clay loams and clays | 309 | 332 | 41.4 |
| **Organic Soils** | | | |
| Organic loams | 5 | 523 | 58.9 |
| Peats | 85 | 579 | 65.4 |

*Source:* Harrison, A. F., *Soil Organic Phosphorus—A World Review of Literature*, CABI International, Wallingford, 1987. With permission.

availability within the rhizosphere (Richardson et al. 2005). The direct hydrolysis of organic-P and the subsequent utilization of released orthophosphate have been demonstrated (McLaughlin et al. 1988). The determination of potentially available P methods therefore need rethinking and cannot afford to neglect the contribution of organic-P. This has become more important in view of the increasing demand for organic foods.

The deficiency in soil P for crop production is met by the application of phosphate fertilizers, of which there are quite a few including finely ground phosphate rock, partially acidulated phosphate rock, ordinary (single) superphosphate, triple super-phosphate, monoammonium phosphate, diammonium phosphate, nitrophosphate, and ammonium polyphosphate (Prasad and Power 1997).

## 4.3   PHOSPHORUS IN PLANTS

### 4.3.1   FUNCTIONS AND DEFICIENCY SYMPTOMS

The essentiality of P for plants was established by Liebig in 1839 and by Ville in 1861 (Prasad et al. 2014b). Phosphorus molecule ATP often called as the *molecular unit of currency of energy transport* (Knowles 1980) is used in photosynthesis and other biological processes, such as cell division, intracellular signaling (Mishra et al. 2006), extracellular signaling (Byrnes 2012), and DNA and RNA synthesis (Stubbe 1990). Another P molecule, nicotinamide adenine dinucleotide phosphate (NADP)/nicotinamide adenine dinucleotide phosphate-oxidase (NADPH), is involved in the Calvin cycle of carbon dioxide assimilation. NADP/NADPH is also used for meta-bolic processes, such as lipid synthesis, cholesterol synthesis, and fatty acid chain elongation. NADPH is also the reducing agent responsible for destruction of reactive oxygen species (Rush et al. 1985).

Phosphorus deficiency leads to reduced photosynthesis (Foyer and Spencer 1986; Rodriguez et al. 1998). Phosphorus is needed for root growth and biological nitrogen fixation in legumes (Israel 1987). Phosphorus is also involved in reproduction, and P fertilization helps in hastening maturity in maize, wheat, cotton, and other crops (Anonymous 1999).

The deficiency symptoms of phosphorus in plants include stunting, darkening of young leaves, and purpling of older leaves (Prasad and Power 1997). In extreme phosphorus deficiency, the older leaves may turn brown and die.

### 4.3.2   ABSORPTION OF SOIL P BY PLANTS

Phosphorus is taken up by plants mostly as $H_2PO_4^-$ and to some extent as $HPO_4^{=}$. The concentration of $H_2PO_4^-$ (abbreviated as Pi) in a soil solution is generally very low (about 10 μM). As a contrast, the Pi concentration in root cells and xylem is a hundred- or thousandfold (5–10 mM) (Lee and Radcliff 1993; Mimura 1995), and this steep concentration gradient helps in the Pi uptake by plants (Mengel et al. 2001). However, most Pi is taken up with the help of a cotransport, which could be $H^+$ or some other source. The review of available literature (Schachtman et al. 1998) brings out that as per the kinetic approach, there appear to be two types of transporters with different affinities (Drew and Saker 1984; Nandi et al. 1987), while the molecular approach suggests that there are at least four genes that encode Pi transporters (Muchhal et al. 1996; Smith et al. 1997).

The absorption of soil P by plants depends upon root characteristics (Dinkelekar et al. 1995; Lynch and Brown 2001; Fan et al. 2003; Rubio et al. 2003; Gahoonia and Nielsen 2004; Hill et al. 2006; Lambers et al. 2006; Banerjee et al. 2010; Postma and Lynch 2011) and their capacity to release organic acids (Bar-Yosef 1991; Gerke 1993; Hissinger and Gikes 1996; Jones 1998; Hissinger 2001; Dakora and Phillips 2002; Jones et al. 2003) and phosphatases (Tadano et al. 1993; Hayes et al. 1999; Richardson et al. 2005). Phosphate application improves root length, volume, and dry matter (Table 4.2). Root–mycorrhizal associations also increase the soil P accessibility to plants (Bolan 1991; Jakobsen et al. 1992; Smith and Read 2008; Richardson et al. 2009). Mycorrhizal plants therefore have two pathways for P uptake: (1) root–soil interface and (2) via vesicular arbuscular mycorrhiza (VAM) hyphae (Smith et al. 2003). VAM is reported to help in the direct use of finely ground phosphate rock (Banerjee et al. 2010) (Table 4.2).

Once absorbed by the roots, Pi is transported in the xylem to the younger leaves. The Pi from the older leaves also moves to the younger leaves and some from the shoots to the roots. The concentration of phosphorus in the xylem may range from 1 mM/L in Pi-starved plants to 7 mM/L in plants grown in solutions containing 125 µM/L Pi (Mimura et al. 1996). In Pi-starved plants, some of the Pi translocated from the shoots to the roots in the phloem is transferred back to the xylem and recycled back to the shoots (Jeschke et al. 1997). In the xylem, phosphorus is almost solely present as Pi, while in the phloem, significant amounts of organic phosphorus compounds are also present (Schachtman et al. 1998). The conversion of Pi to organic compounds is fairly rapid (Jackson and Hager 1980).

## TABLE 4.2
## Effect of P Fertilization on Some Root Parameters of Corn

| Treatment | Root Length (m/plant) | Root Volume (cm³/plant) | Root Dry Weight (g/plant) |
|---|---|---|---|
| 120 kg N/ha | 3.49 | 2.50 | 0.84 |
| 120 kg N + 26 kg P/ha as SSP | 6.24 | 6.25 | 3.42 |
| 120 kg N + 13 kg P/ha as PR + VAM | 8.60 | 10.0 | 4.10 |
| LSD (0.05) | 0.354 | 0.644 | 0.250 |

*Source:* Banerjee, M., Rai, R. K., and Maiti, D., *Arch Agron Soil Sci*, 56(6), 681–695, 2010.

*Note:* PR: phosphate rock, SSP: single superphosphate, VAM: vesicular arbuscular mycorrhyza.

### 4.3.3 Recovery of Applied P by Crop Plants

The recovery of P applied to crop plants (REp) is generally determined by the following expression: REp = 100[(P uptake (kg/ha) in P fertilized plots − P uptake (kg/ha) in check (no P) plots)/P applied (kg/ha)].

This method is known as the difference method. The REp varies from 11 to 46% depending upon the soil, the crop, the source, and the rate of application. In general, the REp decreases as the rate of P application is increased. Some data from recent field experiments in India are in Table 4.3. In rice–wheat cropping system, the REp is lower in rice than in wheat, because rice is grown under submerged conditions. This is partly due to the increased availability of native occluded P (Patrick and Mahapatra 1968), which of course varies from soil to soil (Tomar 2000). This reduces the response of rice to P application (De-Datta 1981). Since wheat is grown under aerobic conditions, the availability of occluded P is reduced. Also during winter–spring, the wheat-growing season soil temperatures are generally lower, which results in reduced oxidation of soil organic matter and thereby lesser release of organic P, forcing the plants to absorb more of the applied fertilizer P.

### 4.3.4 Storage of P in Plant Products

In grains and seeds, phosphorus is mostly stored as phytates (salts of phytic acid [inositol hexakisphosphate]). The phytate content (%) in some grains, seeds, and nuts is wheat (0.39–1.35), maize (0.75–2.22), oat (0.42–1.16), polished rice (0.14–0.60),

## TABLE 4.3
## Recovery Efficiency of P Plants

| Crop | Location (Soil Texture) | P Applied (kg/ha) | REp (%) | P Applied (kg/ha) | REp (%) |
|---|---|---|---|---|---|
| Rice-(wheat)[a] | RS Pura (CL) | 13 | 19.6 (26.9) | 26 | 9.9 (19.9) |
| | Palampur (SL) | 13 | 22.3 (46.2) | 26 | 15.8 (33.7) |
| | Ludhiana (LS) | 13 | 30.0 (24.4) | 26 | 14.2 (24.4) |
| | Pantnagar (CL) | 13 | 28.5 (32.4) | 26 | 17.7 (22.5) |
| | Faizabad (CL) | 13 | 24.4 (41.0) | 26 | 19.5 (32.0) |
| | Ranchi (SiCL) | 13 | 30.4 (22.7) | 26 | 27.1 (19.2) |
| Corn[b] | Udaipur (CL) | 20 | 28.0 | 60 | 12.0 |
| Pigeonpea[c] | Delhi (SL) | 17.2 | 33.0 | 34.4 | 29.0 |
| Groundnut[d] | Delhi (SL) | 17.5 | 12.0 | 35 | 11.0 |
| Cotton[d] | Delhi (SL) | 17/5 | 16.0 | 35 | 18.0 |

*Note:* Soil texture: CL: clay loam, LS: loamy sand, SiCL: silty clay loam, SL: sandy loam.

[a] Singh et al. (2014).

[b] Mehta et al. (2005).

[c] Singh and Ahlawat (2007).

[d] Sepat et al. (2012).

chickpea (0.56), lentils (0.44–0.50), soybean (1.0–2.22), linseed (2.15–2.78), almonds (1.35–3.22), and groundnut (0.95–1.76) (Reddy and Sathe 2001). The phosphorus in phytates is generally not assimilable by humans and nonruminant animals and contributes to the pollution of soil and surface waters. In ruminants, the microorganisms in rumen produce phytase that helps in phytate digestion (Klopfenstein et al. 2002). Adding phytase to the diet can increase the availability of phosphorus from phytates. Phytate reduces the availability of Zn and Fe from the grains, which may lead to their deficiency in humans (Prasad et al. 2014a). Low phytate mutants have been developed in some crops (Guttieri et al. 2006). Sprouting reduces phytate content in grains (Malleshri and Desikachar 1986) and is a technique for increasing the availability of Zn and Fe in grains.

## 4.4 PHOSPHORUS IN HUMAN DIET

### 4.4.1 AMOUNTS AND DISTRIBUTION

Phosphorus nutrition and health has received considerable attention (Calvo et al. 2014). Phosphorus makes about 0.5% of the weight of a newborn infant's body (Fomon and Nelson 1993) and from 0.65 to 1.1% weight in the adult human body (Aloia et al. 1984). About 85% of phosphorus in the human body is in the bones and the teeth, 14% in the soft tissue, and the remaining 1% in the extracellular fluid (Amanzadeh and Reilly 2006). P in bones is mostly present as HA; however, while the Ca:P ratio in HA is 1.67, the bone mineral has a Ca:P ratio ranging from 1.37 to 1.87 due to the presence of additional ions such as Si, Zn, and carbonate (University of Cambridge DoITPoMS via the Internet).

The total P concentration in human blood is 13 mmol/L (40 mg/dL); most of which is as the phospholipids of red blood cells and plasma lipoproteins. Only a fraction (0.1% of the total) is present as inorganic P (Pi), but this tiny fraction is of crucial importance. In adults, the Pi is about 1.5 mmol (4.65 mg/dL) and is mainly located in the blood and extracellular fluids. It is into the Pi compartment that phosphorus is inserted upon absorption from the diet and resorption from the bones. Also most urinary P and HA mineral P is derived from this compartment. This compartment is also the primary source from which the cells of all tissues derive both structural and high-energy phosphate (https:/fnic.nal.usda.gov/.../dri .../calcium-phosphorus-magnesium-vitamin-d).

### 4.4.2 FUNCTIONS

In addition to being a major component of bones and teeth (Takeda et al. 2012), the other functions of P are (1) activation of several catalytic processes by phosphorylation, (2) temporary storage and transfer of energy derived from the metabolic processes, and (3) buffer in maintaining the normal pH in cells. Since phosphorus is not irreversibly consumed in these processes, the first function of dietary P is to support the tissue growth, and the second function is to replace excretory and dermal loss of P (Nordin 1976).

### 4.4.3 ABSORPTION AND RECOMMENDED DIETARY ALLOWANCE

On an average about 800–1400 mg P is ingested daily (Reilly 2005). Phosphorus is absorbed as Pi in the upper small intestine driven by a Na-adopted process, and vitamin D helps in its absorption (Chen et al. 1974; Segawa et al. 2004; Capuano et al. 2005). The recommended dietary allowance (RDA) of P varies from 100 mg/day or less for 0–6 month old babies to 700 mg/day for adults. The RDA for P is most (1250 mg/day) in children and adolescents of 9–18 years of age, because the bone growth is most during this period (Table 4.4) (Standing Committee on the Scientific Evaluation of Dietary Intake, 1997).

### 4.4.4 DEFICIENCY/TOXICITY SYMPTOMS

Phosphorus deficiency can lead to lowered appetite, anemia, muscular pain, improper bone formation (rickets), numbness, and a weakened immune system (Pivnick et al. 1995; Amanzadeh and Reilly 2006; Heaney 2012). The normal serum P concentration is 0.80–1.45 mmol/L (2.5–4.5 mg/dL), and moderate P deficiency (hypophosphatemia) is indicated at serum P concentration of 0.32–0.80 mmol/L (1.0–2.5 mg/dL), while severe hypophosphatemia can occur at serum P concentration of <0.32 mmol/L (<1.0 mg/dL). Since phosphorus is present in most foods, hypophosphatemia is generally not encountered (Lloyd and Johnson 1988), and

## TABLE 4.4
## RDA (Recommended Dietary Allowance) and AI (Adequate Intake) for Phosphorus

| Life Stage | Age | Males (mg/day) | Females (mg/day) |
|---|---|---|---|
| Infants | 0–6 months | 100 (AI) | 100 (AI) |
| Infants | 7–12 months | 275 (AI) | 275 (AI) |
| Children | 1–3 years | 460 | 460 |
| Children | 4–8 years | 500 | 500 |
| Children | 9–13 years | 1250 | 1250 |
| Adolescents | 14–18 years | 1250 | 1250 |
| Adults | 19 years and older | 700 | 700 |
| Pregnancy and Breast-feeding | 18 years and younger | | 1250 |
| Pregnancy and Breast-feeding | 19 years and older | | 700 |
| Breast-feeding | 18 years and younger | | 1250 |
| Breast-feeding | 19 years and older | | 700 |

*Source:*  Standing Committee on the Scientific Evaluation of Dietary Reference Intake. 1997. Chapter 5. Phosphorus pp. 146–189. In *Dietary Reference Intake for Calcium, Phosphorus, Magnesium, Vitamin D and Fluoride.* Institute of Medicine (US), Food & Nutrition Board, Washington, D.C., National Academic Press (http://www.nap.edu/read/5776/Chapter/7).

*Disclaimer:*  The information in this table is of general nature. Individuals must consult nutrition specialists for their specific needs.

severe hypophosphatemia might affect only about 0.43% of hospitalized patients, but it could be higher in alcoholics (0.9%), septic patients (2.4%), malnourished patients (10.4%), and patients with diabetic ketoacidosis (14.6%) (Amanzadeh and Reilly 2006). Severe malnutrition contributes to up to 50% of childhood mortality in developing countries, and it is frequently characterized by electrolyte depletion, including low total body phosphate in children treated at Kenyatta National Hospital with kwashiorkor (protein deficiency). Among children with mild, moderate, and severe hypophosphatemia, 8, 14, and 21% died respectively (Institute of Science in Society 2014). Hypophosphatemia can occur on recovery from alcoholic bouts and from overdose of insulin in diabetics (Matsumma et al. 2000). Excess intake of aluminum-containing antacids also reduces the P intake and leads to hypophosphatemia (Lotz et al. 1988).

On the other hand, having too much phosphorus in the body is actually more common and more worrisome than having too little. Consuming too much phosphorus leads to hyperphosphatemia that causes the body to send calcium from the bones to the blood in an attempt to restore balance. This transfer of calcium weakens the bones, and it can cause calcification of internal organs, increasing risk of heart attack and other vascular diseases. Thus, hyperphosphatemia is central to the development of chronic kidney disease (CKD), mineral bone disorder, and cardiovascular disease (CVD), which are the leading causes of morbidity and mortality worldwide (Martin and Gonzales 2011). In the early stages of CKD, the increase in the phosphate level is relatively small but results in high concentrations of parathyroid hormone (PTH) as well as fibroblast growth factor-23 accompanied with abnormally low 1,2 5-dihydroxyvitamin D (1,25D). (Suki and Massery 1997; Hori 2004; Dhingra

**TABLE 4.5**
**Some High and Low Phosphorus Concentration Foods**

| High Phosphorus Foods | | Low Phosphorus Foods | |
|---|---|---|---|
| Food | Phosphorus (mg) | Food | Phosphorus (mg) |
| 8 oz. milk[a] | 230 | 8 oz. nondairy creamer[a] | 100 |
| 1/2 cup ice cream[a] | 80 | 1/2 cup sherbet or 1 popsicle[a] | 0 |
| 12 oz. can cola[a] | 55 | 12 oz. of ginger ale or lemon soda[a] | 3 |
| 1 1/2 oz. choc-bar[a] | 125 | 1 1/2 oz. hard candy, fruit flavors or jelly beans[a] | 3 |
| 100 g peanuts, roasted, salted[b] | 370 | 100 g chestnut[b] | 70 |
| 100 g brown bread, sliced | 160 | 100 g rice boiled[b] | 35 |
| 100 g wheat bran[b] | 1200 | 100 g wheat, starch reduced[b] | 200 |

[a] National Kidney Foundation, *Phosphorus and Your CKD Diet*, National Kidney Foundation, New York, 2015. Available at https://www.kidney.org/atoz/content/phosphorus.

[b] *Asia Pacific Journal of Clinical Nutrition*, National Institute of Health, Zhunan Township, Taiwan. Available at apjcn.nhri.org.tw. With permission.

et al. 2007; Razzaque 2009). Plasma-low phosphorus concentration suppresses erythrocyte synthesis and storage of 2,3-diphosphoglycerate (2,3-DPG), which plays an important role in the affinity of hemoglobin for oxygen (Shrier 2007; ISIS 2014). These abnormalities, together with reductions in urinary calcium, individually and collectively contribute to renal bone disease, vascular calcification, and CVD. Serum phosphate levels associated with the adverse effects of hyperphosphatemia are >3.5–4.0 mg/dL or higher (de Boer et al. 2009; Tonelli et al. 2005).

The remarks made here are of a very general nature. For more clinical details on hypophosphatemia and hyperphosphatemia, reference may be made to Suki and Massery (1997), Shrier (2007), and Martin and Gonzales (2011).

A list of some high and low phosphorus foods are provided in Table 4.5. Some recent research suggests that phosphorus is more easily absorbed from meat products, and one can only absorb half of the phosphorus contained in plant foods. In a recent study, one week of a vegetarian diet led to lower serum phosphorus levels and decreased FGF-23 levels (Sharon et al. 2011).

## 4.5 SUMMARY AND CONCLUSIONS

Strong and healthy bones are a must for active human life. Bones are made up of calcium and phosphorus-forming mineral HA. Both Ca and P needed for bone formation are obtained from plant and animal food products. In addition to being a component of bones, biological P molecules ATP and NADP are involved in several biochemical processes in plant, animal, and human physiologies. Further P is also a component of DNA and RNA and plays a key role in carrying hereditary characteristics from generation to generation. Plants acquire their P from soils, and animals obtain it from plants. Thus, both native soil P and fertilizer P are the main sources of P for plants, animals, and humans. Plant roots absorb P mostly as $H_2PO_4^-$ and some as $HPO_4^=$. However, $H_2PO_4^-$ is water-soluble and reacts rapidly with Ca, Al, and Fe in a soil solution resulting in insoluble compounds. Most P absorption by plant roots is therefore by diffusion and not by mass flow as in the case of nitrates. Plants develop their root system with increased specific root surface and root volume to forage greater soil volume to procure more P. Root–mycorrhizal associations also help in foraging for P. Plant roots also exude organic acids and phosphatases for increasing P absorption. Phosphorus deficiency in plants leads to poor root development, stunted growth, and dark and purple coloration of leaves. Since soil P (native and applied) is the only source for acquiring P by plants and humans, careful P fertilization of soils and its efficient management should receive utmost attention. Insufficient phosphate could thus be a serious problem that will worsen as the world's supply of phosphate ore shrinks.

## REFERENCES

Aloia, J. F., A. N. Vaswani, J. K. Yeh, K. Ellis, and S. H. Cohn. 1984. Total body phosphorus in post-menopausal women. *Minor Electrolyte Metals* 10: 73–76.
Amanzadeh, J., and R. F. Reilly Jr. 2006. Hypophosphatemia: An evidence-based approach to its clinical consequences and management. *Nat Clin Practice Nephrol* 2(3): 136–148.

Anonymous. 1999. Effects of phosphorus on crop maturity. *Better Crops* 83(1): 14–19.

Ashley, K., D. Cordell, and D. Mavinic. 2011. A brief history of phosphorus: From the philosopher's stone to nutrient recovery and reuse. *Chemosphere* 84: 737–746.

*Asia Pacific Journal of Clinical Nutrition*. National Institute of Health, Zhunan Township, Miaoli. Available at http://apjcn.nhri.org.tw.

Banerjee, M., R. K. Rai, and D. Maiti. 2010. Root characteristics of maize as influenced by various phosphatic chemical fertilizers and biofertilizers. *Arch Agron Soil Sci* 56(6): 681–695.

Bar-Yosef, B. 1991. Root excretions and their environmental effects: Influence of availability of phosphorus. In *Plant Roots: Their Hidden Half*. Eds. Waisal, Y., A. Eshel, and U. Kafkafi. New York, Marcel Dekker, pp. 529–557.

Bleise, A., P. R. Danesi, and W. Burkat. 2003. Properties, use and health effects of depleted uranium (DU): A general overview. *J Environ Radioactiv* 64: 93–112.

Bolan, N. S. 1991. A critical review on the role of mycological fungi in the uptake of phosphorus by plants. *Plant Soil* 133: 189–207.

Byrnes, C. 2012. Treating small fiber neuropathy symptoms with ATP. *Working Towards Wellness* (info@Chris Byrnes.com).

Calvo, M. S., and J. Uribarri. 2013. Public health impact of dietary phosphorus excess on bone and cardiovascular health in general population. *Am J Nutr* 98: 6–15.

Calvo, M. S., A. J. Mospheg, and K. L. Tucker. 2014. Assessing the health impact of phosphorus in food supply: Issues and circumstances. *Adv Nutr* 5(1): 104–113.

Capuano, P., T. Radanovic, C. A. Wagner et al. 2005. Intestinal and renal adaptation to low-Pi diet of type II NaPi cotranspoters in vitamin D receptor- and 1αOHase-deficient mice. *Am J Physiol Cell Physiol* 288: C429–C434.

Chandrajith, R., S. Seneviratna, K. Wickramaarachchi, T. Attanayake, T. N. C. Aturaliya, and C. B. Dissanayke. 2010. Natural radionuclides and trace elements in rice field soils in relation to fertilizer application study of chronic kidney disease (CKD) area in Sri Lanka. *Environ Earth Sci* 60: 193–201.

Chang, S. C., and M. L. Jackson. 1957. Fractionation of soil phosphorus. *Soil Sci* 84(2): 133–144.

Chen, T. C., L. Castillo, M. Korycka-Dahl, and H. F. De Luca. 1974. Role of vitamin D in phosphate transport of rat intestine. *J Nutr* 104: 1056–1060.

Chien, S. H., L. I. Prochnow, S. Tu, and C. S. Snyder. 2011. Agronomic and environmental aspects of phosphate fertilizers varying in source and solubility: An update review. *Nutr Cycl Agroecosystems* 89(2): 229–255.

Cox, M. S. 2001. The Lancaster soil test method as an alternative to the Mehlich 3 soil test method. *Soil Sci* 166(7): 484–489.

Dakora, F. D., and D. A. Phillips. 2002. Root exudates as mediators of mineral acquisition in low nutrient environments. *Plant Soil* 245: 35–47.

de Boer, I. H., T. C. Rue, and B. Kesterbaum. 2009. Serum phosphorus concentrations in the Third National Health and Nutrition Examination Survey (NHANES II). *Am J Kidney Dis* 53: 399–407.

De-Datta, S. K. 1981. *Principles and Practices of Rice Production*. New York, John Wiley.

Dhingra, R., L. M. Sullivan, C. S. Fox et al. 2007. Relation of serum phosphorus and calcium levels to the incidence of cardiovascular diseases in the community. *Arch Int Med* 167(9): 879–885.

Dinkelekar, B., C. Hengeler, and H. Marschner. 1995. Distribution and function of proteoid and other root clusters. *Botanica Acta* 108: 183–200.

Dissanayake, C. B., and R. Chandrajith. 2009. Phosphate mineral fertilizers, trace metals and human health. *J Natn Sci Foundation Sri Lanka* 37: 153–165.

Drew, M. C., and L. R. Saker. 1984. Uptake and long distance transport of phosphate, potassium and chloride in relation to internal ion concentration in barley: Evidence of non-alloteric regulation. *Planta* 60: 500–507.

Epstein, E. 1999. Silicon. *Ann Rev Plant Physiol Plant Mol Biol* 50: 647–664.

Fageria, N. K. 2009. *The Use of Nutrients in Crop Plants*. New York, CRC Press.

Fan, M. S., J. M. Zhu, C. Richards, K. M. Brown, and J. P. Lynch. 2003. Physiological role of aerenchyma in phosphorus stressed roots. *Funct Plant Biol* 30: 493–506.

Fomon, S. J., and S. E. Nelson. 1993. *Calcium, phosphorus, magnesium and sulfur in nutrition of normal infants*. Ed. Fomon, S. J. St. Louis, MO, Mopsby Year Book, pp. 192–216.

Foyer, C., and C. Spencer. 1986. The relationship between phosphorus status and photosynthesis in leaves. *Planta* 167: 369–375.

Gahoonia, T. S., and N. E. Nielsen. 2004. Root traits as tools for creating phosphorus efficient crop varieties. *Plant Soil* 260: 47–57.

Gerke, J. 1993. Solubilizaion of Fe (III) from humic-Fe complexes, humic Fe oxide mixtures and from poorly eroded Fe-oxide by organic acids: Consequences for P-adsorption. *Z Pflanzen Bodenkunde* 156: 253–257.

Graham, R. D., R. M. Welch, D. A. Sanders et al. 2007. Nutritious subsistence food systems. *Adv Agron* 92: 1–74.

Guttieri, M. J., K. M. Peterson, and E. J. Souza. 2006. Milling and baking quality of low phytic acid wheat. *Crop Science* 46(6): 2403–2408.

Harrison, A. F. 1987. *Soil organic Phosphorus—A World Review of Literature*. Wallingford, CABI International.

Hayes, J. E., A. E. Richardson, and R. J. Simpson. 1999. Phytase and acid phytase activities in roots of temperate pasture grasses and legumes. *Aust J Plant Physiol* 26: 801–809.

Heaney, R. P. 2012. Phosphorus. In *Present Knowledge in Nutrition*. Eds. Erdman Jr., J. W., I. A. Macdonald, and S. H. Zeisel. 10th Ed. Ames, IA, Wiley-Blackwell.

Hellal, H. M., and A. Dessler. 1989. Mobilization and turnover of soil phosphorus in the rhizosphere. *Z Pflanz Bodenkunde* 152: 175–180.

Hill, J. O., R. J. Simpson, A. D. Moore, and D. F. Chapman. 2006. Morphology and response of roots of pasture species to phosphorus nd nitrogen nutrition. *Plant Soil* 286: 7–19.

Hissinger, P. 2001. Bioavailability of inorganic P in the rhizosphere as affected by root-induced chemical changes: A review. *Plant Soil* 237: 173–195.

Hissinger, P., and R. J. Gikes. 1996. Mobilization of phosphate from phosphate rock and alumina-sorbed phosphates by the roots of ryegrass and clover as related to rhizosphere pH. *Eur J Soil Sci* 47: 533–544.

Hori, W. H. 2004. The clinical consequences of secondary hyperparathyroidism: Focus on clinical outcomes. *Nephrol Dial Transplant* 19(Suppl. 5): V2–V8.

Institute of Science in Society. 2014. Phosphate starvation threatens the world. British Library, London, Institute of Science in Society. UK Report 22/01/14 (www.i-sis-org.uk).

Israel, D. W. 1987. Investigation on the role of phosphorus in symbiotic nitrogen fixation. *Plant Physiol* 84(3): 835–840.

Jackson, P. C., and C. E. Hager. 1980. Products of orthophosphate absorption by barley roots. *Plant Physiol* 35: 326–332.

Jakobsen, I., L. K. Abbot, and A. D. Robson. 1992. External hyphae of vesicular arbuscular mycorrhyzal fugi associated with *Trifolium subterraneum* L/I: Spread of hyphae and phosphorus inflow into roots. *New Phytol* 120: 371–380.

Jalali, M., and S. Tabar. 2011. Chemical fractionation of phosphorus in calcareous soils of Hamedan, western Iran, under different land use. *J Plant Nutr Soil Sci* 174: 523–531.

Jeschke, W., E. Kirkby, A. Peuke, J. Pate, and W. Hartung. 1997. Effect of phosphorus efficiency on assimilation and transportation of nitrate and phosphate in intact plants of castor bean (*Ricinus communis* L). *J Exp Bot* 48: 75–91.

Jones, D. L. 1998. Organic acids in the rhizosphere: A critical review. *Plant Soil* 205: 25–44.

Jones, D. L., P. G. Dennis, A. G. Owen, and P. A. W. van Hees. 2003. Organic acid behavior in soils—Misconceptions and knowledge gaps. *Plant Soil* 248: 31–41.

Kamprath, E. J., and M. E. Watson. 1980. Conventional soil and tissue tests for assaying the phosphorus status of soils. In *The Role of Phosphorus in Agriculture*. Eds. Khasawneh, F. E., E. C. Sample, and J. E. Kampreth. Madison, WI, American Society of Agronomy and Soil Science Society of America, pp. 433–469.

Khasawneh, F. E., E. C. Sample, and J. E. Kampreth (Eds). 1986. *The Role of Phosphorus in Agriculture*. Second Ed. Madison, WI, American Society of Agronomy and Soil Science Society of America.

Klopfenstein, T. J., A. Rosalina, G. Cromwell et al. 2002. Animal diet modification to decrease the potential for nitrogen and phosphorus pollution. *Faculty Papers and Publications in Animal Science*, Pub. 518. Lincoln, NE, DigitalCommons@University of Nebraska, Lincoln.

Knowles, J. R. 1980. Enzyme-catalyzed phosphoryl transfer reaction. *Annu Rev Biochem* 49: 877–919.

Kuttio, P., A. Auvinen, L. Saloven, H. Saha, J. Pekkanen, I. Makelainen, S. B. Vaisanene, I. M. Penttila, and H. Komulainen. 2002. Renal effects of uranium in drinking water. *Environ Health Perspect* 110(4): 337–342.

Lambers, H., M. W. Shane, M. D. Cramer, S. J. Pearse, and E. J. Veneklass. 2006. Root structure and functioning for efficient acquisition of phosphorus matching morphological and physiological traits. *Ann Bot* 98: 693–713.

Lal, R. 2009. Soils and world food security. *Soil & Tillage Res* 102: 1–4.

Lee, R. B., and R. G. Radcliffe. 1993. Subcellular distribution of major phosphate and levels of nucleoside triphosphate in mature maize roots at low external phosphate concentrations: Measurement with $^{31}$P NMR. *J Exp Bot* 44: 587–598.

Levy, E. C., and W. H. Schlesinger. 1999. A comparison of fractionation methods for soil phosphorus. *Biogeochemistry* 47: 25–38.

Lloyd, C. W., and C. E. Johnson. 1988. Management of hypophosphatemia. *Clinic Pharmacy* 7: 123–128.

Lotz, M., E. Ziman, and F. C. Barter. 1968. Evidence of a phosphorus depletion syndrome in man. *New England J Med* 278: 409–415.

Lynch, J. P., and K. M. Brown. 2001. Top-soil foraging—An architechtural adapatation of plants to low phosphorus availability. *Plant Soil* 237: 225–237.

Malleshri, N. G., and H. S. R. Desikachar. 1986. Nutritive value of malted flour. *Plant Food Human Nutr* 26(3): 191–196.

Marschner, H. 1995. *Mineral Nutrition of Higher Plant*. London, Academic Press.

Martin, K. A., and E. A. Gonzales. 2011. Prevention and control of phosphorus retention/ hyperphosphatemia in CKD.M.D: What is normal, where to start and how to treat. *Clin J Am Soc Nephrol* 6(2): 440–446.

Matsumma, M., A. Nakashima, and Y. Tofuku. 2000. Electrolyte disorder following massive insulin overdose in a patient with Type 2 diabetes. *Intern Med* 39(2): 55–57.

McLaughlin, M. J., A. M. Alsto, and J. K. Martin. 1988. Phosphorus cycling in wheat-pasture rotations: II: The role of the microbial biomass in phosphorus cycling. *Aust J Soil Res* 26: 333–342.

Mehta, Y. K., M. S. Shakavat, and S. M. Singh. 2005. Influence of sulfur, phosphorus and farmyard manure on yield attributes and yield of maize (*Zea mays*) in Southern Rajasthan conditions. *Indian J Agron* 50(3): 203–205.

Mendes, A. M. S., G. P. Duda, G. W. A. Nascimentro, and M. O. Silva. 2006. Bioavailability of cadmium and lead in a soil amended with phosphorus fertilizers. *Scientia Agricola* 63: 328–332.

Mengel, K., E. A. Kirkby, H. Kosegarten, and T. Appel. 2001. *Principles of Plant Nutrition*. Dordrecht, Kluwer Academic Publisher.

Mimura, T. 1995. Homestatis and transport of inorganic P in plants. *Plant Cell Physiol* 36: 1–7.

Mimura, T., K. Sakano, and T. Shimmen. 1996. Studies on the distribution of inorganic phosphate in barley leaves. *Plant Cell Environ* 19: 311–320.

Mishra, N., R. Tuteja, and N. Tuteja. 2006. Signaling through MAP kinase network in plants. *Arch Biochem Biophys* 452(1): 55–68.

Muchhal, U. S., J. M. Pardo, and K. G. Raghothama. 1996. Phosphate transportation in the higher plant *Arabidopsis thaliana*. *Proc Natl Acad Sci USA* 93: 10519–10523.

Nandi, S. K., R. C. Pant, and P. Nissen. 1987. Multiphasic uptake of phosphate by corn roots. *Plant Cell Environ* 10: 463–474.

National Kidney Foundation. 2015. *Phosphorus and Your CKD Diet*. New York, National Kidney Foundation (via internet).

Nordin, B. E. C. 1976. *Calcium, Phosphorus and Magnesium Metabolism*. Edinburgh, Churchill-Livingston.

Oyedele, D. J., C. Asonugho, and O. O. Awotoye. 2006. Heavy metals in soil and accumulation by edible vegetables after phosphate fertilizer application. *Electric J Environ Ag Food Chem* 5(4): 1446–1453.

Patrick, Jr. W. H., and I. C. Mahapatra. 1968. Transformation and availability to rice of nitrogen and phosphorus in waterlogged soils. *Adv Agron* 20: 323–359.

Pivnick, E. K., N. C. Kerr, R. A. Kaufman, D. P. Jones, and R. W. Chesney. 1995. Ricketssecondary to phosphate depletion: A sequella of antacid use in infancy. *Clin Pediatr (phila)* 34: 73–78.

Postma, J. A., and J. P. Lynch. 2011. Root cortical aerenchyma enhances the growth of maize on soils with sub-optimal availability of nitrogen, phosphorus and potassium. *Plant Physiol* 156: 1190–1201.

Prasad, R. 2013. Soil test based fertilizer recommendations in India-how realistic? How reliable? *Indian J Fert* 9(3): 16–19.

Prasad, R., and J. F. Power. 1997. *Soil Fertility Management for Sustainable Agriculture*. Boca Raton, FL, CRC.

Prasad, R., Y. S. Shivay, and D. Kumar. 2014a. Agronomic fortification of cereal grains with iron and zinc. *Adv Agron* 125: 255–339.

Prasad, R., R. K. Tewatia, and K. Majumdar. 2014b. Plant nutrients. In *Textbook of Plant Nutrients*. Eds. Prasad, R., D. Kumar, D. S. Rana, Y. S. Shivay, and R. K. Tewatia. New Delhi, Indian Society of Agronomy, pp. 64–72.

Razzaque, M. S. 2009. Does FGF23 toxicity influence the outcome of chronic kidney disease? *Nephrol Dial Transplant* 24: 4–7.

Reddy, N. R., and S. K. Sathe. 2001. *Food Phytates*. Boca Raton, FL, CRC.

Reilly, R. F. 2005. Disorders of serum phosphorus. In *Nephrology in 30 days*. Eds. Reilly, R. F., and M. A. Prerazella, New York, McGraw Hill, pp. 161–176.

Richardson, A. E., J.-M. Barea, A. M. McNeil, and C. Prigent-Combaret. 2009. Acquisition of phosphorus and nitrogen in the rhizosphere and plant growth promotion by microorganisms. *Plant Soil* 329: 305–339.

Richardson, A. E., T. S. George, M. Hens, and R. J. Simpson. 2005. Utilization of soil organic phosphorus by higher plants. In *Organic Phosphorus in the Environment*. Eds. Turner, B. L., E. Frossard, and B. S. Baldwin. Wallingford, CABI, pp. 165–184.

Rodriguez, D., W. G. Keltjens, and J. Goudriaan. 1998. Plant leaf area expansion and assimilate production in wheat (*Triticum aestivum*) grown under low phosphate. *Plant Soil* 200: 227–240.

Ron-Vaz, M. D., A. C. Edwards, C. A. Shand, and M. S. Cresser. 1993. Phosphorus in soil solutions: Influence of soil acidity and fertilizer additions. *Plant Soil* 148: 175–183.

Rubio, G., H. Liao, X. L. Yan, and J. P. Lynch. 2003. Topsoil foraging and its role in plant competitiveness for phosphorus in common bean. *Crop Sci* 43: 598–607.

Rush, G. F., J. R. Gorshi, M. G. Ripple, J. Sowinski, P. Bugelsky, and W. R. Hewitt. 1985. Organic hyperoxide-induced lipid peroxidation and cell death in isolated hepatocytes. *Toxicol App Pharmcol* 78(3): 473–483.

Sample, E. C., R. J. Sopex, and C. J. Racz. 1986. Reactions of phosphate fertilizers in soils. In *The Role of Phosphorus in Agriculture*. Eds. Khasawneth, F. E., E. C. Sample, and E. J. Kamprath. Madison, WI, American Society of Agronomy, and Soil Science Society of America, pp. 263–310.

Schachtman, D. P., R. J. Reid, and S. M. Ayling. 1998. Phosphorus uptake by plants: From soil to cell. *Plant Physiol* 116(2): 447–453.

Seeling, B., and R. J. Zasoski. 1993. Microbial effects in maintaining organic and inorganic solution phosphorus concentrations in grassland top soil. *Plant Soil* 148: 277–284.

Segawa, H., I. Kaneko, S. Yamanka et al. 2004. Intestinal Na-Pi cotransporter adaptation to dietary Pi content in vitamin D receptor null mice. *Am J Physiol Renal Physiol* 287: F39– F47.

Sepat, S., I. P. S. Ahlawat, and D. S. Rana. 2012. Effect of phosphate sources and levels on Bt cotton (*G. hirsutum*) based cropping systems. *Indian J Agron* 57(3): 235–240.

Shand, C. A., A. E. S. Macklon, A. C. Edwards, and S. Smith. 1994. Inorganic and organic P in soil solutions from three upland soils: I: Effects of soil solution extraction conditions, soil type and season. *Plant Soil* 159: 225–264.

Sharon, M. M., M. P. Zidehsaraj, M. A. Chambers et al. 2011. Vegetarian compared with meat dietary protein source and phosphorus homeostasis in chronic kidney disease. *Clin J Am Soc Nephrol* 6(2): 257–264.

Shrier, R. W. 2007. *Diseases of the Kidney and Urinary Tract*. Eighth Ed., Vol. III. Philadelphia, PA, Lippincott, Williams and Wilkins.

Singh, U., and I. P. S. Ahlawat. 2007. Phosphorus management in pigeonpea (Cajanus cajan)-wheat (*Triricum aestivum*) cropping system. *Indian J Agron* 52(1): 21–26.

Singh, V. K., B. S. Dwivedi, K. N. Tiwai, K. Majumdar, M. Rani, S. K. Singh, and J. Timsina. 2014. Optimizing nutrient management strategies for rice-wheat system in the Indo-Gangetic plains of India and adjacent region for higher productivity, nutrient use efficiency and profits. *Field Crops Res* 164: 30–44.

Smith, S. E., and D. J. Read. 2008. *Mycorrhizal symbiosis*. Third Ed. Amsterdam, Academic Press-Elsevier.

Smith, F. W., P. M. Ealing, B. Dang, and E. Delheize. 1997. The cloning of two Arabidopsis genes belonging to phosphate transporter family. *Plant J* 11: 83–92.

Smith, S. E., F. A. Smith, and I. Jakobsen. 2003. Mycorrhizal fungi can dominate phosphate suuply to plants irrespective of growth responses. *Plant Physiol* 133: 16–20.

Standing Committee on the Scientific Evaluation of Dietary Reference Intake. 1997. Chapter 5. Phosphorus pp. 146–189. In *Dietary Reference Intake for Calcium, Phosphorus, Magnesium, Vitamin D and Fluoride*. Institute of Medicine (US), Food & Nutrition Board, Washington, D.C., National Academic Press (http://www.nap.edu/read/5776 /Chapter/7).

Stein, A. J. 2010. Global impacts of human mineral nutrition. *Plant Soil* 335: 133–154.

Stubbe, J. 1990. Ribonucleotide reductase: Amazing and confusing. *J Biol Chem* 265(10): 5329–5332.

Suki, W. N., and S. G. Massery. 1997. *Suki's and Massery's Therapy of Renal Diseases and Related Disorders*. Third Ed. New York, Springer Science & Business Media.

Tadano, T., K. Ozawa, H. Sakai, M. Osaki, and H. Matsui. 1993. Secretion of acid phosphatase by roots of crop plants under phosphorus-deficient conditions and some properties of the enzyme secreted by lupin roots. *Plant Soil* 155/156: 95–98.

Takeda, A., H. Tsukuda, Y. Takaku, S. Hisamastu, and M. Nanzyo. 2006. Accumulation of uranium derived from long term fertilizer applications in a cultivated Andisol. *Sci Total Environ* 367: 924–931.

Takeda, E., H. Yamamoto, H. Yamanka-Okumura, and Y. Taketani. 2012. Dietary phosphorus in bone health and quality of life. *Nutr Rev* 5(1): 104–113.

Tomar, N. K. 2000. Dynamics of phosphorus in soils. *J Indian Soc Soil Sci* 48: 640–668.

Tonelli, M., F. Sacks, M. Pfeffer, Z. Gao, and G. Curhan. 2005. Relation between serum phosphate level and cardiovascular event rate in people ith coronary disease. *Circulation* 112: 2627–2633.

Van Wazer, J. R. 1958. *Phosphorus and its Compounds*. New York, Interscience.

White, P. J., and P. H. Brown. 2010. Plant overview for sustainable development and global health. *Annals of Botany* 105: 1073–1080.

Williams, C. H. 1950. Studies on soil phosphorus: I: A method for partial fractionation of soil phosphorus. *J Agric Sci* 40: 233–242.

# 5 Phosphorus Management

*Rajendra Prasad, Yashbir Singh Shivay,*
*Kaushik Majumdar, and Samendra Prasad*

## CONTENTS

## 5.1  INTRODUCTION

The world population was merely 10 million nearly 10,000 years ago and took almost 8000 years to reach 100 million and another 3804 years to reach the 1 billion mark in 1804. The next billion was added in 123 years (1927) and another billion in the next 33 years (1964). However, since then, an additional billion has been added each 12–14 years and the seven billionth baby was born on 31 October 2011, near Lucknow in India. According to the United Nations, the global population may level off at 9 billion by 2050 (Prasad 2013). As a consequence of this boom in human population, the cereal demand will increase from 2.1 t year$^{-1}$ to 3.0 billion t year$^{-1}$, and the demand for meat from 200 million t year$^{-1}$ to 470 million t year$^{-1}$ by 2050 (Food and Agriculture Organization 2009; Alexandros and Bruinsma 2012; Lal 2014). The global phosphorus demand has increased from about 2 Tg in 1890 to 4 Tg in 1950 and 15 Tg in 2000 (Murray 2003) to cope with increasing food demand, and it will continue to grow. In spite of its importance in biology, plants strive hard to obtain this essential nutrient from the rhizosphere. This is primarily due to low availability of P in many natural ecosystems (Barber et al. 1963). P deficiency is considered to be one of the major limitations for crop production particularly in low-input agricultural systems around the world. It is estimated that globally, 5.7 billion hectares of cultivated land is deficient in P for achieving optimal crop production (Lynch 2007). It is a significant concern in the highly weathered and volcanic soils of the humid tropics and sandy soils of the semiarid tropics (Sanchez et al. 1997). Phosphate fixation significantly increases under acid soil conditions, which affects nearly 26% of the world's soils (Eswaran et al. 1997). All phosphorus in P fertilizers is obtained from the phosphate rock (PR) deposits, which are nonrenewable. The known PR deposits are sufficient for 100–1000 years, depending on the efficiency of the resource use during the manufacture of phosphate fertilizers (Smil 2002; Zhang et al. 2008) and their efficient management on the farm fields. Some efforts have been made in the recent years to extract magnesium phosphates struvite and dittmarite from municipal and industrial waste waters, and Massey et al. (2009) reported that these were as effective as triple superphosphate for wheat (*Triticum aestivum* L.). Nevertheless efficient phosphorus management is critical for the augmented food production.

## 5.2  PHOSPHORUS AS AN ELEMENT

Phosphorus with an atomic number 15 and an average atomic mass of 30.97 occurs in group 15 of the periodic table and has an electronic configuration of [Ne] $3s^23p^3$

(Rao 1999). The highest and most important oxidation state (valence) is +5 as in phosphoric acid ($H_3PO_4$). It also has a −3 oxidation state as in $PH_3$ (phosphine gas). Elemental phosphorus could be white or red. When exposed to oxygen, white phosphorus emits a glow; hence, it was known as *lightening bear* in Greek mythology referring to Venus the *morning star.* The term *phosphorescence* is derived from this property of phosphorus. Its property of getting excited by solar radiation is the main pathway of converting solar energy into chemical energy as carbohydrates in photosynthesis, which is the only physiological process that produces food. Phosphorus is thus the key element in global food production. Phosphorus has radioactive isomer $^{32}P$, which has been extensively used in determining the pathways of P transformation in soils and plants.

Phosphorus is a highly reactive element, and it is never found as a free element in the earth's crust and it mostly occurs as apatite in PRs, which are the main source for manufacturing phosphate fertilizers. The high reactivity of phosphorus makes its management on farm fields difficult and depends upon how well it is managed for different farming systems on a variety of soils and under varying climates across the globe.

## 5.3   PHOSPHORUS IN THE EARTH'S CRUST

Both sedimentary and igneous deposits of phosphorus occur in earth's crust. Sedimentary deposits make up 80% of the world's production of PRs; generally called *phosphorites*, they are used for manufacturing the phosphate fertilizers. Phosphate ores from igneous deposits are often low in grade but can be beneficiated to high-grade concentrates (from about 157 to >175 g P $kg^{-1}$ [36 to more than 40% $P_2O_5$]). Igneous deposits typically contain apatites of the FA ($Ca_{10}(PO_4)_6F_2$) and HA ($Ca_{10}(PO_4)_6OH_2$) varieties or intermediate compositions (Hogan 2011). The igneous apatite varieties are relatively unreactive, and the PR from these deposits may have to be ground very finely for use in fertilizer processing (Stewart et al. 2005). Most of the sedimentary deposits contain varieties of carbonate-FAP that are collectively called *francolite*. Francolites form a continuous series from end members, which contain practically no carbonate substitution, to end members with 60–70 g $kg^{-1}$ (6–7%) carbonate for phosphate substitution. The higher the carbonate substitution, the higher is the reactivity of the sedimentary PR. This factor controls the ultimate grade to which francolite-containing PR could be beneficiated (McConnell, 1938; Stewart et al. 2005). The global estimates of PR are at 67,000 billion tons (Bt). The top ten countries/regions holding these deposits are Morocco and Western Sahara (50,000 Bt), China (3700 Bt), Algeria (2200 Bt), Syria (1800 Bt), Jordan (1500 Bt), South Africa (1500 Bt), United States (1400 Bt), Russia (1300 Bt), Peru (820 Bt), and South Africa (750 Bt) (van Kauwenbergh et al. 2013).

## 5.4   PHOSPHORUS IN SOIL

The total phosphorus content of soils is generally one-tenth to one-fourth that of nitrogen and one-twentieth that of K (Brady 1990). The total P content in surface soil and subsoil may vary from a few milligrams to over 1 g $kg^{-1}$. Phosphorus is fairly well distributed in soil profile, and the P content in subsoil may be less than, equal to, or greater than that in the surface soil (Prasad and Power 1997). The two major forms of phosphorus in soils are organic-P and inorganic-P.

## 5.4.1 ORGANIC-P

The amount of P present in organic form in soils varies from a few milligrams to about 0.5 g kg$^{-1}$ soil (20–80% of total P). Organic matter content, texture, land use pattern, fertilizer practices, drainage and irrigation, etc., affect the organic-P content in soil. In general, the organic-P in surface soils is a smaller fraction of the total soil P in the warm regions than it is in the cool regions of the world, about 35.2% as organic-P in equatorial regions (countries between 40° parallels) and 48.6% in colder regions (Harrison 1987). Also more soil P is present as organic-P in organic soils and peats (histosols) than in mineral soils. Again in mineral soils, high clay soils have a greater percentage of their P in organic form than do sandy soils. Several authors have observed a positive correlation between the clay content and the organic-P contents (Sood and Minhas 1988; Singh and Dutta 1987). Clay minerals can inhibit hydrolysis of phosphate esters by adsorbing the enzyme phosphatase. For instance, the smectites can provide high surface areas for the adsorption of substrates and enzyme phosphatase and exchangeable cations and may serve as the most effective soil constituent for the accumulation of organic-P (Sanyal and De Datta 1991; Tomar 2000). The large variations in the organic-P content of soils have been attributed to several factors, namely, type of parent materials, climate, vegetation, land use, soil properties such as content and composition of humus, texture, pH, drainage, and supply of inorganic-P and other nutrients. Tomar (2000), while reviewing the effect of parent materials on the organic-P content of soils, found that the soils derived from basalts and basic igneous parent materials contained higher amounts of organic-P than did those derived from granite, while soils derived from parent materials of higher P content also contained higher organic P. The effect of temperature on soil organic-P is a consequence of its effect on the growth of soil microorganisms responsible for immobilization and mineralization of P in soils. This is evident from the fact that the soils of Himachal Pradesh, Assam, hilly areas of Haryana and Uttar Pradesh with colder climates, have higher organic-P than soils of the Indo-Gangetic plains of Uttar Pradesh, Haryana, Punjab, and Rajasthan (Tomar 2000). Saxena (1979) observed that the organic-P content of soils is generally higher in humid regions.

In general, organic-P tends to accumulate in the surface soil because organic-P is a part of the soil organic matter. However, there are soils such as uncultivated pine bogs where the subsoil is richer in P than the surface soil (Prasad and Power 1997).

About half of the soil organic-P is phytic acid (inositol hexakisphosphate), while the other half includes mono- and diesters of aliphatic acids, phospholipids, phosphoproteins, sugar phosphates, and phosphonates (Tate 1984). The lack of suitable analytical techniques for assessing the contribution of soil organic-P has led to phosphate management practices that often ignore the contribution of organic-P in crop production. The need for this has risen due to the recent increase in the area under organic farming. A procedure involving citrate, phytase, and phosphatase for predicting the bioavailability of organic-P is now being developed (Helmke et al. 2000).

## 5.4.2 Inorganic-P

The inorganic-P in soil is mostly present as compounds of Ca, Fe, and Al; Ca phosphates dominate in neutral to alkaline soils, while Fe- and Al-phosphates dominate in acidic soils. At any specific time, a very small amount of phosphorus is present in a soil solution in equilibrium with the solid inorganic phase. The concentration of P in the soil solution is commonly about 0.05 mg $L^{-1}$ and seldom exceeds 0.3 mg $L^{-1}$ in unfertilized soils. The ionic forms of inorganic P are pH dependent. Between the pH values of 4.0 and 6.0, most of the P in the soil solution is present as the $H_2PO_4^-$, the form in which it can be readily absorbed by plant roots because this ionic form is soluble in water. Between pH 6.5 and 7.5, the P in the soil solution is present partly as $H_2PO_4^-$ and partly as the $HPO_4^=$. $HPO_4^=$ ions can also be taken up by the plant roots but not as readily as $H_2PO_4^-$. The $HPO_4$ ion is dominant between pH 8.0 and 10.0. Under such conditions, $Na^+$ ions dominate the soil cation exchange complex, and some phosphate is present as sodium phosphate, making $H_2PO_4$ ions available on hydrolysis. Beyond pH 10.0, the dominant ionic form of P is $PO_4^{3-}$, and unless present as sodium phosphate, P is not available to crop plants. At the other extreme, that is, below pH 3.0, which is generally not found in cultivated soils, P would be present in $H_3PO_4$ (phosphoric acid), a very reactive form. Under such conditions, there is plenty of soluble Fe and Al in the soil that reacts with the P to form insoluble phosphates, and the process is referred to as *phosphate fixation* or *phosphate retention or reversion* of applied P (Lindsay 1979; Prasad and Power 1997). That is why in highly acidic ultisols and oxisols, the phosphate fixation or reversion of applied P is rapid, and large amounts of phosphate fertilizers are required to obtain good crop growth.

## 5.4.3 Fixation/Reversion of Applied P

Once water-soluble phosphorus (WSP) is applied to soil, it starts reacting with the soil constituents and is precipitated or retained by the soil. Phosphorus is retained in soils by hydroxides and oxyhydroxides of Fe and Al, silicate minerals, carbonates, and soil organic matter. Researchers using dilute P solutions, usually in the millimolar range, have developed several adsorption equations, while those working with concentrated P solution, usually in the molar range, observed precipitates forming separately at the surfaces of the soil constituents (Prasad and Power 1997). There are four steps in the precipitation process that occur when dissolved P reacts with soil: (1) formation of a surface-adsorbed P complex; (2) dissolution of clay minerals, which increases the P-reactive metal ion concentration in the solution; (3) slow desorption of surface-adsorbed P compounds; and (4) slow nucleation, crystallization, and recrystallization of P compounds (Talibuddin 1981). If the P retention processes are viewed throughout the entire zone influenced by the fertilizer application and over an entire growing season or over a longer period, the P retention in soil should be considered as a continuum embodying precipitation, chemisorption, and adsorption (Sample et al. 1986). Phosphorus sorption is affected by the type of surfaces contacted by P in the soil. As mentioned earlier, amorphous iron and aluminum oxides are the most effective P sorbents due to their high specific surface area. Sorption is also affected by the quantity and the mineralogy of the clay fractions and has been extensively studied (Sanyal and Majumdar

2009). Other factors that significantly influence the P sorption are species and concentration of cations in the system, competing anions (Deb and Datta 1967), amount, and reactivity of calcium carbonate present in the soil (Brar and Vig 1988), ionic nature and composition of the supporting medium (Bidappa and Rao 1973), and temperature. In the late 1950s and the early 1960s, a large number of reaction products of phosphate fertilizers based on solubility product principle and some other techniques were reported, which include: $Ca_{10}(PO_4)_6(OH)_2$ HA; $Ca(NH_4)_2P_2O_7H_2O$; $Ca_{10}(PO_4)_6F_2$ (FA); $Ca_3(NH_4)_2(P_2O_7)_2 \cdot 6H_2O$; $CaAlH(PO_4)_2 \cdot 6H_2O$; $Ca_5(NH_4)_2(P_2O_7)_3 \cdot 6H_2O$; $CaAlH_4(PO_4)_3 \cdot 2H_2O$; $CaNH_4HP_2O_7$; $Ca_2NH_4H_3(P_2O_7)_2 \cdot 3H_2O$; $Ca(NH_4)_2(HPO_4) \cdot 2H_2O$; $CaK_2P2O_7$; $Ca_2NH_4H_7(PO_4)_4 \cdot 2H_2O$; $NH4 \cdot Ca_3K_2(P_2O_7)_2 \cdot 2H_2O$; $Ca_2(NH_4)_2(HPO_4)_3 \cdot 2H_2O$; $CaKPO_4 \cdot H_2O$; $Ca_2KH_3 \cdot (P_2O_7)_2 \cdot 3H_2O$; $CaK_3H(PO_4)_2$; $CaNa_2P_2O_7 \cdot 4H_2O$; $Ca_2KH_7 \cdot (PO_4)_4 \cdot 2H_2O$; $KFe(NH_4)_2P_2O_72H_2O$; $CaFe_2H_4 \cdot (PO_4)_4 \cdot 5H_2O$; $Mg(NH_4)_2P_2O_7 \cdot 4H_2O$; $CaFe_2H_4(PO4)_4 \cdot 5H_2O$; $Mg(NH_4)_6(P_2O_7)_2 \cdot 6H_2O$; $Ca_3Mg_3(PO_4)_4$; $Mg(NH_4)_2H_4 (P_2O_7)_2 2H_2O$; $FePO_4 2H_2O$ (strengite); $Ca(NH_4)P_3O_{10} \cdot 2H_2O$, and $AlPO_4 2H_2O$ (variscite), $AlFePO_4 2H_2O$ (Lindsay and Stephenson 1959; Lindsay et al. 1959; Lindsay and Moreno 1960; Wright and Peech 1960; Lindsay et al. 1962; Yadav and Mistry 1984; Sample et al. 1986). Little attention, however, was paid to the subsequent bioavailability of fixed or retained phosphorus during this early phase of P fixation research. The need to apply more P than was removed in the harvested crop raised the question as to what happened to the residual phosphate. It had commonly been assumed that the failure of a crop to respond to fertilizer P was because of the rapid fixation of P by the soil. In this context, the study by Coleman (1942) is particularly interesting where the author showed that this could also be due to a sufficiency of plant-available P already in the soil and that large amounts of P formerly considered fixed were available to plants (Syers et al. 2008). Evidence of early recognition of the reversibility of the P fixation reaction in the soil was noted by Kurtz (1953) who observed that "contrary to the apparent belief of two decades ago, more recent evidence indicates that the reactions of phosphate with soil are not entirely irreversible and that for most soils the term fixation is an exaggeration." The pioneering work in Australia in the late 1960s and 1970s (Posner and Barrow 1982; Barrow 1983) on P adsorption in soils and its reversibility (desorption) led to better understanding of P reactions in soils. Syers et al. (2008) and Johnston and Syers (2009) have therefore proposed that there are basically four pools of P in soil, namely, (1) soil solution pool, where P is soluble and immediately available and accessible to plants; (2) surface-adsorbed P, which is readily extractable and available to plants; (3) strongly bonded or adsorbed P, which has low extractability, low availability and low accessibility to plants; and (4) very strongly bonded, precipitated, and mineral P, which has very low extractability, availability, and accessibility to plants (Table 5.1). The transformation of P from pool 1 to pool 2 and vice versa is fairly fast, while the transformation from pools 3 and 4 is slow. The quantification of soil P in different pools is a difficult task and methodology for this is not available at the moment. Sanyal and Majumdar (2009) reviewed several studies that have described the desorption of phosphate from soils using the Langmuir equation and the first-order kinetics. Parabolic and radial diffusion equations have also been successfully used to describe the release of surface-adsorbed P from soil suggesting that the relationship of cumulative P desorbed to $t^{+1/2}$ is diffusion controlled (Brar and Vig 1988).

**TABLE 5.1**
**Soil P Pools 1, 2, 3, and 4**

| Immediately Accessible | | High Accessibility | | Low Accessibility | | Very Low Accessibility |
|---|---|---|---|---|---|---|
| In solution | | Readily extractable | | Low extractability | | Very low extractability |
| Soil solution P | $\rightarrow$ | Surface- | $\rightarrow$ | Strongly bonded or | $\rightarrow$ | Very strongly bonded or |
| | $\leftarrow$ | adsorbed P | $\leftarrow$ | absorbed P | $\leftarrow$ | inaccessible or mineral or precipitated P |
| Pool 1 | | Pool 2 | | Pool 3 | | Pool 4 |
| Immediately available | | Readily available | | Low availability | | Very low availability |

*Source:* Johnston, A. E. and J. K. Syers, *Better Crops*, 93(3), 14–16, 2009.

### 5.4.4 Phosphorus Use Efficiency

The efficiency in plant mineral nutrients has been defined based on the process by which plants acquire, transport, store, and use the nutrient in order to produce dry matter or grain, at low or high nutrient supply (Ciarelli et al. 1998). Nutrient acquisition efficiency, used in the sense of plant nutrient acquired from the soil, and nutrient internal utilization efficiency, defined as a plant's internal ability to produce yield units per unit of nutrient in the plant, have been considered as the two major components of plant nutrient use efficiency (Good et al. 2004). Improving the efficiency of phosphorus fertilizer used for crop growth requires enhanced P acquisition by plants from the soil (P-acquisition efficiency) and enhanced use of P in processes that lead to faster growth and greater allocation of biomass to the harvestable parts (internal P-use efficiency [PUE]). As only 15–30% of applied fertilizer P is taken up by crops in the year of its application (Syers et al. 2008), potentially large gains in efficiency can be made by improving P acquisition. This aspect of P efficiency has received significant attention and has been recently reviewed (White and Hammond 2008; Ramaekers et al. 2010; Wang et al. 2010; Richardson et al. 2011).

The new concept of P pools in soil has allowed the development of a balance method for calculating the PUE where it can be measured by expressing total P uptake ($U_p$) as a percentage of the P applied ($F_p$), i.e., the total P in the crop divided by the P applied ($U_p/F_p$), expressed as a percentage. This was initially proposed by Johnston and Poulton (1977) and later developed by Syers et al. (2008). This method has the advantage that the recovery of P from soil reserves is allowed for, and there is no need for a control or a check plot. Syers et al. (2008) cited several examples, which showed that the recovery (efficiency) of the applied fertilizer P plus the residual soil P frequently ranged from about 50–90% when measured by a suitable method and over an appropriate timescale.

## 5.5  PHOSPHORUS NUTRITION OF PLANTS

### 5.5.1  Phosphorus as an Essential Plant Nutrient

The credit for establishing the essentiality of phosphorus as a plant nutrient goes to Justus von Liebig in 1839 and G. Ville in 1861 (Glass 1989; Fageria and Baligar 2005).

### 5.5.2  Accessibility of Soil P to Plants

Accessibility of soil P to plants depends upon the root characteristics (Hill et al. 2006) and their capacity to release organic acids and phosphatases. Root–mycorrhizal associations also increase soil P accessibility to plants.

#### 5.5.2.1  Root Characteristics

Since the P uptake by plant roots is governed by diffusion rather than by mass flow of soil water, the root characteristics facilitate the exploration of soil P and increase its uptake (Gahoonia and Nielsen 2004). Some of these characteristics are specific root length (Silberbush and Barber 1983), increased root branching (Lynch and Brown 2001; Rubio et al. 2003), increased root volume and root dry weight (Banerjee et al. 2010), and more root hair (Gahoonia and Nielsen 1997, 1998). Some plant species produce specialized roots for P absorption, such as aerenchyma in corn (Fan et al. 2003; Postma and Lynch 2011) and proteoid roots or cluster roots in lupins (*Lupinus* spp.) (Dinkelaker et al. 1995; Lambers et al. 2006). Braum and Helmke (1995) reported that under P-deficient conditions, lupin (*Lupinus albus* L.) roots develop proteoids that excrete large quantities of citrate, phytase, and phosphatase, supporting the findings that lupin has access to a part of the soil phosphorus pool, which is unavailable to soybean. Phosphorus fertilization improves the root characteristics and helps in increasing the P uptake by plants and some data are in Table 5.2.

#### 5.5.2.2  Capacity to Secrete Organic Acids

The roots of a number of plant species, such as white lupins (*Lupinus alba* L.), secrete organic acids, such as citric, oxalic, and maleic (Hissinger 2001; Dakora and Phillips

---

**TABLE 5.2**
**Effect of P Fertilization on Some Root Parameters of Maize**

| Treatment | Root Length (m/plant) | Root Volume (cm³/plant) | Root Dry Weight (g/plant) |
|---|---|---|---|
| 120 kg N/ha | 3.49 | 2.50 | 0.84 |
| 120 kg N + 26 kg P/ha as SSP | 6.24 | 6.25 | 3.42 |
| 120 kg N + 13 kg P/ha as PR + VAM | 8.60 | 10.0 | 4.10 |
| Least Significant Difference (LSD) (0.05) | 0.354 | 0.644 | 0.250 |

*Source:* Banerjee, M., R. K. Rai, and D. Maiti, *Arch Agron Soil Sci*, 56, 681–695, 2010. With permission.
*Note:* PR: phosphate rock, SSP: single superphosphate, VAM: vesicular arbuscular mycorrhyza.

2002), which increase the solubility of sparingly soluble inorganic-P compounds, especially calcium phosphates (Gahoonia and Nielsen 1992; Hissinger and Gikes 1996; Hissinger 2001). Anions in organic acids also help in the desorption of P from adsorption sites (Jones and Darrah 1994; Jones 1998). Organic acids also mobilize the P bonded in humic–metal complexes (Gerke 1993). As would be expected, the concentration of organic acids is reported to be greater (about tenfold) in the rhizosphere than in the bulk soil (Jones et al. 2003). The effectiveness of different organic acids differs; citrate and oxalate are more effective than malate, malonate, tartarate, succinate, fumate, acetate, and lactate (Bar-Yosef 1991).

### 5.5.2.3 Capacity to Exude Phosphatases

Sharpley (1999) suggested that enzyme phosphatase and phytase may play a role in the cycling of phosphorus in soils and in the nutrition of plants. The secretion of phosphatases by some plant species in an under deficient-P condition has been reported (Dinkelaker and Marschner 1992; Tadano et al. 1993; He 1998; Hayes et al. 1999). Phosphatase helps in the hydrolysis of organic-P and hence in its release for uptake by the roots (Richardson et al. 2005).

### 5.5.2.4 Root–Mycorrhizal Associations

The role of the root–mycorrhizal association in increasing the accessibility of soil P to plants has received considerable attention and has been reviewed (Bolan 1991; Smith and Read 2008; Richardson et al. 2009). Both ectomycorrhiza and endomycorrhiza (VAM) associations are known. Ectomycorrhiza are generally associated with forest trees, such as members of Fagaceae, Betulaceae, and Pinaceae families and *Eucalyptus*, while VAM have associations with agricultural crops (Smith and Read 2008). The increased P uptake by plants with roots having VAM association is mainly due to increased length and surface area created by fungal hyphae that permit the extraction of P from a greater soil volume (Jakobsen et al. 1992), although VAM hyphae may also be able to tap P from P pools that are inaccessible to the root hair (Cardoso et al. 2006). Mycorrhizal plants therefore have two pathways for P uptake: (1) root–soil interface and (2) via VAM hyphae (Smith et al. 2003). VAM is reported to permit the direct use of finely ground PR (Banerjee et al. 2010) (Table 5.2).

### 5.5.3 Absorption of P by Plants

Plants face a considerable problem in the absorption of phosphorus, which, as already stated, is mostly taken up as $H_2PO_4^-$ (abbreviated as Pi); the concentration of which in a soil solution is generally very low (about 10 µM). As a contrast, the Pi concentration in root cells and xylem is a hundred- or thousandfold (5–10 mM) (Lee and Radcliff 1993; Mimura 1995), and this steep concentration gradient helps in the Pi uptake by plants (Mengel and Kirkby 1987). However, most Pi is taken up with the help of a cotransport, which could be $H^+$ or some other source. The review of available literature (Schachtman et al. 1998) brings out that as per kinetic approach, there appear to be two types of transporters with different affinities (Drew and Saker 1984; Nandi et al. 1987), while the molecular approach suggests that there are at least four genes that encode the Pi transporters (Muchhal et al. 1996; Smith et al. 1997).

### 5.5.4 Transportation of P in Plants

Once absorbed by the roots, the Pi is transported in the xylem to the younger leaves. The Pi from older leaves also moves to younger leaves and some from the shoots to the roots. The concentration in the xylem may range from 1 mM/L in Pi-starved plants to 7 mM/L in plants grown in solutions containing 125 µM/L Pi (Mimura et al. 1996). In Pi-starved plants, some of the Pi translocated from the shoots to the roots in the phloem is transferred back to the xylem and recycled back to the shoots (Jeschke et al. 1997). In the xylem, phosphorus is almost solely present as Pi, while in the phloem, significant amounts of organic phosphorus compounds are also present (Schachtman et al. 1998). The conversion of Pi to organic compounds is fairly rapid (Jackson and Hager 1980).

### 5.5.5 Functions

Phosphorus as ATP plays the key role in photosynthesis. In its conversion from ATP to ADP (adenosine diphosphate), the energy released ($\Delta G^0$) by the cleaving of one unit of ATP at standard state of 1 M is −7.3 kcal/mole, However, in the cells, the conditions do not conform to the standard state, primarily because the reactant and product concentrations differ from 1 M. Under actual conditions, the $\Delta G^0$ for the transformation ATP to ADP is −13 kcal/mole, 78% more than that under standard conditions (Reece et al. 2011). There are some analogs to ATP, namely, uridine triphosphate, cytidine triphosphate (CTP), and guanosine triphosphate, which also function as a source of energy and act as a substrate for the synthesis of RNA (Berg et al. 2002; Blackburn 2006). CTP is also involved in the synthesis of phospholipids. Another biological phosphorus molecule, nucleotide adenosine diphosphate, is involved in the electron transfer in photosystem I of photosynthesis (Salisbury and Ross 1986). Phosphorus is an integral part of the DNA and the RNA. Phosphorus helps in the development of a better root system (Drew 1975; Zhang and Barber 1992). Phosphorus has a major role in reproduction in plants and is reported to hasten maturity in a number of crops including corn, sorghum (*Sorghum vulgare* Pers.), wheat (*Triticum aestivum* L.), cotton (*Gossypium arboretum* L.), and some vegetables (Anonymous 1999). In a study in Kansas, the application of 19.1 kg P/ha reduced the thermal units (degree days) from emergence to mid-silk in corn by 110 units (Gordon 2003). A recent study on plant species diversity in Eurasian wetlands and grasslands (Fujita et al. 2014) showed that plants in phosphorus-deficient communities had shorter flowering periods and lesser seed production, and therefore, endangered species were more frequent in phosphorus-deficient environments.

### 5.5.6 Deficiency Symptoms

Phosphorus deficiency limits leaf expansion and surface area as well as leaf number. Phosphorus-deficient plants generally turn dark green due to poor carbohydrate utilization, and they may be stunted. The older leaves turn purple due to the accumulation of anthocyanins (Hamy 1983; Yin et al. 2012; Prasad et al. 2014). Phosphorus deficiency reduces the quality of fruits and vegetables. Deficiency symptoms of P in some crop plants are shown in Figure 5.1.

(a)

(b)

**FIGURE 5.1** **(See color insert.)** Phosphate deficiency symptoms in some crops: (a) symptoms of P deficiency in lettuce (*Lactuca sativa*) and (b) symptoms of P deficiency in barley (*Hordeum vulgare*). *(Continued)*

**FIGURE 5.1 (CONTINUED)    (See color insert.)** Phosphate deficiency symptoms in some crops: (c) symptoms of P deficiency in corn (*Zea mays*) and (d) P deficiency in guava (*Psidium gujava*). (Courtesy of International Plant Nutrition Institute, Peachtree Corners, Georgia.)

## 5.6   CROP RESPONSE TO PHOSPHORUS

A volume of data exists all over the world on the response of various crops to P. Some data from India on crop responses on irrigated as well as dryland soils are presented in Tables 5.3 through 5.5. Rice and wheat are the major crops in India where irrigation is available. Data from a large number of trials carried out in different states practicing rice–wheat cropping system (RWCS) in India, by the Directorate of Farming Systems Research, Modipuram, have shown that both rice and wheat respond well to P application and in the states of Punjab, Uttar Pradesh, and Bihar,

**TABLE 5.3**
**Check Yield (t/ha) and Response (kg grain/kg $P_2O_5$) of Rice and Wheat to P Fertilization in the Indo-Gangetic Plains on Farmers' Fields**

| | Rice | | | Wheat | | |
|---|---|---|---|---|---|---|
| State | Trials | Check Yield | Response | Trials | Check Yield | Response |
| Punjab | 48 | 4.1 | 16.3 | 48 | 3.1 | 11.7 |
| Haryana | 24 | 3.4 | 11.3 | 70 | 2.0 | 14.1 |
| Uttar Pradesh | 147 | 1.2 | 5.9 | 137 | 0.9 | 5.3 |
| Bihar | 22 | 1.6 | 15.5 | 46 | 1.0 | 12.9 |

*Source:* Prasad, R., *Better Crops-India*, 1, 8–11, 2007. With permission.
*Note:* Average over 3 years (1999–2002).

**TABLE 5.4**
**Response of Oilseeds to P Fertilization**

| Crop | $P_2O_5$ Applied (kg/ha) | Yield Increase due to $P_2O_5$ (kg/ha) (±SE) | Net Return due to $P_2O_5$ (Rs/ha) | Net Return Rs/Re Invested on $P_2O_5$ | Response per kg of $P_2O_5$ Applied (kg/kg) |
|---|---|---|---|---|---|
| Soybean (30)[a] | 101 | 633 (±60) | 17714 | 7.08 | 8.1 |
| Pigeon pea (2)[a] | 79 | 605 (±15) | 16940 | 6.67 | 7.7 |
| Groundnut (7)[a] | 64 | 401 (±55.4) | 11216 | 6.34 | 7.3 |
| Mustard (32)[a] | 140 | 611 (±75.82) | 17098 | 4.85 | 5.6 |
| Sunflower (2)[a] | 80 | 73 (±30) | 2044 | 0.79 | 0.9 |
| Raya (4)[a] | 60 | 720 (±95.31) | 20160 | 10.4 | 12.0 |

*Source:* Majumdar, K., and V. Govil, *Better Crops-South Asia*, 9, 15–18, 2015. With permission.
*Note:* Rs/Re: Indian Rupees/Indian Rupee.
[a] Number of trials.

the response of rice to P was better than that of wheat (Prasad 2007a) (Table 5.2). On the other hand, oilseeds and pulses (beans) are mostly grown under dryland agriculture conditions either during rainy season or on stored moisture after the cessation of rains. In general, the response of oilseeds to P application was higher than that of pulses (Tables 5.4 and 5.5). The lesser response of pulses to P application could be partly due to their deep tap root system that forages deeper layers of soil. Both under irrigation and dryland conditions application of P was profitable.

## 5.7 AGRONOMIC MANAGEMENT

There is an urgent need to seek strategies by which P fertilizers can be more effectively used in those farming systems where P is currently deficient and where its use

**TABLE 5.5**
**Response of Pulses to P Fertilization**

| Crop | $P_2O_5$ Applied (kg/ha) | Yield Increase due to $P_2O_5$ (kg/ha) (±SE) | Net Return due to $P_2O_5$ (Rs/ha) | Net Return Rs/Re Invested on $P_2O_5$ | Response per kg of $P_2O_5$ Applied (kg/kg) |
|---|---|---|---|---|---|
| Blackgram (7)[a] | 90 | 106 (±19.8) | 3821 | 1.9 | 1.7 |
| Gram (5)[a] | 147 | 445 (±91.01) | 16,020 | 3.7 | 3.3 |
| Greengram (6)[a] | 58 | 221 (±19.1) | 7956 | 4.9 | 4.4 |
| Pigeonpea (9)[a] | 111 | 460 (±40.9) | 16,572 | 5.1 | 4.6 |
| Urdbean (5)[a] | 72 | 129 (±32.9) | 4658 | 2.1 | 1.9 |
| Cowpea (1)[a] | 20 | 139 | 5004 | 7.8 | 7.0 |
| Chickpea (26)[a] | 68 | 640 (±60.2) | 23,051 | 13.0 | 11.5 |
| Mungbean (3)[a] | 59 | 127 (±3.33) | 4560 | 3.8 | 3.4 |

*Source:* Majumdar, K., and V. Govil, *Better Crops-South Asia*, 9, 15–18, 2015. With permission.
*Note:* Rs/Re: Indian Rupees/Indian Rupee.
[a] Number of trials.

is economically feasible (Raghothama 2005). There is also a need to more efficiently use fertilizer P in areas where plant-available P levels are well above the appropriate critical value for the soil and the farming system under consideration. The ideal phosphorus management involves the choice of appropriate source, application rate and timing, and method of application.

## 5.7.1 RATE OF APPLICATION

The determination of the appropriate P application rate in crops is a critical issue as farm economics and environmental stewardship of the nutrient depend significantly on the application rate. The amount of P to be applied depends upon the available P in the soil, the crop to be grown, and several other factors. A key scientific principle to selecting the right fertilizer rate is matching the nutrient supply with the plant nutrient demand. Nutrient demand refers to the total amount of nutrients that will need to be taken up by the crop during the growing season. The nutrient demand differs among crops. The phosphorus requirement for cereals is generally lower than for pulses and oilseeds. Along with the nutrient demand, the targeted yield of a crop needs to be factored in while deciding the P rates. One of the major challenges in using a yield-based approach for determining the fertilizer rates is that the yield levels are known to vary widely in a given environment from year to year, as well as among growing seasons within a year where multiple cropping is practiced. Crop responsiveness to fertilizer also fluctuates as a result of the environment, independent of the crop yield potential. Both yield potential and crop responsiveness affect the annual fertilizer rate requirement (Bruulsema et al. 2012). A portion of the plant nutrient demand is met by the soil. The soil's capacity to supply nutrients to a growing crop needs to be assessed before deciding external application rates. A number of chemical extractants have been suggested for determining the available P in soil. However, Bray's P1, Mehlich III, and Olsen's extract are most often used (Fixen and Grove 1990). Bray's P1 extract is 0.03 M $NH_4F$ + 0.025 M HCl, while Mehlich III is 0.015 M $NH_4F$ + 0.2 M $CH_3COOH$ + 0.25 M $NH_4NO_3$ + 0.013 M $HNO_3$. Both these extractants are suited for acidic soils. Olsen's extracting solution is 0.5 M $NaHCO_3$ of pH 8.5 and is best suited for neutral, calcareous, and alkaline soils. For soil with large reserves of residual P in Norway, extraction with ammonium lactate (Egner et al. 1960) and just with water (Singh et al. 2005) are being practiced. Predicting phosphorus requirements based on sorption isotherms has been considered more accurate than the one determined by using conventional chemical extractants (Klages 1988; Patiram 1994; Singh et al. 2005). According to these researchers, sorption isotherm takes into account both intensity and capacity factors. However good these methods may be, they are not likely to become rapid soil tests. When selecting the right fertilizer P rate, the contribution toward meeting crop nutrient requirements coming from all available nutrient sources needs to be considered. Some of these sources include indigenous nutrient supplies (those not applied to the land such as crop residues and green manures), animal manures, composts, biosolids, atmospheric deposition, and irrigation water. PUE is another major factor in determining the fertilizer P rate needed. Even with the best management practices, the amount of the applied fertilizer utilized by the crop will be less than 100%. While growers strive to minimize losses and increase efficiencies, some applied nutrients may also be utilized by soil organisms, particularly while soil organic matter levels are being built up. The inherent sinks and loss mechanisms that exist in every field can also adversely affect the efficiency of fertilizer nutrient uptake. That is why the adjustments for efficiency should be included when determining the fertilizer P rate requirements (Bruulsema et al. 2012).

### 5.7.2 SOURCE, TIME, AND METHOD OF P APPLICATION

Most phosphorus fertilizers are made by the acidulation of PR with sulfuric and phosphoric acids and some with nitric acid. Phosphorus fertilizers include single or ordinary superphosphate (SSP or OSP), triple superphosphate (TSP), monoammonium phosphate (MAP), diammonium phosphate (DAP), and a number of NP-/NPK-granulated products (Prasad 2007b). Most of these contain WSP. Nitrophosphates generally contain part WSP and part citrate-soluble P, although some nitrophosphates with 100% WSP are also manufactured. Solution NP and NPK fertilizers are also marketed. Ground PR is also being utilized, especially on acid soils. Some efforts have been made to develop polymer-coated slow-release P fertilizers (Pauly et al. 2002; Chien et al. 2009), but results from a large number of on-farm trials are yet not available.

Keeping in view the fact that plants often absorb about 50% of their seasonal P requirements by the time they have accumulated 25% of their total seasonal dry mass (Black 1968), most fertilizer P is applied to crops at sowing. However, assessing whether there are issues with immobilization or other processes (fixation of P) that might disrupt the nutrient supply or compromise the availability of added nutrients over time is critical for determining the timing as well as the fertilizer source and the method of application. Phosphorus may be broadcast and incorporated in soil or be band placed near and below the seed. The latter is highly desirable for widely spaced crops, such as corn, sorghum (*Sorghum vulgare* Pers.), cotton (*Gossypium arboretum* L.), and sugarcane (*Saccharum officinarum* Linn.). This is also possible in wheat, barley, and oats, which are sown using a seed-cum-fertilizer drill. Only surface broadcast application is possible on established pastures. Phosphorus cannot be top dressed as nitrogen, because of its rapid fixation and low mobility in soils. In view of these problems and increasing cost, especially in those countries that do not have large deposits of PRs, there have been three major innovations in phosphorus management, which are discussed in the following sections.

### 5.7.3   FOLIAR APPLICATION

Foliar application of plant nutrients has been practiced for a long time (Boyton 1954). In recent years, there have been several studies on the foliar application of P to crops, especially cereals. McBeath et al. (2011) reported that the foliar application of P increased the grain yield and the P uptake of wheat. Mosali et al. (2006) found that delaying the P application in wheat until complete emergence of head (GS58 stage) increased the PUE by 8% as compared to its application at GS32 stage. The stage of growth in wheat is important in determining the foliar applied P. Romer and Schilling (1986) reported that P applied at GS31–39 (flag leaf ligule and collar visible stages) at 1 ppm rate increased the grain yield as compared to its application at GS75 (milk stage). Ali et al. (2014) reported that the application of Nutri-Phite, a phosphite ($PO_3$)-based product containing 3% N, 8.7% P, and 5.8% K, increased the yield of winter wheat at one of the four locations in Oklahoma, while the P uptake was increased at three out of the four locations. Torres (2011), from Oklahoma reported an increase of PUE by 28% in wheat with foliar application than its application to soil. Girma et al. (2007) reported a greater PUE at 2 kg/ha than at 4 or 8 kg P/ha, when applied at the V8 stage in corn. Light, humidity, and temperature at the time of the foliar application of P affect its absorption (Kannan 1986). Only WSP fertilizers can be used for the foliar P application.

### 5.7.4   PARTIALLY ACIDULATED ROCK PHOSPHATE

In countries where native sulfur deposits are not present and the manufacture of sulfuric acid or phosphoric acid involves costs and foreign exchange, partial acidulation with sulfuric acid or other acids can be done (McLay et al. 2000). In partially acidulated rock phosphate (PARP), generally about half the quantity of acid is used. On a highly acidic, high P-fixing, dark red oxisol, PARP was as good as the TSP for rice (*Oryza sativa* L.) and beans (Fageria et al. 1991). Sometimes PARPs may even perform better than ordinary or concentrated superphosphate (Marwaha 1989). It is possible that the drop in pH in the soil surrounding a dissolving PARP granule is less than that around the fully acidulated fertilizer. In acidic soils, this would result in less dissolution of Fe and Al and consequently less fixation of P. The residues around the PARP granule and the reaction products in the soil may thus sustain a higher solution P concentration than for ordinary or concentrated superphosphate (Sharpley et al. 1992).

### 5.7.5  DIRECT USE OF GROUND PHOSPHATE ROCK

The rise in the price of the P fertilizers containing WSP has forced the direct use of ground PR (Bolan et al. 1990; White et al. 1999; Akintokun et al. 2003). Since the P in PR is not water soluble, its direct use is restricted to acid soils. However, some researchers have reported PR to be useful even on nonacid soils (Tiwari 1979). Phosphate availability from PR is reported to be related to its citrate-soluble P content. Chien and Friesen (1992) showed that a high reactive ground North Carolina PR containing 4.2% citrate-soluble P was as good as TSP for corn grown on an acid soil. The less reactive Jordan PR (2.9% P) and Togo PR were less effective than TSP. Similarly, Msolla et al. (2005) reported that Mnjingu PR was fairly good for corn production. Among crop species, oilseed rape (*Brassica campestris*) is reported to be highly efficient in utilizing PR due to its ability to exude malic and citric acids that help in the utilization of P in PR (Habib et al. 1999). Lin et al. (2003) have developed a process called mechanomilling at high energy to enhance the chemical and physical reactivities of PRs. The direct use of PR has gained importance in view of the increasing demand for organic food (Nelson and Mikkelsen 2008; Prasad and Nagarajan 2008). Organic farming permits the use of PR but not the industry-made phosphate fertilizers.

A number of innovations have been attempted to increase the efficiency of directly applied PR. These include the following:

#### 5.7.5.1  Mixing Rock Phosphate with Soluble P Fertilizers

Several researchers have experimented with mixtures of finely ground rock phosphate and soluble P fertilizers such as superphosphate. The proportion of the two will, of course, depend upon the soil conditions and the crop. Fageria et al. (1990) suggested the use of rock phosphate in combination with a soluble P source as a strategy for improving the P status of acidic soils; they recommended the broadcast application of rock phosphate and the band application of soluble P sources. Mashori et al. (2013) suggested the use of 25% PR and 75% SSP mixture with 10 t/ha of farmyard manure for corn in Pakistan.

#### 5.7.5.2  Mixing Rock Phosphate with Elemental S
####          or S-Producing Compounds

The mixing of rock phosphate with elemental S was suggested in the early part of the twentieth century (Lipman et al. 1916) for increasing the availability of P in PR. The principle involved is that the S in the mixture is oxidized to sulfuric acid by chemoautotrophic bacteria (*Thiobacillus thioxidans* and *Thiobacillus thioparus*), which solubilizes P in the rock phosphate (Prasad and Power 1997). This has been demonstrated on a number of soils. The efficiency of rock phosphate–sulfur mixtures appears to be greatly influenced by a number of factors such as soil pH, soil temperature, soil water, and particle size of the mixture. Swaby (1975) inoculated the rock phosphate–sulfur mixtures with *T. thiooxidans* and *T. thioparus* and called them *biosupers*; these mixtures were found to be superior to uninoculated mixtures for pasture production in tropical soils.

Iron pyrites as a source of sulfur have also been suggested for mixing with rock phosphate. This strategy has been used for utilizing low-analysis PR (18–20% $P_2O_5$), which cannot be used for making soluble phosphate fertilizers (Tiwari 1979). Sulfur–PR mixtures have an advantage of supplying S, which is becoming deficient in more and more soils (Mitchell and Mullins 1990; Messick 2003; Singh 2001) due to the large-scale use of MAP, DAP, and TSP in place of SSP or OSP.

### 5.7.5.3  Use of Phosphate-Solubilizing Microorganisms

A number of phosphorus-solubilizing organisms (PSO), both bacteria and fungi, have been reported to help in solubilizing P from PR (Didiek et al. 2000; Richardson 2001; Richardson et al. 2009, 2011; Alzoubi and Gaibore 2012). Viveganandan and Jauhri (2002) reported an increase in seed yield and P uptake by soybean by a granular formulation of low-grade PR and phosphate-solubilizing bacteria (PSB) (*Pseudomonas striata* P27 and *Bacillus polymyxa* H5). Panhwar et al. (2011) reported significantly more P solubilization from Christmas Island PR due to PSB (*Bacillus* spp. PSB9 and PSB16) resulting in improved growth and increased P uptake by aerobic rice. In some studies, the effect of the combined inoculation with PSB and VAM was studied, and it was found that their individual effects were an additive for solubilizing P from rock phosphate (Azcon et al. 1976). In a study in India, the relative agronomic efficiency of a combination of local Mussoorie PR, a low-grade and low-reactive PR, with PSB and VAM was 53–65% in the first cycle of rice–rapeseed (*Brassica campestris*)–mung bean (*Phaseolus radiates* L.) cropping system but reached 69–106% in the third cycle (Sharma et al. 2010).

## 5.8  PHOSPHORUS MANAGEMENT IN CROPPING SYSTEMS

### 5.8.1  Rice–Wheat Cropping System (RWCS)

The RWCS occupies about 26.5 million ha of which 13 million ha are in China, 10 millon ha in India, 2.2 million ha in Pakistan, 0.8 million ha in Bangladesh, and 0.5 million ha in Nepal (Singh et al. 2004). These five countries practicing RWCS represent 43% of the world's population on 20% of the world's arable land (Paroda et al. 1994; Prasad 2005). This cropping system is unique, since rice is grown under complete or partial submergence (Prasad 2011), while wheat is grown under well-drained conditions. The two crops are grown in the same year. Wheat succeeds rice, and the time interval between rice harvest and wheat seeding is 3–4 weeks. The chemistry of P in rice soils has been reviewed (Sanyal and DeDatta 1991). Submergence in a rice field leads to a flush of available P (Kirk et al. 1990), which could be due to the release of P from occluded P fraction (Patrick and Mahapatra 1968), and this has led to the general belief that on the same soil, rice responds less well to fertilizer P than wheat. Gill and Meelu (1983) suggested that if 26 kg P/ha is applied to wheat, rice could be grown without application of P. However, Singh et al. (2000) recommended the application of 32 kg P/ha to wheat and 15 kg P/ha to rice in the RWCS. Kumar and Yadav (2001) on the basis of a long-term experiment at Faizabad, India, on RWCS also reported that both rice and wheat responded well to P application. Singh et al. (2014) in a multilocation on-farm study observed that the application of 13 kg P ha[-1] increased the average yield of rice by 0.87 t ha$^{-1}$, whereas the yield increase at 26 kg P ha$^{-1}$ was 1.23 t ha$^{-1}$ grain at different locations. In wheat, the yield response to 13 kg P ha$^{-1}$ varied from 0.55 to 1.11 t ha$^{-1}$ at different locations with a mean of 0.79 t ha$^{-1}$. At 26 kg P ha$^{-1}$, the mean yield increase due to P application was 1.32 t ha$^{-1}$. Nevertheless, in general, the response of dwarf high-yielding wheat varieties to P application is fairly good. This can be attributed to their high yields, shorter root system, and lower temperatures during the winter–spring growing season (Prasad 2007a). Further, Sah and Mikkelsen (1989) and Sah et al. (1989a) have shown that drying of soil during wheat reduces P availability. Thus, in RWCS in India, both rice and wheat have to be adequately fertilized with phosphorus. The rates of P application to rice in China have been generally high (19.6–78.4 kg P/ha) (Bi et al. 2014), and there are suggestions to reduce P application in rice to prevent pollution (Guan et al. 2014).

### 5.8.2 CORN–SOYBEAN CROPPING SYSTEM

Corn–soybean (grown in alternate years) is an important cropping system in the United States and several other countries. In a 42-year long-term experiment at North Central Kansas Experimental Field near Scandia, the application of phosphorus (14.6 kg P/ha) to corn increased both corn and soybean yields. The corn yield obtained with 55 kg N and 13 kg P/acre was equal to that obtained with 160 or 200 kg N/acre without P (Gordon 2003). Without P application, Bray-1 available P in soil declined from 80 ppm in 1960 to less than 40 ppm in 2000. However, with P application, there was very little decline in availability P. Similarly in a 30-year long-term experiment at the Northeast Research and Demonstration farm, in Iowa, combinations of 0, 22.4, and 45 kg P/ha and 0, 60, and 120 kg K/acre were applied to both corn and soybean. The average increase in corn yield (1979–2008) was from 9980 kg/ha in no-P plots to 10,859 kg/ha in plots receiving 45 kg/ha; the corresponding increase in soybean yield was from 3430 kg/ha in check plots to 3712 kg/ha with 45 kg/ha (Mallarino and Pecinovsky 2013). In this study, Bray-1 available P increased from 9 ppm in check plots to 105 ppm in plots receiving 45 kg/ha each year by 2008.

## 5.9 INTERACTION WITH ZINC

The P × Zn interaction is antagonistic and has received considerable attention from researchers. Negative effects of high rates of P application on absorption of Zn and vice versa have been reported in corn (Adriano et al. 1971; Christensen and Jackson 1981; Sharma et al. 1986; Bukovic et al. 2003), wheat (Webb and Loneragan 1988; Nayak and Gupta 1995), rice (Haldar and Mandal 1981), potato (*Solanum tuberosum* L.) (Soltanpour 1969; Christensen and Jackson 1981; Barben et al. 2007), soybean (Shittu and Ogunwale 2012), oilseed rape (Hu et al. 1996), and groundnut (*Arachis hypogaea*) (Mirvat et al. 2006), and the subect has been reviewed (Alloway 2008; Mousavi et al. 2012). As regard to the causes for this antagonistic interaction, some researchers (Nair and Babu 1975; Khan and Zende 1977; Rupa et al. 2003) pointed out that high levels of P inhibit the translocation of Zn from the roots to the metabolic sites in leaves. The formation of zinc phosphate on root surfaces (Sarret et al. 2001) and in leaves (Kupper et al. 2000) has also been suggested. Loneragan and Webb (1993) showed that when plants are supplied with high concentrations of P and low concentrations of Zn, they accumulate high P in their leaves, which precipitates Zn.

## 5.10   PHOSPHORUS AND POLLUTION

The pollution of soil and water due to phosphate has received considerable attention in recent years. There are two major areas, namely, the eutrophication of surface water with phosphorus and the contamination of soil with the heavy metals cadmium and lead.

### 5.10.1   Eutrophication of Surface Waters

*Eutrophication* refers to the enrichment of surface waters with plant nutrients and could be natural or anthropogenic, and phosphorus eutrophication of waters has been reported (Reed et al. 2000). An Olsen extractable P value of 100 mg P/kg soil could represent a critical level to distinguish soils that may be a potential source of surface water eutrophication (Indiati and Rossi 1999). Phosphorus is perhaps more critical than nitrogen for some species of algae, such as blue green algae (Cyanobacteria) (Chien et al. 2011). Algal blooms lead to the depletion of dissolved oxygen in water (hypoxia), which can lead to fish mortality, and their death and decomposition impair such waters for drinking, recreation, and industry (Shumway 1990; Martin and Cooke 1994; Smith et al. 1999; Foy 2005). The use of WSP is one of the potential sources of P responsible for P eutrophication of surface waters (Alexander et al. 2008). A number of factors can be responsible for this. These include P runoff after application, which is affected by landscape and rainfall characteristics, WSP of fertilizer, method of P application, and crops and cropping systems.

#### 5.10.1.1   Landscape and Rainfall Characteristics

Landscape characteristics strongly affect the runoff of P. Duan et al. (2005) reported that in Yunnan Province, in China, P loss was most from farmlands with 6°–12° of slope. The amount of P loss also depends on soil erodibility, land cover, and cultivation practices (Sharpley et al. 2000). Phosphorus loss is more in the start of a rainy season and gradually decreases as the rainy season advances (Jiao et al. 2007; Xu et al. 2007).

#### 5.10.1.2   Method and Source and Rate of P Application

The incorporation of P in soil reduces the loss of applied P (Withers et al. 2005). Fresh surface application of P fertilizer leads to elevated P loss due to inadequate time for P to react with other elements in soil (Haygarth et al. 1998). Kimmel et al. (2001) reported that broadcast P application increased P losses from ridge-till to no-till system in a grain sorghum–soybean system on a Woodson silt loam having 1.0–1.5% slope. It is reported that P fertilizers having no WSP, such as ground PR, can minimize the runoff losses of P as compared to WSP fertilizers (Hart et al. 2004; Shigaki et al. 2006, 2007). As would be expected, higher rates of P application result in higher loss of P. In rice, the application of 0, 25, 60, 120, and 240 kg P/ha resulted in a loss of 0.13, 0.50, 0.94, 3.02, and 5.97 kg P/ha to runoff water (Xia et al. 2008). The period of most risk of P loss from rice field was within 1–2 months after the application (Zhang et al. 2007). In pastures, the application of P prior to irrigation can result in increased P loss in runoff, and large P losses (5.5–9.0 kg/ha) under such conditions have been reported (Bush and Austin 2001).

### 5.10.1.3 Farming Systems and Vegetation

The losses of applied P are more from grasslands or forests than from farmlands. According to Wu et al. (2008), P runoff losses were influenced by the vegetation in the following order: turf grass → secondary forest → sloping land → bamboo. The lack of soil inversion in grasslands permits a greater soil–rainfall interaction leading to higher fertilizer P runoff (Haygarth et al. 1998; Turtola and Yi-halla 1999).

### 5.10.1.4 Crops and Cropping Systems

Cao et al. (2005) reported that P losses were most in mulberry fields (1.1 kg/ha in 4 months), followed by wheat and rice (0.84 kg P/ha) and vegetables (0.6 kg P/ha). Zhu et al. (2007) reported that the P losses were higher in rice–rape rotation than in rice–wheat or rice–green gram rotation due to more irrigations in the rice–rape rotation. Zhang et al. (2003) reported that in rice–wheat rotation, the losses were more during the wheat season.

## 5.10.2 Pollution of Soils with Cadmium

The application of phosphate fertilizers leads to the accumulation of cadmium (Cd) in soils and crop plants grown on them, and the subject has been thoroughly reviewed by several researchers (Alloway and Steinnes 1999; Grant et al. 1999; Mortvedt 2005; Chien et al. 2011; Roberts 2014). Only the main findings are brought out here. Cd in P fertilizers comes from PRs, where it gets in by substituting Ca in apatite structure (Iretskaya and Chien 1999). Cd in PRs can vary from 0.2 mg Cd/kg PR in deposits from Kola, in Russia, to 100 mg Cd/kg PR in deposits from Nauru (Alloway and Steinnes 1999). The chemical forms of Cd in acidulated P fertilizers are $Cd(H_2PO_4)_2$ and $CdHPO_4$, the analogs of calcium monophosphate and calcium diphosphate, and the availability of these compounds is also similar to calcium phosphates and is more in acid soils (Iretskaya and Chien 1999). Since the amounts of Cd added through the P fertilizers in a crop season are too small, it takes considerable time before the Cd levels reach toxic limits.

Kukier et al. (2010) from the Virginia Polytechnic Institute and State University, Blacksburg have suggested that chemical extractability and phytoavailability of Cd in biosolids amended soils over a long period is reduced.

Bolan et al. (2003) calculated that at an annual application of 20 kg P/ha, it would take 214 years with Gafsa PR containing 70 mg Cd/kg to 2250 years with DAP containing only 10 mg Cd/Kg to exceed the threshold limit of 3 mg C/kg soil, set for Cd for sewage sludge application in New Zealand. The U.S. EPA (2007) has set the national cumulative ceiling concentration limits for land application of biosolids at 85 mg Cd/kg soil and a maximum loading rate of 1.9 kg Cd/ha/year. The Association of American Plant Food Control Officials has suggested the standard Cd limit in

P fertilizers at 4.4 mg Cd/kg per percentage of P content in fertilizer (Mortvedt 2005). More details are available in the paper by Iretskaya and Chien (1999). Recently conducted scientific risk assessments showed that the use of P fertilizer containing current levels of Cd is generally safe and does not pose a risk to the farmers that use them or the general public (Roberts 2014).

## 5.11  CONCLUSION

To meet the food demand of the ever-increasing world population, phosphorus consumption, which has already increased from 4 Tg in 1890 to 15 Tg in 2000, will have to be further increased. However, since phosphorus is a nonrenewable resource and the present PR reserves may not last more than 1000 years, it has to be carefully and economically managed. Careful management of phosphorus has also become important because in some regions of the world it is being overused, resulting in the eutrophication of surface waters and increased concentration of cadmium in soil. Phosphorus is taken up by the plants mostly as $H_2PO_4^-$, and partly as $HPO_4^=$. The $H_2PO_4^-$ is the water-soluble form (WSP) and is present in most P fertilizers made by the acidulation of PR, which requires sulfur and energy, making it an expensive input in crop production. Once applied to soil, the WSP reacts fairly quickly with Ca, Al, and Fe and reverts back to less-soluble or insoluble forms. This high reactivity of WSP also reduces its mobility in soil and makes P management on farm fields a difficult task because it cannot be top dressed as nitrogen. Most phosphorus is therefore applied at seeding as broadcast and incorporated in the soil or is drilled below the seed. In the case of established grasslands and pastures, it can be only surface broadcast, which leads to high surface runoff losses. Attempts have been made toward the foliar application of P in cereals with mixed success. Efforts have also been made to directly use ground PR alone or mixed with WSP fertilizers, sulfur, or organic manures especially on acid soils. Cultures of PSO and VAM have also been experimented upon to increase the efficiency of ground PR with some success. Efficient phosphorus management for crop production is a must for countries, such as India, which have very little PR deposits and no sulfur deposits, forcing lower P applications to crops, as well as for countries where there are plenty of PR deposits but excessive P application is leading to soil and water pollution.

## REFERENCES

Adriano, D. C., G. M. Paulsen, and L. S. Murphy. 1971. Phosphorus-iron and phosphorus-zinc relation-ship in corn (*Zea mays* L.) seedlings as affected by mineral nutrition. *Agron J* 63:36–39.

Akintokun, M. O., E. A. Makinde, F. I. Oluwatoyinbo, and M. T. Adetunji. 2003. Effects of phosphate rock application on dry matter yield and phosphorus recovery of maize and cowpea grown in sequence. *African J Environ Sci Tech* 4:293–303.

Alexander, R. B., R. Smith, G. Schwartz, E. Boyer, J. Nolan, and J. Brakebill. 2008. Differences in phosphorus and nitrogen delivery to the Gulf of Mexico from the Mississippi River Basin. *Environ Sci Technol* 42(3):822–830.

Alexandros, A., and J. Bruinsima. 2012. *World Agriculture Towards 2030/2050*. Rome, Food and Agriculture Organization of the United Nations.

Ali, M. S., A. Sutradhar, M. L. Edano et al. 2014. Response of winter wheat grain yield and phosphorus uptake to foliar phosphite fertilization. *Int J Agron.* Article ID 801626. Vol. 2014.10.1155/2014/80126. Available at http://dx.doi.org/10.1155/2014/801626.

Alloway, B. J. 2008. *Zinc in Soils and Crop Nutrition.* Second Ed. Brussels, International Zinc Association; Paris, International Fertilizer Industry Association.

Alloway, B. J., and E. Steinnes. 1999. Anthropogenic additions of cadmium to sols. In *Cadmium in Soils and plants.* McLaughlin, M. J., and B. R. Singh, eds. pp. 97–123, Dordrecht, Kluwer Academic.

Alzoubi, M. M., and M. Gaibore. 2012. The effect of phosphate solubilizing bacteria and organic fertilization on availability of Syrian PR and increase of triple superphosphate efficiency. *World J Agric Sci* 8(5):473–478.

Anonymous. 1999. Effects of phosphorus on crop maturity. *Better Crops* 83(1):14–19.

Azcon, R., J. M. Barea, and D. S. Hayman. 1976. Utilization of rock phosphate in alkaline soils by plant inoculated with mycorrizal fungi and phosphate solubilizing bacteria. *Soil Biol Biochem* 8(2):L135–138.

Banerjee, M., R. K. Rai, and D. Maiti. 2010. Root characteristic of maize as influenced by various phosphatic chemical fertilizers and biofertilzers. *Arch Agron Soil Sci* 56(6):681–695.

Barben, S. A., B. A. Nicholas, B. G. Hopkins, V. D. Jolley, J. W. Ellsworth, and B. L. Webb. 2007. Phosphorus and zinc interactions in potato. *Water Nutrient Management Conference, Salt Lake City, UT* 7:219–223.

Barber, S. A., J. M. Walker, and E. H. Vasey. 1963. Mechanisms for the movement of plant nutrients from the soil and fertilizers to the plant root. *J Agric Food Chem* 11:204–207.

Barrow, N. J. 1983. On the reversibility of phosphate sorption by soils. *J Soil Sci* 34:751–758.

Bar-Yosef, B. 1991. Root excretions and their environmental effects: Influence of availability of phosphorus. In *Plant Roots: Their Hidden Half.* Waisal, Y., A. Eshel, and U. Kafkafi, eds. pp. 529–557, New York, Marcel Dekker.

Berg, J. M., J. L. Tymoczko, and L. Stryer. 2002. *Biochemistry.* Fifth Ed. W. H. Freeman and Co.

Bi, L., J. Xia, K. Liu, D. Li, and X. Yu. 2014. Effect of long-term chemical fertilization on trends of rice yield and nutrient use efficiency under double rice cultivation in subtropical China. *Plant Soil Environ* 60(12):537–543.

Bidappa, C. C., and B. V. V. Rao. 1973. Studies on the relationship between sesquioxides, phosphorus contents and phosphorus fixing capacity of coffee soils of South India. *J Indian Soc Soil Sci* 21:155–159.

Black, C. A. 1968. *Soil-Plant Relationships.* New York, John Wiley.

Blackburn, G. M. 2006. *Nucleic Acids in Chemistry and Biology.* London, The Royal Society.

Bolan N. S. 1991. A critical review on the role of mycological fungi in the uptake of phosphorus by plants. *Plant Soil* 133:189–207.

Bolan, N. S., D. C. Adriano, and R. Naidu. 2003. Role of phosphorus in immobilization and bioavailability of heavy metals in the soil-plant system. *Rev Environ Contam Toxicol* 177:1–44.

Bolan, N. S., R. E. White, and M. J. Hedley. 1990. A review of the use of phosphate rocks as fertilizers for direct application in Australia and New Zealand. *Australian J Exp Agric* 30:297–313.

Boyton, D. 1954. Nutrition by foliar application. *Annu Rev Plant Physiol* 5:31–54.

Brady, N. C. 1990. *The Nature and Properties of Soils*, Tenth Ed. New York, John Wiley.

Brar, B. S., and A. C. Vig. 1988. Kinetics of phosphate release from soils and its uptake by wheat. *J Agric Sci (Cambridge)* 110:505–513.

Braum, S. M., and P. A. Helmke. 1995. White lupin utilizes soil phosphorus that is unavailable to soybean. *Plant Soil* 176:95–100.

Bruulsema, T. W., P. E. Fixen, and G. D. Sulewski, eds. 2012. *4R Plant Nutrition: A Manual for Improving the Management of Nutrition.* Norcross, GA, International Plant Nutrition Institute.

Bukovic, G., M. Antunovic, C. Popvic, and M. Rastija. 2003. Effect of P and Zn fertilization on biomass yield and its uptake by maize lines (*Zea mays* L.). *Plant Soil Environ* 49:505–510.

Bush, R. T., and N. R. Austin. 2001. Timing of phosphorus fertilizer application within an irrigation cycle for perennial pasture. *J Environ Qual* 30:939–946.

Cao, Z. H., X. G. Lin, L. Z. Yang et al. 2005. Ecological function of "paddy field ring" to urban and rural environment: I: Characteristics of soil P losses from paddy field to water bodies with runoff. *Acta Ped Sinica* 42:799–804 (in Chinese).

Cardoso, I. M., C. L. Boddington, B. H. Janssen, O. Oenema, and T. W. Kuyper. 2006. Differential access to phosphorus pools of an oxisol by mycorrhyzal and non-mycorrhyzal maize. *Commun Soil Sci Plant Anal* 37:1537–1551.

Chien, S. H., and D. K. Friesen. 1992. Phosphate rock for direct application. In *Future Directions for Agricultural Research*, TVA/NFERC Bulletin Y-224. Sikora, F. J., ed. pp. 47–52, Muscle Shoals, AL, Tennessee Valley Authority.

Chien, S. H., L. I. Prochnow, and H. Cantarella. 2009. Recent developments of fertilizer production and use to increase nutrient efficiency and minimize environmental impacts. *Adv Agron* 102:261–316.

Chien, S. H., L. I. Prochnow, S. Tu, and C. S. Synder. 2011. Agronomic and environmental aspects of phosphate fertilizers varying in source and solubility: An update review. *Nutr Cycl Agroecocyst* 89:229–255.

Christensen, N. W., and T. L. Jackson. 1981. Potential for phosphorus toxicity in zinc stressed corn and potato. *Soil Sci Soc Am J* 45:904–909.

Ciarelli, D. M., A. M. C. Furlani, A. R. Dechen, and G. Lima. 1998. Genetic variation among maize genotypes for phosphorus uptake and phosphorus use efficiency in nutrient solution. *J Plant Nutr* 21:2219–2229.

Coleman, R. 1942. Utilization of adsorbed phosphate by cotton and oats. *Soil Sci* 54:237–246.

Dakora, F. D., and D. A. Phillips. 2002. Root exudates as mediators of mineral acquisition in low nutrient environments. *Plant Soil* 245:35–47.

Deb, D. L., and N. P. Datta. 1967. Effect of association of ions on phosphorus retention in soils under variable anion concentrations. *Adv Plant Sci* 26(1):432–444.

Didiek, G., H. Siswanto, and Y. Sugiato. 2000. Bioactivation of poorly soluble phosphate rocks with a phosphate solubilizing fungus. *Soil Sci Soc Am J* 64:927–937.

Dinkelaker, B., and H. Marschner. 1992. In vivo demonstration of acid phosphatase activity in the rhizosphere of soil-grown plants. *Plant Soil* 144:199–205.

Dinkelaker, B., C. Hengeler, and H. Marschner. 1995. Distribution and function of proteoid and other root clusters. *Bot Acta* 108:183–200.

Drew, M. C. 1975. Comparison of the effects of a localized supply of phosphate, nitrate, ammonium and potassium on the growth of the seminal root system, and the shoot in barley. *New Phytologist* 75(3):479–490.

Drew, M. C., and L. R. Saker. 1984. Uptake and long distance transport of phosphate, potassium and chloride in relation to internal ion concentration in barley: Evidence of non-alloteric regulation. *Planta* 60:500–507.

Duan, Y. H., N. M. Zhang, B. Hong, and J. J. Chen. 2005. Factors influencing the N and P loss from farmland runoff in Dianchi watershed. *Chin J Econ-Agric* 13:116–118 (in Chinese).

Egner, H., H. Riehm, and W. R. Domingo. 1960. Untersuchungen uber die chemische Bodenanalyse als Grundlage fur den Beurteilung des Narstoffzustandes der Boden. *Kungl Lantbrukshogskolans Annaler* 26:199–215 (in Norwegian).

Eswaran, H., P. Reich, and F. Beinroth. 1997. Global distribution of soils with acidity. In *Plant-Soil Interactions at Low pH*. São Paulo, Brazilian Soil Science Society.

Fageria, N. K., and V. C. Baligar. 2005. Nutrient availability. In *Encyclopedia of Soils and Environment*. Hillel, D., ed. pp. 63–72. San Diego, CA, Elsevier.

Fageria, N. K., V. C. Baligar, and D. G. Edwards. 1990. Soil-plant nutrient relationships at low pH stress. In *Crops and Enhancers of Nutrient Use*. Baligar, V. C., and R. R. Duncan, eds. pp. 475–508, New York, Academic Press.

Fageria, N. K., B. C. Baligar, and R. J. Wright. 1991. Influence of phosphate rock sources and rates on rice and common bean production in an oxisol. *Plant Soil* 134:137–144.

Fan, M. S., J. M. Zhu, C. Richards, K. M. Brown, and J. P. Lynch. 2003. Physiological role of aerenchyma in phosphorus stressed roots. *Funct Plant Biol* 30:493–506.

Fixen, P. E., and J. H. Grove. 1990. Testing soil for phosphorus. In *Soil Testing and Plant Analysis*, Book Ser. 3. Westerman, R. L., ed. pp. 141–180, Madison, WI, Soil Science Society of America.

Food and Agriculture Organization. 2009. *How to Feed the World in 2050*. Rome, Food and Agriculture Organization of the United Nations.

Foy, R. H. 2005. The return of phosphorus paradigm: Agricultural phosphorus and eutrophication. In *Phosphorus in Agriculture and the Environment*, Agronomy Monograph No. 46. Sims, J. T., and A. N. Sharpley, eds. pp. 911–939, Madison, WI, American Society of Agronomy, Crop Science Society of America, and Soil Science Society of America.

Fujita, Y., H. O. Venterink, P. M. van Bodegom et al. 2014. Lower investment in sexual reproduction threatens plants adapted to phosphorus limitation. *Nature* 505:82–86.

Gahoonia, T. S., and N. E. Nielsen. 1992. The effect of root induced pH changes on the depletion of inorganic and organic P in the rhizosphere. *Plant Soil* 143:185–191.

Gahoonia, T. S., and N. E. Nielsen. 1997. Variation in root hairs of barley cultivars doubled phosphorus uptake from soil. *Euphytica* 98:177–182.

Gahoonia, T. S., and N. E. Nielsen. 1998. Direct evidence on participation of root hairs in phosphorus ($^{32}$P) uptake from soil. *Plant Soil* 198:147–152.

Gahoonia, T. S., and N. E. Nielsen. 2004. Root traits as tools for creating phosphorus efficient crop varieties. *Plant Soil* 260:47–57.

Gerke, J. 1993. Solubilizaion of Fe (III) from humic-Fe complexes, humic Fe oxide mixtures and from poorly eroded Fe-oxide by organic acids: Consequences for P-adsorption. *Z Pflanzen Bodenkunde* 156:253–257.

Gill, H. S., and O. P. Meelu. 1983. Studies on the utilization of phosphorus and causes for its differential response in rice-wheat rotation. *Plant Soil* 74:211–222.

Girma, K., K. L. Martin, K. W. Freeman et al. 2007. Determination of optimum rate and growth stage for foliar-applied phosphorus in corn. *Commun Soil Sci Plant Anal* 3899:1137–1154.

Glass, A. D. M. 1989. *Plant Nutrition—An Introduction to Current Concepts*. London, Jones & Bartlett.

Good, A. G., A. K. Sahrawat, and D. G. Muench. 2004. Can less yield more? Is reducing nutrient input into the environment compatible with maintaining crop production? *Trends Plant Sci* 9:597–605.

Gordon, W. B. 2003. Nitrogen and phosphorus management for corn and soybean in rotation. *Better Crops* 87(1):4–5.

Grant, C. A., L. D. Bailey, M. J. McLaughlin, and B. R. Singh. 1999. Management factors which influence cadmium concentrations in crops. In *Cadmium in Soils and Plants*. McLaughlin, M. J., and B. R. Singh, eds. pp. 151–198, Dordrecht, Kluwer Academic.

Guan, G., S. Tu, H. Li, J. Yang, T. Zhang, S. Wen, and L. Wang. 2014. Phosphorus fertilization modes affect crop yield, nutrient uptake and soil biological properties in rice-wheat cropping system. *Soil Sci Soc Am J* 77(1):166–172.

Habib, L., S. H. Chien, G. Carmona, and J. Henao. 1999. Rape response to a Syrian phosphate ock and its mixture with triple superphosphate on a limed alkaline soil. Commun *Soil Sci Plant Anal* 30:449–456.

Haldar, M., and L. N. Mandal. 1981. Effect of phosphorus and zinc on the growth and phosphorus, zinc, copper, iron and manganese nutrition of rice. *Plant Soil* 59:415–425.

Hamy, A. 1983. Effect of phosphorus deficiency on pigmentation of barley leaves. *CR Seances l'Acad d'Agric France* 69:935–943.

Harrison, A. F. 1987. *Soil Organic Phosphorus—A Review of World Literature*. Wallingford, CAB International.

Hart, M. R., B. Quin, and M. L. Nguyen. 2004. Phosphorus runoff from agricultural land and direct fertilizer effects: A review. *J Environ Qual* 33:1954–1972.

Hayes, J. E., A. E. Richardson, and R. J. Simpson. 1999. Phytase and acid phytase activities in roots of temperate pasture grasses and legumes. *Aust J Plant Physiol* 26:801–809.

Haygarth, P. M., L. Hepworth, and C. C. Jarvis. 1998. Forms of phosphorus transfer in hydrological pathways from soil under grazed grassland. *Eur J Soil Sci* 49:65–72.

He, X. 1998. Mineralization and bioavailability of phosphorus bound to soil organic matter by enzymes from *Lupinus albus*. PhD Thesis, University of Wisconsin, Madison, WI.

Helmke, P. A., T. J. Boerth, and X. He. 2000. Bioavailability of organically bound soil phosphorus. *Proc. 2000 Wisconsin Fertilizer, Aglime and Pest Manage Conf* January 18–20, 2000, Madison, WI.

Hill, J. O., R. J. Simpson, A. D. Moore, and D. F. Chapman. 2006. Morphology and response of roots of pasture species to phosphorus nd nitrogen nutrition. *Plant Soil* 286:7–19.

Hissinger, P. 2001. Bioavailability of inorganic P in the rhizosphere as affected by root-induced chemical changes: A review. *Plant Soil* 237:173–195.

Hissinger, P., and R. J. Gikes. 1996. Mobilization of phosphate from phosphate rock and alumina-sorbed phosphates by the roots of ryegrass and clover as related to rhizosphere PH. *Eur J Soil Sci* 47:533–544.

Hogan, C. M. 2011. *Phosphate* In *Encyclopedia of Earth*. Cleveland, C. J., ed. Washington, DC, National Council for Science and the Environment.

Hu, D., R. W. Bell, and Z. Xie. 1996. Zinc and phosphorus responses in transplanted oilseed rape. *Soil Sci Plant Nutr* 42:333–344.

Indiati, R., and N. Rossi. 1999. Extractability of residual phosphorus from highly manured soil. *Italian J Agron* 3:63–67.

International Plant Nutrition Institute (IPNI), 3500 Parkway Lane, Suite 550, Peachtree Corner, GA 30092-2844, USA.

Iretskaya, S. I., and C. H. Chien. 1999. Comparison of cadmium uptake by five different food grain crops on three soils of varying pH. *Commun Soil Sci Plant Anal* 30:441–448.

Jackson, P. C., and C. E. Hager. 1980. Products of orthophosphate absorption by barley roots. *Plant Physiol* 35:326–332.

Jakobsen, I., L. K. Abbot, and A. D. Robson. 1992. External hyphae of vesicular arbuscular mycorrhyzal fugi associated with Trifolium subterraneum L/I: Spread of hyphae and phosphorus inflow into roots. *New Phytolologist* 120:371–380.

Jeschke, W., E. Kirkby, A. Peuke, J. Pate, and W. Hartung. 1997. Effect of phosphorus efficiency on assimilation and transportation of nitrate and phosphate in intact plants of castor bean (*Ricinus communis* L.). *J Exp Bot* 48:75–91.

Jiao, S. J., X. M. Hu, G. X. Pan, H. J. Zhou, and X. D. Xu. 2007. Effects of fertilization on nitrogen and phosphorus runoff from Qingzini paddy soil in Taihu lake region during rice growth season. *Chin J Ecol* 26:495–500 (in Chinese).

Johnston, A. E., and P. R. Poulton. 1977. Yields on the exhaustion land and changes in the NPK contents of the soils due to cropping and manuring. *Rothamsted Exp Sta Rept for 1976*. Part 2:53–85.

Johnston, A. E., and J. K. Syers. 2009. A new approach to assessing phosphorus use efficiency in agriculture. *Better Crops* 93(3):14–15.

Jones, D. L. 1998. Organic acids in the rhizosphere: A critical review. *Plant Soil* 205:25–44.

Jones, D. L., and P. R. Darrah. 1994. Role of root derived organic acids in the mobilzaion of nutrients in the rhizosphere. *Plant Soil* 166:247–257.

Jones, D. L., P. G. Dennis, A. G. Owen, and P. A. W. van Hees. 2003. Organic acid behavior in soils-misconceptions and knowledge gaps. *Plant Soil* 248:31–41.

Kannan, S. 1986. Foliar absorption and transport of inorganic nutrients. *Critical Rev Plant Sciences* 4(4):347–376.

Khan, A., and G. K. Zende. 1977. The site of Zn-P interaction in plants. *Plant Soil* 46:259–262.

Klages, M. G., R. A. Olsen, and V. A. Haby. 1988. Relationship f phosphorus ieotherms to NaHCO₃ extractable phosphorus as affected by soil properties. *Soil Sci* 146:85–91.

Kimmell, R. J., G. M. Pierzynski, K. A. Janssen, and P. L. Barn. 2001. Effects of tillage and phosphorus placement on phosphorus runoff losses in a grain sorghum-soybean rotation. *J Environ Qual* 30:1324–1330.

Kirk, J. G. D., Y. Tian-ren, and F. A. Chaudhary. 1990. Phosphorus chemistry in relation to water regimes. In *Phosphorus Requirements for Sustainable Agriculture in Asia and Oceania*. pp. 211–223. Los Baños, International Rice Research Institute.

Kukier, U., R. L. Chaney, T. A. Ryan, W. L. Daniels, R. H. Dowdy, and T. C. Grants. 2010. Phytoavailability of cadmium in long-term biosolids amended soils. *J Environ Qual* 39(2):59–530.

Kumar, A., and R. L. Yadav. 2001. Long-term effects of fertilizers on soil fertility and productivity of a rice-wheat system. *J Agron Crop Sci* 186:47–54.

Kupper, H., E., Lombi, F. G. Zhao, and S. P. McGarth. 2000. Cellular compartmentation of cadmium and zinc in relation to other elements in the hyperaccumulator *Arabidopsis halleri*. *Planta* 212:75–84.

Kurtz, L. T. 1953. Phosphorus in acid and neutral soils. In *Phosphorus in Crop Nutrition*, Agronomy Monograph No. 4. Pierre, W. H., and A. G. Norman, eds. pp. 59–85, Madison, WI, American Society of Agronomy; New York, Academic Press.

Lal, R. 2014. Societal value of soil carbon. *J Soil Water Conserv* 69(6):186A–192A.

Lambers, H., M. W. Shane, M. D. Cramer. S. J. Pearse, and E. J. Veneklass. 2006. Root structure and functioning for efficient acquisition of phosphorus matching morphological and physiological traits. *Ann Bot* 98:693–713.

Lee, R. B., and R. G. Radcliffe. 1993. Subcellular distribution of major phosphate and levels of nucleoside triphosphate in mature maize roots at low external phosphate concentrations: Measurement with ³¹P NMR. *J Exp Bot* 44:587–598.

Lin, H. H., R. J. Gikes, and P. G. McCormick. 2003. Benefication of rock phosphate fertilizers by mechano-milling. *Nutr Cycl Agroecosyst* 67:177–186.

Lindsay, W. L. 1979. *Chemical Equilibria in Soils*. New York, John Wiley.

Lindsay, W. L., and E. C. Moreno. 1960. Phosphate phase equilibria in soils. *Soil Sci Soc Am Proc*. 24:177–182.

Lindsay, W. L., and H. F. Stephenson. 1959. Nature of the reactions of monocalcium phosphate monohydrate in soils: II: Dissolution and precipitation reactions involving iron, aluminum, manganese and calcium. *Soil Sci Soc Am Proc* 23:18–22.

Lindsay, W. L., A. W. Frazierand, and H. F. Stephenson. 1962. Identification of reaction products from phosphate fertilizers in soils. *Soil Sci Soc Am Proc* 26:446–452.

Lindsay, W. L., M. Peech, and J. S. Clark. 1959. Solubility criteria for the existence of variscite in soils. *Soil Sci Soc Am Proc* 23:357–360.

Lipman, J. G., H. C. McLean, and H. C. Lont. 1916. Sulfur oxidation in soils and its effect on availability of mineral phosphates. *Soil Sci* 2:499–538.

Loneragan, J. F., and M. J. Webb. 1993. Interaction between zinc and other nutrients affecting the growth of plants. In *Zinc in Soils and Plants*. Robson, A. D., ed. Dordrecht, Kluwer Academics.

Lynch, J. P. 2007. Roots of the second green revolution. *Aust J Bot* 55:1–20.

Lynch, J. P., and K. M. Brown. 2001. Top-soil foraging—An architechtural adapatation of plants to low phosphorus availability. *Plant Soil* 237:225–237.

Majumdar, K., and V. Govil. 2015. Phosphorus responses of oilseeds and pulses in India and profitability of phosphorus fertilizer application. *Better Crops-South Asia* 9:15–18.

Mallarino, A. and K. Pecinovsky. 2013. Phosphorus and potassium fertilization for corn and soybean in rotation for 30 years. *Iowa State University Research Farm Progress Reports*. Paper 2048. Available at http://lib.dr.iastate.edu/farm_reports/2048.

Martin, A., and G. D. Cooke. 1994. Health risks in eutrophic water supplies. *Lake Line* 14:24–26.

Marwaha, B. C. 1989. Rock phosphate holds the key to productivity in acid soils. *Fertil News* 34:23–29.

Mashori, N. M., M. Memon, K. S. Memon, and H. Kakar. 2013. Maize dry matter yield and uptake as influenced by rock phosphate and single superphosphate treated with farm manure. *Soil Environ* 32(3):130–134.

Massey, M. S., J. G. Davis, J. A. Ippolito, and R. E. Sheffield. 2009. Effectiveness of recovered magnesium phosphates as fertilizers in neutral and slightly alkaline soils. *Agron J* 101:323–329.

McBeath, T. M., M. J. McLaughlin, and S. R. Noack. 2011. Wheat grain yield response to and tranloction of foliar applied phosphorus. *Crop & Pasture Sci* 62(1):58–65.

McConnell, D. 1938. A structural investigation of the isomorphism of the apatite group. *Am Mineral* 54:1379–1391.

McLay, C. D. A., S. S. S. Rajan, and Q. Liu. 2000. Agronomc effectiveness of partially acidulated phosphate rock fertilzers in an allophanic soil at near neutral pH. *Commun Soil Sci Plant Anal* 31:423–435.

Mengel, K., and E. A. Kirkby. 1987. *Principles of Plant Nutrition*. New Delhi, Panima Educational Book Agency.

Messick, D. L. 2003. Sulfur fertitilizers: A global perspective. *Proc TSI-FAI-IFA Workshop on Balanced Fertilization, New Delhi, February 25–26, 2003*. Sarkar, M. C., B. C. Biswas, S. Das, S. P. Kalwe, and S. K. Maity, eds. pp. 1–7. New Delhi, The Fertilizer Association of India.

Mimura, T. 1995. Homestatis and transport of inorganic P in plants. *Plant Cell Physiol* 36:1–7.

Mimura, T., K. Sakano, and T. Shimmen. 1996. Studies on the distribution of inorganic phosphate in barley leaves. *Plant Cell Environ* 19:311–320.

Mirvat, E. G., M. H. Mohamed, and M. M. Tawfiq. 2006. Effect of phosphorus fertilization and foliar spraying with zinc on growth, yield and quality of groundnut under reclaimed sandy soils. *J Appl Sci Res* 2:491–496.

Mitchell, C. C., and G. L. Mullins. 1990. Sources, rates and time of fertilizer application to wheat. *Sulphur Agric* 14:20–24.

Mortvedt, J. J. 2005. Heavy metals in fertilizers: Their effects on soil and plant health. *The International Fertilizer Society Proceedings No. 575*. p. 23

Mosali, J., K. Desta, R. K. Teal et al. 2006. Effect of foliar application of phosphorus on winter wheat grain yield, phosphorus uptake and use efficiency. *J Plant Nutr* 29(12):2147–2163.

Mousavi, S. R., M. Galavi, and M. Rezae. 2012. The interaction of zinc with other elements in plants—A review. *Int J Agric Crop Sci* 24:1881–1884.

Msolla, M. M., J. M. R. Semoka, and O. K. Boggard. 2005. Comparison of phosphate rock and triple superphosphate on a phosphorus-deficient Kenyan soil. *Commun Soil Sci Plant Anal* 72:299–308.

Muchhal, U. S., J. M. Pardo, and K. G. Raghothama. 1996. Phosphate transportation in the higher plant *Arabidopsis thaliana*. *Proc Natl Acad Sci USA* 93:10519–10523.

Murray, P. 2003. *The Fertilizer Industry*. Beijing, Publishing House of the Chinese Customs (in Chinese).

Nair, K. P. P., and R. G. Babu. 1975. Zinc-phosphorus interaction studies in maize. *Plant Soil* 42:517–536.

Nandi, S. K., R. C. Pant, and P. Nissen. 1987. Multiphasic uptake of phosphate by corn roots. *Plant Cell Environ* 10:463–474.

Nayak, A. K., and M. L. Gupta. 1995. Phosphoru, Zn and organic matter interaction in relation to uptake, tissue concentration absorption rate of P in wheat. *J Indian Soc Soil Sci* 43:633–636.

Nelson, N., and R. Mikkelsen. 2008. Meeting the phosphorus requirement on organic farms. *Better Crops* 92(1):12–14.

Panhwar, Q. A., O. Radziah, A. R. Zaharah, M. Saraiah, and I. M. Razi. 2011. Role of phosphate solubilizing bacteria on rock phosphate solubility and growth of aerobic rice. *J Environ Biol* 32(5):607–612.

Paroda, R. S., T. Woodhed, and R. B. Singh. 1994. *Sustainability of the Rice-Wheat Cropping System in Asia*. Bangkok, Food and Agriculture Organization- Regional Office for Asia and the Pacific. Pub. 1994/11.

Patiram, R. 1994. Phosphate sorption parameters in relation to phosphate desorption in acid soil. *J Indian Soc Soil Sci* 42:22–25.

Patrick, W. H. Jr., and I. C. Mahapatra. 1968. Transformations and availability to rice of nitrogen and phosphorus. *Adv Agron* 20:323–359.

Pauly, D. G., S. S. Malhi, and M. Nyeborg. 2002. Controlled-release fertilizer concept evaluation using growth and P uptake of barley from three soils in greenhouse. *Can J Soil Sci* 82:201–210.

Posner, A. M., and N. J. Barrow. 1982. Simplification of a model for ion adsorption on oxide surfaces. *J Soil Sci* 33:211–217.

Postma, J. A., and J. P. Lynch. 2011. Root cortical aerenchyma enhances the growth of maize on soils with sub-optimal availability of nitrogen, phosphorus and potassium. *Plant Physiol* 156:1190–1201.

Prasad, R. 2005. Rice-wheat cropping systems. *Adv Agron* 86:255–339.

Prasad, R. 2007a. Phosphorus management in the rice-wheat cropping system of the Indo-Gangetic plain. *Better Crops-India* 1:8–11.

Prasad, R. 2007b. *Crop Nutrition—Principles and Practices*. New Delhi, New Vishal Publications.

Prasad, R. 2011. Aerobic rice systems. *Adv Agron* 111:207–247.

Prasad, R. 2013. Population growth, food shortages and sways to alleviate hunger. *Curr Sci* 105(1):32–36.

Prasad, R., and S. Nagarajan. 2008. Organic vis-à-vis modern agriculture. *Curr Sci* 89:252–254.

Prasad, R., and J. F. Power. 1997. *Soil Fertility Management for Sustainable Agriculture*. Boca Raton, FL, CRC-Lewis.

Prasad, R., R. K. Tewatia, and K. Majumdar. 2014. Plant nutrients. In *Textbook of Plant Nutrients*. Prasad, R., D. Kumar, D. S. Rana, Y. S. Shivay, and R. K. Tewatia, eds. pp. 64–72, New Delhi, Indian Society of Agronomy.

Raghothama, K. G. 2005. Phosphorus and plant nutrition: An overview. In *Phosphorus, Agriculture and the Environment*, Agronomy Monograph No. 46. Sims J. T., and A. H. Sharpley, eds. pp. 355–378, Madison, WI, American Society of Agronomy, Crop Science Society of America and Soil Science Society of America.

Ramaekers, L., R. Remans, I. M. Rao, M. W. Blair, and J. Vanderleyden. 2010. Strategies for improving phosphorus acquisition efficiency of crop plants. *Field Crops Res* 117:169–176.

Rao, C. N. R. 1999. *Understanding Chemistry*. Hyderabad, University Press.

Reece, J. B., L. A. Urry, M. L. Cain et al. 2011. *Campbell Biology*, Ninth Ed. Boston, Benjamin Cummings.

Reed, A. T., S. R. Carpenter, and R. C. Lathrop. 2000. Phosphorus flow in a watershed-lake ecosystem. *Ecocyst* 3:561–573.

Richardson, A. E. 2001. Prospects of using soil microorganisms to improve the acquisition by plants. *Aust J Plant Physiol* 28:8797–8906.

Richardson, A. E., J.-M. Barea, A. M. McNeil, and C. Prigent-Combaret. 2009. Acquisition of phosphorus and nitrogen in the rhizosphere and plant growth promotion by microorganisms. *Plant Soil* 329:305–339.

Richardson, A. E., T. S. George, M. Hens, and R. J. Simpson. 2005. Utilization of soil organic phosphorus by higher plants In *Orghanic Phosphorus in the Environment*. Turner, B. L., E. Frossard, and B. S. Baldwin, eds. pp. 165–184, Wallingford, CAB International.

Richardson, A. E., J. P. Lynch, P. R. Ryan et al. 2011. Plant and microbial strategies to improve the phosphorus efficiency of agriculture. *Plant Soil* 349:121–156.

Roberts, T. L. 2014. Cadmium and phosphorus fertilizers: The issues and the science. *Procedia Engineering* 83:52–59.

Romer, W., and G. Schilling. 1986. Phosphorus requirements of the wheat plant in various stages of its life cycle. *Plant and Soil* 91(2):221–229.

Rubio, G., H. Liao, X. L. Yan, and J. P. Lynch. 2003. Topsoil foraging and its role in plant competitiveness for phosphorus in common bean. *Crop Sci* 43:598–607.

Rupa, T. R., Ch. Srinivasa Rao, A. Subba Rao, and M. Singh. 2003. Effect of FYM and phosphorus on Zn transformations and phyto-availability in two Alfisols of India. *Bioresource Tech* 87:279–288.

Sah, R. N., and D. S. Mikkelsen. 1989. Phosphorus behavior in flooded soils: I: Effects on P sorption. *Soil Sci Soc Am J* 53:1718–1722.

Sah, R. N., D. S. Mikkelsen, and A. A. Hafeez. 1989a. Phosphorus behavior in flooded soils. II: Iron transformations and phosphorus sorption. *Soil Sci Soc Am J* 53:1723–1729.

Sah, R. N., D. S. Mikkelsen, and A. A. Hafeez. 1989b. Phosphorus behavior in flooded soils: I: Phosphorus desorption and availability. *Soil Sci Soc Am J* 53:1729–1732.

Salisbury, F. B., and C. W. Ross. 1992. *Plant Physiology*. New Delhi, CBS Publishers and Distributors.

Sample, E. C., R. J. Sopex, and C. J. Racz 1986. Reactions of phosphate fertilizers in soils. In *The Role of Phosphorus in Agriculture*. Khasawneth, F. E., E. C. Sample, and E. J. Kamprath, eds. Madison, WI, American Society of Agronomy and Soil Science Society of America.

Sanchez, P. A., K. D. Shephard, M. J. Soule et al. 1997. Soil fertility replenishment in Africa: An investment in natural resource capital. In *Replenishing Soil Fertility in Africa*. Buresh, R. J., P. A. Sanchez, and F. Calhoun, eds. Madison, WI, Soil Science Society of America.

Sanyal, S. K., and S. K. DeDatta. 1991. Chemistry of phosphorus transformation in soil. *Adv Soil Sci* 16:1–120.

Sanyal, S. K., and K. Majumdar. 2009. Nutrient dynamics in soil. *J Ind Soc Soil Sci* 57(4):477–493.

Sarret, G., J. Vangronsveld, A. Manceau, M. Musso, J. D'Haen, J. J. Menthonnex, and J. L. Hazelmann. 2001. Accumulation forms of Zn and Pb in *Phaseolus vulgaris* in the presence of EDTA. *Environ Sci Tech* 35:2854–2859.

Saxena, S. N. 1979. Biochemistry of soil phosphorus. *Bull Indian Soc Soil Sci* 12:42–46.

Schachtman, D. P., R. J. Reid, and S. M. Ayling. 1998. Phosphorus uptake by plants: From soil to cell. *Plant Physiol* 116(2):447–453.

Sharma, K. B., A. Krantz, A. L. Brown, and S. Quick. 1986. Interaction of zinc and phosphorus in topand root of corn and tomato. *Agron J* 60:453–456.

Sharma, S. N., R. Prasad, Y. S. Shivay et al. 2010. Relative efficiency of diammonium phosphate and mussoorie phosphate rock on productivity and phosphorus balance in a rice–rape seed–mungbean cropping system. *Nutr Cycl Agroecocyst* 86:199–209.

Sharpley, A. 1999. Phosphorus availability. In *Handbook of Soil Science*. Sumner, M. E., ed. pp. D18–D38, Boca Raton, FL, CRC Press.

Sharpley, A. N., R. Foy, and P. Withers. 2000. Practical and innovative measures for control of agricultural phosphorus losses to water: An overview. *J Environ Qual* 21:30–35.

Sharpley, A. N., S. J. Smith, O. K. Jones, W. A. Berg, and G. A. Coleman. 1992. The transport of bioavailable phosphorus in agricultural runoff. *J Environ Qual* 29:1–9.

Shigaki, F., A. Sharpley, and L. I. Pochnow. 2006. Source related transport of phosphorus in surface runoff. *J Environ Qual* 35:2229–2235.

Shigaki, F., A. Sharpley, and L. I. Pochnow. 2007. Rainfall ntensity and phosphorus source effects on phosphorus transport in surface runoff from soil trays. *Soil Total Environ* 373:334–343.

Shittu, O. S., and J. A. Ogunwale. 2012. Phosphorus-zinc interaction for soybean production in soil developed on charnockite in Ekiti State. *J Emerging Trends Engg Appl Sci* 3:938–942.

Shumway, S. E. 1990. A review of the effects of algal bloom on shellfish and aquaculture. *J World Aquaculture Soc* 21(2):65–104.

Silberbush, M., and S. A. Barber. 1983. Sensitivity of simulated phosphorus uptake to parameters used by a mechanistic-mathematical model. *Plant Soil* 74:93–100.

Singh, M. V. 2001. Importance of sulfur in balanced fertilizer use in India. *Fertil News* 46(10):13–35.

Singh, O. P., and B. Dutta. 1987. Phosphorus status of some hill soils of Mizoram in relation to pedogenic properties. *J Indian Soc Soil Sci* 35:699–705.

Singh, Y., A. Dobermann, B. Singh, K. F. Bronson, and C. S. Khind. 2000. Optimal phosphorus management strategies for wheat-rice cropping on a loamy sand. *Soil Sci Soc Am J* 64:1413–1422.

Singh, V. K., B. S. Dwivedi, K. N. Tiwari et al. 2014. Optimizing nutrient management strategies for rice–wheat system in the Indo-Gangetic Plains of India and adjacent region for higher productivity, nutrient use efficiency and profits. *Field Crops Res* 164:30–44.

Singh, B., Y. Singh, P. Imes, and X. Jian-chang. 2004. Potassium nutrition of the rice-wheat cropping system. *Adv Agron* 81:203–259.

Singh, B. R., T. Krogstad, Y. S. Shivay, B. G. Shivakumar, and M. Bakkegard. 2005. Phosphorus fractionation and sorption in P-enriched soils of Norway. *Nutr Cycl Agroecosyst* 73:245–256.

Smil, V. 2002. Phosphorus in the environment: Natural flows and the human interferences. *Annu Rev Energy Environ* 25:53–88.

Smith, S. E., and D. J. Read. 2008. *Mycorrhizal symbiosis*, Third Ed. Amsterdam, Academic Press-Elsevier.

Smith, F. W., P. M. Ealing, B. Dang, and E. Delheize. 1997. The cloning of two Arabidopsis genes belonging to phosphate transporter family. *Plant J* 11:83–92.

Smith, S. E., F. A. Smith, and I. Jakobsen. 2003. Mycorrhizal fungi can dominate phosphate suuply to plants irrespective of growth responses. *Plant Physiol* 133:16–20.

Smith, V. H., G. D. Tilman, and J. C. Nekola. 1999. Eutrophication: Impacts of excess nutrient inputs on fresh water, marine and terrestrial ecocystems. *Environ Pollu* 100(1–3): 179–186.

Soltanpour, P. N. 1969. Effect of N, P and Zn placement on yield and composition of potatoes. *Agron J* 61:288–289.

Sood, R. D., and R. S. Minhas. 1988. Organic phosphorus fractions in some soil profiles of humid temperate highland zone of Himachal Pradesh. *J Indian Soc Soil Sci* 36:660–665.

Stewart, W. M., L. L. Hammond, and S. J. van Kauwenbergh. 2005. Phosphorus as natural resource. In *Phosphorus: Agriculture and the Environment*, Agronomy Monograph No. 46. Sims, T., and A. H. Sharpley, eds. pp. 3–22, Madison, WI, American Society of Agronomy, Crop Science Society of America and Soil Science Society of America.

Swaby, R. J. 1975. Biosuper-Biological superphosphate. Tn *Sulfur in Australian Agriculture*. McLacklan, K. D., ed. pp. 213–220, Sydney, Sydney University Press.

Syers, J. K., A. E. Johnston, and D. Curtin. 2008. Efficiency of soil and fertilizer phosphorus. FAO Fertilizer and Plant Nutrient Bull. 18. Rome, Food and Agricultural Organization of the United Nations.

Tadano, T., K. Ozawa, H. Sakai, M. Osaki, and H. Matsui. 1993. Secretion of acid phosphatase by roots of crop plants under phosphorus-deficient conditions and some properties of the enzyme secreted by lupin roots. *Plant Soil* 155/156:95–98.

Talibuddin, O. 1981. Precipitation. In *The Chemistry of Soil Processes*. Greenland, D. J., and M. H. B. Hayes, eds. pp. 81–116, Chichester, John Wiley & Sons.

Tate, K. R. 1984. The biological transformation of phosphorus in soil. *Plant Soil* 76:245–256.

Tiwari, K. N. 1979. Efficiency of Mussorie rock phosphate, pyrites, and their mixtures in some soils of Uttar Pradesh. *Indian Soc Soil Sci Bull* 12:519–526.

Tomar, N. K. 2000. Dynamics of phosphorus in soils. *J Indian Soc Soil Sci* 48:640–668.

Torres, G. M. 2011. Foliar phosphorus fertilization and the effect of surfactants on winter wheat (*Triticum aestivum* L.). MS Thesis, Oklahoma State University, Stillwater, OK.

Turtola, E., and M. Yi-halla. 1999. Fate of phosphorus applied in slurry and mineral fertilizer: Accumulation in soil and release to surface runoff water. *Nutr Cycl Agroecosyst* 55:156–174.

U.S. Environmental Protection Agency (EPA). 2007. Electronic code of federal regulations: Title 40: Protection of environment: Part 503—Standards for the use of disposal of sewage sludge: Part B:Land Application. Available at http://yosemite.epa.gov/r10/water.nsf /NPDES+Permits/Sewage+S825/4FILE/503-032007.pdf.

Van-Kauwenbergh, S. J., M. Stewart, and R. Mikkelsen. 2013. World reserves of phosphate rocks and unfolding a dynamic. Story. *Better Crops* 97(3):18–20.

Viveganandan, G., and K. S. Jauhari. 2002. Efficiency of rock phosphate base soil implant formulation of phosphbacteria in soybean (Glycine max Merrill). *Indian J Biotech* 1:180–187.

Wang, X., J. Shen, and H. Liao. 2010. Acquisition or utilization, which is more critical for enhancing phosphorus efficiency in modern crops. *Plant Sci* 179:302–306.

Webb, M. J., and, J. F. Loneragan. 1988. Effect of zinc deficiency on growth, phosphorus concentration and phosphorus toxicity of wheat plants. *Soil Sci Soc Am J* 52:1676–1680.

White, P. J., and J. P. Hammond. 2008. *The Ecophysiology of Plant-Phosphorus Interactions*. Dordrech, Springer.

White, P. F., H. J. Nesbitt, C. Ross, V. Seng, and B. Lor. 1999. Local rock phosphate deposits are a good source of phosphorus fertilizer for rice production in Cambodia. *Soil Sci Plant Nutr* 45:51–63.

Withers, P. J. A., D. Nash, and C. A. M. Laboski. 2005. Environmental management of phosphorus fertilizers. In *Phosphorus: Agriculture and the Environment*, Agronomy Monograph No. 46. Sims, J. T., and A. N. Sharpley, eds. pp. 782–827, Madison, WI, American Society of Agronomy, Crop Science Society of America and Soil Science Society of America.

Wright, B. C., and M. Peech. 1960. Characterization of phosphate reaction products in acid soils by the application of solubility criteria. *Soil Sci* 90:32–43.

Wu, X. Y., L. P. Zhang, H. B. Ni., and X. M. Zhu. 2008. Researech on characteristics of nitrogen and phosphorus loss under different coverage in Qingshan Lake Valley. *J Soil Water Conserv* 22:56–59.

Xia, T. X., W. C. Li, and J. Z. Pan. 2008. Risk assessment on soil environment quality and losses of nitrogen and phosphorus for the gravel soils under different farming practices in the watershed of Lake Fuxian, *J Lake Sci* 20:110–116.

Xu, Q. G., H. L. Liu, Z. Y. Shen, and B. D. Xi. 2007. Characteristics on nitrogen and phosphorus losses in the typical small watershed of the three Georges Reservoir area. *Acta Sci Circumstantiae* 27:326–331 (in Chinese).

Yadav, V. V., and K. B. Mistry. 1984. Reaction products of tri-ammonium pyrophosphate in different Indian soils. *Fert Res* 5:423–434.

Yin, Y., G. Berges, M. Sakuta. A. Crozier, and A. Ashihara. 2012. Effect of phosphate deficiency on the content and biosynthesis of anthocyanin and the expression of related genes in suppression cultured grapes (*Vitis* sp.) cells. *Plant Physiol Biochem* 55:77–84.

Zhang, J., and S. A. Barber. 1992. Maize root distribution between phosphorus fertilized and unfertilized soil. *Soil Sci Soc Am J* 56(3):819–822.

Zhang, H. C., Z. H. Cao, Q. R. Shen, and M. H. Wong. 2003. Effect of phosphate fertilizer application on phosphorus (P) losses from paddy soils in Tai Lake Region: I: Effect of phosphate fertilizer rate on P losses from paddy soil. *Chemosph* 50:695–701.

Zhang, W., W. Ma, Y. Ji, M. Fan, O. Oenema, and F. Zhang. 2008. Efficiency, economics, and environmental implications of phosphorus resource use and the fertilizer industry in China. *Nutr Cycl Agroecosyst* 80:131–144.

Zhang, Z. J., J. Y. Zhang, R. He, Z. D. Wang, and Y. M. Zhu. 2007. Phosphorus interception in floodwater of paddy field during the rice growing season in Tai Lake Basin. *Environ Pollut* 145:425–433.

Zhu, J. P., Z. Z. Chang, J. C. Zheng, and L. G. Chen. 2007. Analysis on nitrogen and phosphorus losses and economic returns of major cropping systems in Tai Lake region. *Jiangsu Agri Sci* 2007(3):612–613 (in Chinese).

# 6 Phosphorus Effluxes from Lake Sediments

*Sergei Katsev*

## CONTENTS

## 6.1 INTRODUCTION

Phosphorus inputs into lakes and reservoirs are one of the primary factors that affect water quality, biological productivity, and health of aquatic ecosystems. In addition to the external inputs from the watershed, water bodies receive large amounts of phosphorus from their sediments, as the materials deposited to the lake floor are recycled by a suite of biogeochemical processes there. This internal loading can dominate P budgets and frustrate lake-remediation efforts aimed at the reduction of bioavailable P. This chapter reviews the biogeochemical processes that regulate phosphorus effluxes from sediments on timescales from seasonal to decadal and the implications for the management of aquatic resources.

Phosphorus plays a crucial role in terrestrial aquatic ecosystems. Being commonly a productivity-controlling nutrient (Schindler et al. 2008; Paterson et al. 2011), it regulates the primary production in the water columns of lakes, and its supply often determines the trophic status of water bodies. While nitrogen—another important nutrient—can be fixed from the atmosphere by nitrogen-fixing bacteria, phosphorus can be supplied to lakes only from the outside (Gächter et al. 2004), primarily with inflows and surface runoff and with a minor contribution from aeolian dust deposition. Consequently, increases in the external P supplies (also referred to as *loading*) may lead to increased productivity and eutrophication (Correll 1998; Smith et al. 2006), causing potentially harmful algal blooms and decreases in water quality.

Throughout most of the developed world, the increased uses of P-based fertilizers and particularly the use of P-based detergents by the 1950s and the 1960s have led to problems with the eutrophication of lakes and reservoirs. Phosphorus from inadequately treated sewage effluents and agricultural lands was entering water streams, ending up in lakes, many of which underwent transitions from oligotrophic or mesotrophic to eutrophic in a time span of less than two decades. The resulting widespread problems demanded legislative and engineering actions. In North America, two crucial pieces of legislation came in the form of the Clean Water Act of 1972 (Adler et al. 1993) and the Safe Drinking Water Act of 1974. Similar pieces of legislation were adopted elsewhere. Investments were made into better sewage treatment facilities, and the usage of P-based detergents was curtailed. Subsequent decades saw the decline in both P inputs and P levels in lakes (Dolan and Chapra 2012; Chapra and Dolan 2012), and the primary production levels responded accordingly, decreasing in most impacted lakes (National Research Council 1992). Nevertheless, the recovery of these inland waters took several decades, in many cases significantly longer than anticipated. In the mid-1980s, significant attention was given to the processes that recycle and recirculate phosphorus within the lakes, particularly to the recycling of P from sediments back into the water column. These fluxes of P from sediments were termed *internal loading*, in contrast to the *external loading* from the atmosphere and watershed. Processes that regulate the efficiency of P recycling in lacustrine sediments have been investigated, and engineering solutions aimed at decreasing the internal loading were suggested. After more than a decade of the application of these methods, however, the community learned their limitations (Gächter and Wehrli 1998), as not all of them were effective, and some of the improvements were only short term. Based on the gained experience, adjustments to the management practices were made.

Since the 1990s, a progressive shift, advancing from mere correlation studies toward a mechanistic understanding of the processes of P mobilization, has been an important development. This progress has been aided by a proliferation of detailed reactive-transport geochemical models that simulate a plethora of interconnected biogeochemical reactions within the sediment (Katsev et al. 2004, 2006a; Paraska et al. 2014) and allow investigation of the interactions and the contributions of the individual processes that are normally difficult to tackle experimentally. Important advances also came from refined analytical procedures (Anagnostou and Sherrell 2008), development of sequential extraction protocols (Ruttenberg 1992; Roden and Wetzel 2002; Poulton and Canfield 2005), and improvements in instrumentation. The debate on the principles of P management in inland waters, however, is far from being settled. Although our current knowledge allows us to understand, given a sufficient quantity of analyses, the internal versus external loadings into a lake and the response of the lake to interventions, outstanding problems remain at larger scales. For example, evidence accumulates for a "cleaner lakes are dirtier lakes" (Bernhardt 2013) paradigm where decreasing the amounts of P in inland lakes may inadvertently lead to the stronger accumulation of nitrogen (Finlay et al. 2013; Li and Katsev 2014) and its export downstream and eventually to the coastal ocean where it contributes to the proliferation of dead zones (Diaz and Rosenberg 2008). Whereas P runoffs from agricultural lands and urban sources have been regulated for decades, N outputs have not, and arguments are being made for a dual-nutrient regulation (Conley et al. 2009).

This chapter focuses on the processes that take place in the upper decimeters of lake sediments and several centimeters of water column above them. Despite the seemingly small size, this zone has a deciding influence on the recycling of P in aquatic ecosystems and on the timescales on which the lake responds to eutrophication and remediation. After reviewing the main forms of P in the sediment and the processes that determine its dynamics and ultimate fate, we investigate how these processes interact to affect the P effluxes from the sediment into the water column on a variety of timescales.

## 6.2 CYCLING OF P IN LAKE SEDIMENTS

In most lacustrine environments, phosphorus is deposited into the sediments from the overlying water column primarily in the form of organic P, as part of settling organic matter (Figure 6.1). As phosphorus is commonly the nutrient that limits the productivity in the photic zone, the uptake of dissolved bioavailable phosphorus in the water column is rapid, and the concentrations of dissolved inorganic phosphorus (phosphate) are low, often below detection (<1 μM for commonly used methods). The C:P ratios in the seston vary within a substantial range, depending on the type of organic material and the strength of nutrient limitation, but the averages are on the order of 166:1 (Hecky et al. 1993). This is higher than the corresponding ratio of 106:1 (the Redfield ratio) that is commonly observed for marine phytoplankton, reflecting stronger dependence on phosphorus. In contrast to the marine environments, the P

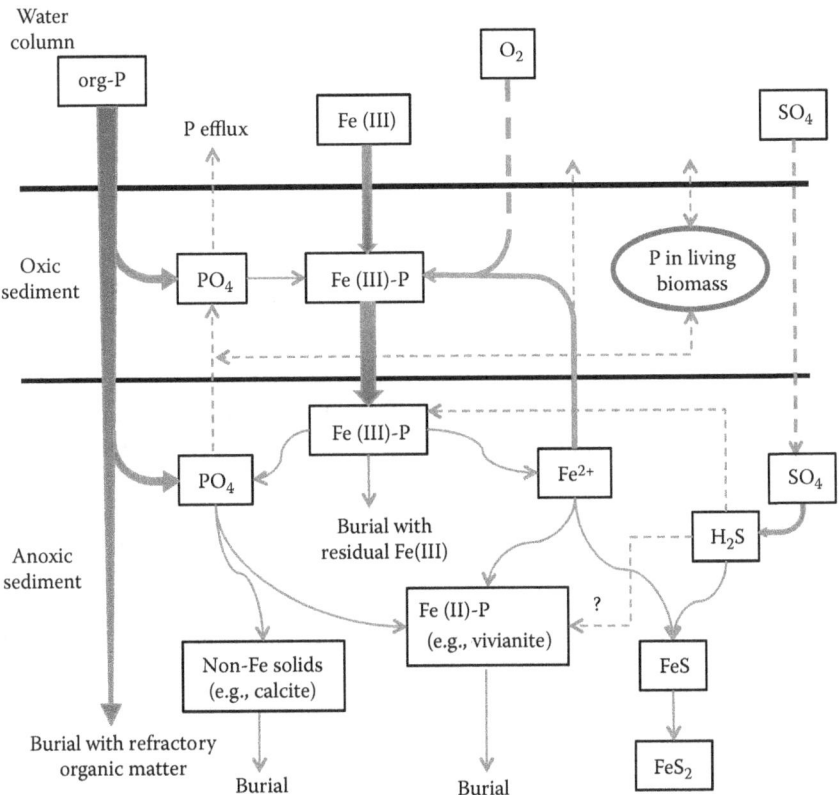

**FIGURE 6.1**    Simplified diagram of P cycling in sediments.

content of freshwater organic material is significantly more variable. Between 20 and 90% of the organic matter deposited to the sediment surface is typically mineralized within the sediment, and the fate of the mobilized soluble phosphorus depends on whether it is resequestered within the sediment or diffuses upward into the water column. Mineral phosphorus can be deposited to the sediment surface as detrital apatite (Williams et al. 1976) or as P adsorbed to calcite, iron, or aluminum minerals or clays (Søndergaard et al. 2001).

Within the sediment, several additional pools of P may authigenically form (within the sediment) that may be important for the P geochemical cycling. Sediment microbial communities may accumulate P within their cells (Gächter et al. 1988), particularly as polyphosphates, which can be used as the cells' energy reserves. This live biomass P may be exchanged with the environment, contributing to P releases from sediments under certain conditions. Phosphate ($PO_4$) released during the process of organic matter mineralization may become adsorbed to a variety of sediment surfaces. Of primary importance are freshly precipitated iron oxyhydroxides (ferric iron phases), as they have large specific surface areas and have a strong affinity for orthophosphate (Spiteri et al. 2008). Besides adsorption, P association with Fe oxyhydroxides can also occur via the inclusion of P in the mineral structure upon the Fe mineral precipitation (Hyacinthe and Van Cappellen 2004; Thibault et al. 2009). Of the possible authigenic P-bearing minerals, the most important in lake environments is probably vivianite ($Fe_3(PO_4)_2$) (Williams et al. 1976; Fagel et al. 2005; Rothe et al. 2014), a ferrous phosphate mineral that can form under reducing conditions and may become a significant long-term sink of phosphorus. Blue vivianite crystals are readily observed under a microscope in many lake sediments. In calcareous lakes, biologically induced precipitation of calcite ($CaCO_3$) may also be important (Boström et al. 1988; Dittrich and Koschel 2002).

The sediment geochemical cycling of P is tightly coupled to the cycling and redox dynamics of iron (Figure 6.1). When the water column overlying the sediment contains oxygen, the oxidizing conditions at the sediment surface favor the precipitation and the preservation of ferric (Fe (III)) particles in the oxidized upper layer of the sediment. Fe (III) phases include iron oxides and, importantly, oxyhydroxides (ferrihydrite, goethite, lepidocrosite). The latter are typically nanophases (van der Zee et al. 2003; Waychunas et al. 2005) characterized by large specific surface areas, high bioavailability, and strong capacity for binding P. Ferric iron phases are largely insoluble under oxic conditions and circumneutral pH. Upon their burial into the deeper (anoxic) sediment, they can undergo microbially mediated reduction to Fe (II). A number of microbial species (e.g., from genera *Geobacter* and *Shewanella*) can catalyze this reaction, coupling it to the oxidation of organic carbon in their dissimilatory or sometimes fermentative metabolisms (Jørgensen 2000; Weber et al. 2006; Blöthe and Roden 2009):

$$[CH_2O] + 7CO_2 + 4Fe(OH)_3 \rightarrow 4Fe^{2+} + 8HCO_3^- + 3H_2O \quad (\Delta G^0 = -114 \text{ kJ mol}^{-1})$$

This process can mobilize iron, releasing it as soluble $Fe^{2+}$ together with the Fe-bound phosphorus, into the sediment porewaters (Figure 6.1). In most sediments, oxygen penetrates into the sediment by a few millimeters (Katsev et al. 2006b; Li et al. 2012), and the zone immediately below this oxygen penetration becomes the locus of intense redox cycling. The dissolved $Fe^{2+}$ released in the anoxic sediment can diffuse upward where it can oxidatively reprecipitate near the oxic–anoxic boundary, forming Fe (III) oxyhydroxides again and binding porewater phosphate (Gunnars et al. 2002):

$$4Fe^{2+} + O_2 + 10H_2O \rightarrow 4Fe(OH)_3 + 8H^+$$

In certain cases, the oxidation of $Fe^{2+}$ can be coupled to the reduction of nitrate $\left(NO_3^-\right)$ instead of oxygen (Straub et al. 1996; Senko et al. 2005), but the contribution of this reaction is typically minor (Li and Katsev 2014). The redox cycling of Fe is often mediated by the redox cycling of Mn, a less abundant but more reactive redox metal. Like the ferric Fe minerals, Mn oxides $(MnO_2)$ are used in the microbially mediated oxidation of organic matter $(\Delta G^0 = -349 \text{ kJ mol}^{-1})$, but they can also serve as electron acceptors to reoxidize $Fe^{2+}$ (Davison 1993):

$$2Fe^{2+} + MnO_2 + 4H_2O \rightarrow 2Fe(OH)_3 + Mn^{2+} + 2H^+$$

The soluble $Mn^{2+}$ produced in this reaction is in turn oxidized by oxygen, making the diagenetic cycling of Mn analogous to the cycling of Fe and effectively coupling the oxidation of $Fe^{2+}$ to the reduction of oxygen with Mn as an intermediary.

A continued cycling and recycling of Fe across the sediment redox boundary (at or slightly below the depth of oxygen penetration) results in the accumulation of Fe there, producing Fe-rich layers that are typically several millimeters thick (Katsev and Dittrich 2013). In some cases, e.g., in the Fe-rich carbon-poor Lake Superior (Li et al. 2012; Dittrich et al. 2015) or in Lake Baikal (Torres et al. 2014), these enrichments can take the form of dense Fe-rich crusts, millimeter to centimeters thick. The concentrations of iron within them can reach several percent by weight. Mn layers are typically found immediately above the Fe-rich layers (e.g., Li 2014), in accordance with the thermodynamically determined sequence in which different electron acceptors are utilized by resident microbial populations as they heterotrophically oxidize organic matter (in order of the respective Gibbs free energy yields $\Delta G$; Froelich et al. 1979).

The produced Fe-rich layers serve as a barrier to the diffusion of phosphate, which is produced upon the decomposition of organic matter in deeper sediment. Their redox dynamics then becomes important for moderating the P efflux from sediments. Within the oxic sediment layer, the concentrations of dissolved phosphate are significantly lower than in the anoxic porewaters below (sometimes below detection), and the P effluxes across the sediment–water interface are largely determined by the diffusive vertical gradients of P in that oxidized near-interface zone.

Whereas this conceptual model captures the essential features of the diagenetic P transformations, several additional factors can significantly alter the P dynamics. As the adsorption of phosphate (Krom and Berner 1980), as a charged molecule, onto the surfaces of iron oxyhydroxides is pH dependent (Kuo and Lotse 1974), changes in the water column pH may induce changes in the sediment P-binding capacity

(Jensen and Andersen 1992). For instance, the increases in water pH (especially above 9) due to photosynthetic activity in shallow lakes may cause desorption and release of phosphate. The incorporation of phosphate into solid particles, however, is a multistep process, and the dependence on pH may be complex (Froelich 1988). Under oxic conditions, the sediment–water exchange fluxes of phosphorus can be significantly affected by the actions of benthic organisms, which burrow (bioturbate) and pump water (bioirrigate) through the upper centimeters of sediment and facilitate exchanges with the water column (Holdren and Armstrong 1980; Matisoff et al. 1985; Boström et al. 1988; Søndergaard et al. 2003). In contrast to marine sediments where bioturbation zone commonly extends 10–20 cm below the sediment–water interface, bioturbation in freshwater sediments is shallower (1–2 cm), as it is carried out by smaller organisms (Matisoff 1995). The intensities of these biologically enhanced exchanges, nevertheless, seem to be comparable between the freshwater and marine environments with similar rates of supply of organic matter (Li et al. 2012). Bioturbation and especially bioirrigation cause the total sediment–water exchange fluxes to exceed the purely diffusive fluxes, often by more than a factor of two (Matisoff and Wang 1998; Meile and Van Cappellen 2003). In shallow lakes (Scheffer 2004), one may also need to consider the effects of sediment resuspension (Søndergaard et al. 2003), epipelic photosynthesis (Carlton and Wetzel 1988), macrophyte senescence (Welch and Cooke 1995), and potentially blooms of cyanobacteria that may access sediment P (Cottingham et al. 2015).

Sulfate $\left(SO_4^{2-}\right)$ reduction and resultant production of hydrogen sulfide ($H_2S$) may significantly interfere with the cycle of Fe, as Fe (III) phases can be reductively dissolved by sulfide (Figure 6.1). Sulfate, predominantly supplied from the overlying waters, is used by sulfate-reducing microbes below the zone of iron reduction as an electron acceptor to oxidize organic matter (Jørgensen 2000):

$$2[CH_2O] + SO_4^{2-} \rightarrow H_2S + 2HCO_3^- \quad (\Delta G^0 = -77 \text{ kJ mol}^{-1})$$

The produced hydrogen sulfide can diffuse within the anoxic sediment porewater toward the Fe (III)-rich layer where it can contribute to its dissolution (Morse et al. 1987; Golterman 1995):

$$H_2S + 4CO_2 + 2Fe(OH)_3 \rightarrow 2Fe^{2+} + S^0 + 4HCO_3^- + 2H_2O$$

The hydrogen sulfide can also react with the mobilized $Fe^{2+}$ in the sediment porewater, binding iron to a solid phase through the precipitation of iron sulfides (FeS, operationally defined as acid-volatile sulfide (Rickard and Morse 2005), or under highly sulfidic conditions as pyrite [$FeS_2$]):

$$Fe^{2+} + HS^- \rightarrow FeS + H^+$$

This sequestering of $Fe^{2+}$ from the anoxic sediment porewaters prevents the precipitation of iron as Fe oxides at the sediment redox boundary. Hence, high concentrations of sulfate in lakes have been linked to a decreased capacity for P retention and higher sediment P effluxes (Caraco et al. 1989, 1993; Katsev et al. 2006a). Likewise, high concentrations of sulfate in marine and brackish systems must be one of the main reasons why phosphorus is released from such sediments with a significantly higher efficiency than from freshwater lake sediments (Caraco et al. 1990).

The earlier description has been so far limited to nearly steady-state conditions. Seasonal and longer-term changes in the lacustrine environment can strongly affect the dynamics of phosphorus exchanges between the sediments and the water column. Summer increases in temperature, in particular, can have a strong effect on the rates of chemical and biological processes in the sediment, increasing P effluxes (e.g., Jensen and Andersen 1992). The changes in organic sedimentation or chemistry of the overlying waters can bring about changes in the sediment P retention on a variety of timescales, as discussed in the following sections. In some cases, the sediment can even become "locked" in a particular regime. For example, an increase in the organic matter sedimentation can initiate a positive feedback in the internal P loading (Gächter and Imboden 1985; Carpenter 2005; Katsev et al. 2006a; Jeppesen et al. 2007), which can further support biological productivity in the lake. The possibility of such hysteresis effects implies that restoring the lake's oligotrophic status may require strong measures beyond the reduction in the external P inputs. Besides coming from a purposeful management intervention, such strong regime-switching interference may also come from natural causes or human-caused unintended processes. Most famously, the trophic status and the phosphorus cycling of the lower Great Lakes seem to have been strongly modified (toward oligotrophication) by the invasion of dreissenid mussels (Higgins and Zanden 2010; Ozersky et al. 2015). Besides removing a large quantity of P from the lake's water column and storing it in biomass, these suspension feeders reengineer lake ecosystems by shifting the biological production from the pelagic areas (which become more oligotrophic) to the littoral areas that may become more productive and even experience more frequent algal blooms (Hecky et al. 2004; Higgins et al. 2008). Shallow and small lakes typically respond to changes in the external conditions faster than large lakes, although the exact timescale of their responses is also affected by the effective rate of phosphorus removal from the water column into the sediments and the efficiency of P recycling within the sediments (e.g., Chapra and Canale 1991). Slow rates of permanent P removal, for example, in unstratified shallow lakes where phosphorus released from sediments can readily enter the trophogenic zone, lead to longer response times and delayed recovery after the reductions in the external P inputs (Welch and Cooke 1995).

## 6.3 BURIAL OF P INTO DEEP SEDIMENT

The long-term removal of P from circulation requires its effective burial into the deep sediment, below the diagenetically active upper sediment layers. In the deep reduced sediment, P is typically present as organic-P that is associated with the unmineralized organic matter and the dead and the living sediment biomass, as P associated with ferrous phases such as vivianite or as Ca-bound P (e.g., with detrital apatite or authigenic calcite; Dittrich and Koschel 2002). In marine environments, the characterization of P-bearing mineral phases mostly focused on detrital and authigenic fluorapatite (Slomp et al. 1996; Jilbert and Slomp 2013), which is not regularly reported in freshwater. Recent findings from the Baltic Sea, however, suggest that vivianite may also be important (Slomp et al. 2013). In limnological investigations, vivianite was assumed to be unstable in the presence of hydrogen sulfide (Roden and Edmonds 1997; Gächter and Müller 2003; Katsev et al. 2006a):

$$Fe_3(PO_4)_2 \cdot H_2O + 3H_2S \rightarrow 3FeS + 2H_3PO_4 + 8H_2O$$

Observations of vivianite (and potentially other authigenic P-bearing phases; Jilbert and Slomp 2013; Dijkstra et al. 2014) in high-sulfate marine sediments may thus indicate the need for a reevaluation of the factors that regulate the long-term burial of P in lakes.

Probably the largest pool with which P can be buried into the deep sediment is the pool of refractory organic material, which remains unmineralized for decades or centuries. Depending on the conditions, the fraction of the deposited organic material that escapes mineralization can be between 10 and 80% (Sobek et al. 2009; Katsev and Crowe 2015). The stoichiometric ratio with which organic phosphorus is buried with this refractory organic matter (relative to organic carbon) is approximately equal to or lower than the P:C ratio in the settling matter (Reitzel et al. 2007). The burial efficiencies for P in lake sediments (Søndergaard et al. 2001) have not been comprehensively systematized. As a matter of argument, however, it is reasonable to assume that the trends for organic phosphorus there are approximately similar to the respective trends for organic carbon. The percentage of the deposited organic carbon that escapes mineralization (Burdige 2007) depends on factors such as the type of organic material, the mineralogy of inorganic sediment phases, the sedimentation rate, and the oxygen content of overlying waters. For example, terrigenous organic matter (soil matter, leaf matter, etc.) is generally considered less reactive than the autochthonous aquatic material such as phytoplankton (Sobek et al. 2009). In small- and medium-sized lakes, in particular, the bulk of allochthonous organic matter is dominated by recalcitrant compounds such as humic substances (Fenchel et al. 1998). Organic compounds such as lignins and certain pigments that are more abundant in the terrestrial matter are more difficult for the aquatic microbes to degrade and make the allochthonous organic matter more refractory. In lakes where organic inputs are dominated by such materials, organic matter is buried into the sediments with high efficiency, with between 40 and 90% of the amount deposited to the sediment surface being buried into the deep sediment (Sobek et al. 2009). Lower efficiencies correspond to the environments where the organic matter spends more time in the oxidized uppermost sediment layer. The exposure to oxygen generally speeds up the mineralization process and makes it more complete (Canfield 1994; Katsev and Crowe 2015). Anaerobic microbes lack enzymes to break down some of the more complex organic molecules (Jørgensen 2000), and oxygen affects the mineralization rates in several additional ways, e.g., by stimulating mixing by bioturbation (Burdige 2007). As a consequence of this enhanced degradation, sediments in cold unproductive well-oxygenated large lakes (with predominantly autochthonous organic inputs) may bury as little as 10% of the deposited organic carbon (Sobek et al. 2009; Li 2014; Katsev and Crowe 2015). This is the case, for example, in slowly accumulating and well-oxygenated sediments of Lake Superior (Li et al. 2012) and Lake Baikal (Maerki et al. 2006; Li 2014), and perhaps to some degree in Lake Michigan (Thomsen et al. 2004; Li 2014). In contrast, the organic carbon burial efficiencies in permanently anoxic large lakes, such as the East African Lake Malawi, are comparable to those in anoxic marine basins: on the order of 60–80% (Canfield 1994; Li 2014). Interestingly, the mineralization kinetics for the decomposition of organic material appears to be similar between the freshwater and marine sediments. For example, despite the obvious differences in water chemistries, the phenomenological relationship that links the reactivity of organic material to its age (Middelburg 1989) has been shown to be approximately the same in both types of environments (Katsev and Crowe 2015). Higher temperatures likely accelerate the mineralization rates (Gudasz et al. 2010), although the effect on sediment mineralization appears to be not as strong as that of oxygen (Katsev and Crowe 2015). A more complete decomposition in a warmer water column (Katsev et al. in press) can also render the material reaching the sediment surface more refractory, effectively decreasing the mineralization efficiency when it is measured relative to the amount at the sediment surface (Katsev and Crowe 2015).

In contrast to the refractory organic-P, phosphorus that is released from organic matter upon its decomposition can experience a very different fate than the respective organic carbon (which is predominantly oxidized to $CO_2$ and diffuses out of the sediment). The pathways of the mobilized P can strongly differ between the well-oxygenated and anoxic environments. For example, in well-oxygenated and iron-rich sediments of Lake Superior (Li et al. 2012), the released phosphorus is efficiently immobilized by iron oxides (Dittrich et al. 2015). Despite the high (~90%) efficiency of mineralization for organic carbon, only about 10–20% of the deposited phosphorus is returned into the water column (Li 2014). This low recycling efficiency of sediment P in Lake Superior leads to a strong phosphorus limitation of phytoplankton (C:P of 301:1–455:1), to the point that the microbial flora cells utilize S-based lipids (sulfolipids) and N-based lipids in place of regular P-based lipids (phospholipids) (Bellinger et al. 2014). The rate of P release from such strongly P-binding sediments is approximately proportional to the rate of iron reduction (Li 2014). This is a relatively rare situation, however. The percentage of the recycled P in more productive (or less oxygenated) systems is higher, while the relative proportions of the regenerated carbon versus the phosphorus are closer to the C:P ratios in the organic sedimentation (Moore et al. 1998; Li 2014).

## 6.4 EVOLUTION OF UNDERSTANDING OF P RELEASES FROM SEDIMENTS

The classical paradigm of P releases from sediments and the phosphorus–iron connection were born out of the pioneering works of Einsele (1936) and Mortimer (1941, 1942). The key observation was that, upon the onset of summer anoxia in stratified lakes, the phosphorus concentrations in the anoxic hypolimnia sharply and concomitantly increased with the concentrations of dissolved iron. Understanding of the P–Fe binding association and the redox chemistry of iron had led to a paradigm that stood as a pillar of P management for decades: the releases of sediment P are tightly linked to the presence of oxygen in the bottom waters. Maintaining the oxic conditions at the sediment–water interface, in particular, came to be seen as a means of maintaining the effluxes of P at a low level. This paradigm has recently undergone revision (Golterman 2001; Gächter and Müller 2003; Katsev et al. 2006a; Hupfer and Lewandowski 2008). Large-scale oxygenation projects where the hypolimnia of seasonally stagnant lakes were flushed with air or even pure oxygen demonstrated that these measures were ineffective in stemming the internal loading of P from sediments (Gächter and Wehrli 1998). Whereas the induced precipitation of oxides could briefly immobilize some amount of phosphate, the long-term effect often could not be sustained as the precipitated iron oxyhydroxides were redissolving after being buried into the anoxic sediment zone, releasing the adsorbed phosphorus. Likewise, under the high-phosphorus conditions in eutrophic lakes, the surfaces of the available iron oxides can become saturated with phosphate (with the Fe:P molar ratios falling below the typical threshold values of 7–15; Jensen et al. 1992; Roden and Edmonds 1997), ceasing to serve as barriers to the phosphate diffusion. In some shallow lakes, for example, fluxes of sediment phosphorus were observed into the oxic overlying waters (Jensen and Andersen 1992). Interestingly, a situation where phosphate can flux undeterred across the Fe (III) barrier into the oxic waters was also described in the oceanographic literature (Sundby et al. 1992) and observed in marine sediment incubations (Katsev et al. 2007) but was not readily accepted in limnology. Given the often short-term nature of the Fe (III) phosphorus sink, a long-term solution for the reduction of sediment P effluxes requires a long-term sink for phosphorus that would allow its burial into the deep reduced sediment (Katsev et al. 2006a). Unless the reduction of Fe oxyhydroxides is limited by the available amounts of organic matter (a rare situation in mesotrophic and eutrophic lakes), there are only a few pools of P that can remain immobile over the long term. For example, mineral P can be formed and buried as vivianite or other authigenic P-bearing minerals. The conditions for the formation and the long-term preservation of such phases have not been comprehensively reviewed to the best of the author's knowledge.

The effectiveness of the ferric iron sink near the oxidized sediment surface is largely limited by the duration of time it takes for the Fe oxyhydroxides to be buried into the anoxic sediment, whereas long-term phosphorus retention requires a deep-sediment sink. The differential roles of these two sinks point to the following time-scale argument (Katsev et al. 2006a; Katsev and Dittrich 2013). The classical paradigm of Einsele–Mortimer can be applicable to processes that occur on timescales that are significantly faster than the burial and the reduction of iron oxyhydroxides (typically the time that it takes a particular sediment layer to be buried by about a centimeter, the spatial extent of the Fe reduction/reoxidation zone). Such processes include, for example, P releases from sediments of seasonally anoxic lakes. Provided that the Fe (III) phases in such sediments are not at the limit of their capacity to adsorb P, their reductive dissolution and/or precipitation in response to varying redox conditions should have a noticeable effect on P concentrations. Another example may be the relatively short-term decreases in the lake's P levels in response to hypolimnetic oxygenation (Katsev and Dittrich 2013). The internal loading on those timescales (less than several years) becomes regulated by factors that are internal to the lake, such as redox conditions at the sediment–water interface. On longer timescales, which are typically of interest for environmental management applications, the issue of P burial into deep sediment becomes central: What does not become buried becomes released (Katsev et al. 2006a; Katsev and Dittrich 2013). The burial timescale is approximately the depth of the diagenetically active zone (10–20 cm) divided by the sediment burial velocity (on the order of several millimeters in most lakes). On those multidecadal and centurial timescales, the internal loading from sediments can be considered as part of the internal P cycling in the lake, not affecting the mass balance considerations that determine the total amount of P that remains in circulation (Katsev et al. 2006a). This total P amount becomes regulated by the balance between the burial of P into the deep sediment and the net external P inputs into the lake. As the burial fluxes of P are only weakly variable, the external inputs of P become the controlling factor. Sensitivity analyses performed with sediment diagenesis models (Katsev et al. 2006a) indicate that on the decadal and longer timescales, sediment phosphorus effluxes are primarily determined by the sedimentation fluxes of reactive organic matter, fluxes of iron oxyhydroxides, and the concentrations of oxygen and sulfate at the sediment–water interface. These factors do not act separately and independently, however, with the immobilization efficiency of phosphorus being determined by their combinations and interplays (Katsev et al. 2006a).

Management strategies are currently faced with an adaptation to this new understanding. Remediation projects aimed at binding the excess phosphorus into the sediment have shifted from oxygenation and treatments with ferric iron to methods that do not depend on the redox dynamics. Aluminum compounds (Huser and Pilgrim 2014; Huser et al. 2015), such as alum ($Al(OH)_3$), in particular are redox insensitive. Their presence, as well as that of unreducible Fe (III) minerals, can prevent P releases even under anoxic conditions (Hupfer and Lewandowski 2008). Biological methods for P removal by restructuring the biological communities of the lakes (biomanipulation) (Shapiro et al. 1975; Nurnberg et al. 2012), as well as sediment dredging, also remain as options. The choice of the treatment, however, should not be overshadowed by a single dominant paradigm: lake-specific conditions such as sediment types or land use in the catchment may be important. For example, alternative P mobilization mechanisms, such as dissolution of calcium-bound P may be more important in certain environments than the redox-driven cycling of Fe and P (Hupfer and Lewandowski 2008). Most importantly, restrictions in the external P inputs from the watershed (Gächter 1987; Sharpley et al. 1994) remain as probably the most effective method for combating eutrophication.

## REFERENCES

Adler, R. W., Landman, J. C., and Cameron, D. M. 1993. *The Clean Water Act 20 Years Later.* Island Press, Washington, DC.

Anagnostou, E., and Sherrell, R. M. 2008. MAGIC method for subnanomolar orthophosphate determination in freshwater. *Limnology and Oceanography: Methods*, 6(1), 64–74.

Bellinger, B. J., Van Mooy, B. A., Cotner, J. B., Fredricks, H. F., Benitez-Nelson, C. R., Thompson, J., Cotter, A., Knuth, M. L., and Godwin, C. M. 2014. Physiological modifications of seston in response to physicochemical gradients within Lake Superior. *Limnology and Oceanography*, 59(3), 1011–1026.

Bernhardt, E. S. 2013. Cleaner lakes are dirtier lakes. *Science, 342*(6155), 205–206.

Blöthe, M., and Roden, E. E. 2009. Microbial iron redox cycling in a circumneutral-pH groundwater seep. *Applied and Environmental Microbiology, 75*(2), 468–473.

Boström, B., Andersen, J. M., Fleischer, S., and Jansson, M. 1988. Exchange of phosphorus across the sediment-water interface. In *Phosphorus in Freshwater Ecosystems*, pp. 229–244. Springer, Netherlands.

Burdige, D. J. 2007. Preservation of organic matter in marine sediments: Controls, mechanisms, and an imbalance in sediment organic carbon budgets? *Chemical Reviews, 107*(2), 467–485.

Canfield, D. E. 1994. Factors influencing organic carbon preservation in marine sediments. *Chemical Geology, 114*(3), 315–329.

Caraco, N. F., Cole, J. J., and Likens, G. E. 1989. Evidence for sulphate-controlled phosphorus release from sediments of aquatic systems. *Nature, 341*, 316–318.

Caraco, N. F., Cole, J. J., and Likens, G. E. 1993, January. Sulfate control of phosphorus availability in lakes. In *Proceedings of the Third International Workshop on Phosphorus in Sediments*, pp. 275–280. Springer, Netherlands.

Caraco, N., Cole, J., and Likens, G. E. 1990. A comparison of phosphorus immobilization in sediments of freshwater and coastal marine systems. *Biogeochemistry, 9*(3), 277–290.

Carlton, R. G., and Wetzel, R. G. 1988. Phosphorus flux from lake sediments: Effect of epipelic algal oxygen production. *Limnology and Oceanography, 33*(4), 562–570.

Carpenter, S. R. 2005. Eutrophication of aquatic ecosystems: Bistability and soil phosphorus. *Proceedings of the National Academy of Sciences of the United States of America, 102*(29), 10002–10005.

Chapra, S. C., and Canale, R. P. 1991. Long-term phenomenological model of phosphorus and oxygen for stratified lakes. *Water Research, 25*(6), 707–715.

Chapra, S. C., and Dolan, D. M. 2012. Great Lakes total phosphorus revisited: 2: Mass balance modeling. *Journal of Great Lakes Research, 38*(4), 741–754.

Conley, D. J., Paerl, H. W., Howarth, R. W., Boesch, D. F., Seitzinger, S. P., Havens, K. E., Lancelot, C., and Likens, G. E. 2009. Controlling eutrophication: Nitrogen and phosphorus. *Science, 323*(5917), 1014–1015.

Correll, D. L. 1998. The role of phosphorus in the eutrophication of receiving waters: A review. *Journal of Environmental Quality, 27*(2), 261–266.

Cottingham, K. L., Ewing, H. A., Greer, M. L., Carey, C. C., and Weathers, K. C. 2015. Cyanobacteria as biological drivers of lake nitrogen and phosphorus cycling. *Ecosphere, 6*(1), art1.

Davison, W. 1993. Iron and manganese in lakes. *Earth-Science Reviews, 34*, 119–163.

Diaz, R. J., and Rosenberg, R. 2008. Spreading dead zones and consequences for marine ecosystems. *Science, 321*(5891), 926–929.

Dijkstra, N., Kraal, P., Kuypers, M. M. M., Schnetger, B., Slomp, C. P. 2014. Are iron-phosphate minerals a sink for phosphorus in anoxic Black Sea sediments? *PLoS ONE, 9*(7), e101139.

Dittrich, M., and Koschel, R. 2002. Interactions between calcite precipitation (natural and artificial) and phosphorus cycle in the hardwater lake. *Hydrobiologia, 469*(1–3), 49–57.

Dittrich, M., Moreau, L., Gordon, J., Quazi, S., Palermo, C., Fulthorpe, R., Katsev, S., Bollmann, J., and Chesnyuk, A. 2015. Geomicrobiology of iron layers in the sediment of Lake Superior. *Aquatic Geochemistry, 21*(2–4), 123–140.

Dolan, D. M., and Chapra, S. C. 2012. Great Lakes total phosphorus revisited: 1: Loading analysis and update (1994–2008). *Journal of Great Lakes Research, 38*(4), 730–740.

Einsele, W. 1936. Über die Beziehungen des Eisenkreislaufs zum Phosphatkreislauf im eutrophen See. *Archiv für Hydrobiologie, 29*(6), 664–686.

Fagel, N., Alleman, L. Y., Granina, L., Hatert, F., Thamo-Bozso, E., Cloots, R., and André, L. 2005. Vivianite formation and distribution in Lake Baikal sediments. *Global and Planetary Change, 46*(1), 315–336.

Fenchel, T., King, G. M., Blackburn, T. H. 1998. *Bacterial Biogeochemestry: The Ecophysiology of Mineral Cycling*, Second edn, pp. 48–50. Academic Press, New York.

Finlay, J. C., Small, G. E., and Sterner, R. W. 2013. Human influences on nitrogen removal in lakes. *Science*, *342*(6155), 247–250.

Froelich, P. N. 1988. Kinetic control of dissolved phosphate in natural rivers and estuaries: A primer on the phosphate buffer mechanism1. *Limnology and Oceanography*, *33*(4part2), 649–668.

Froelich, P., Klinkhammer, G. P., Bender, M. A. A., Luedtke, N. A., Heath, G. R., Cullen, D., Dauphin, P., Hummond, D., and Maynard, V. 1979. Early oxidation of organic matter in pelagic sediments of the eastern equatorial Atlantic: Suboxic diagenesis. *Geochimica et Cosmochimica Acta*, *43*(7), 1075–1090.

Gächter, R. 1987. Lake restoration: Why oxygenation and artificial mixing cannot substitute for a decrease in the external phosphorus loading. *Aquatic Sciences-Research Across Boundaries*, *49*(2), 170–185.

Gächter, R., and Imboden, D. 1985. Lake restoration. In Stumm, E. (Ed.), *Chemical Processes in Lakes*, pp. 365–388. John Wiley & Sons, Hoboken, NJ.

Gächter, R., and Müller, B. 2003. Why the phosphorus retention of lakes does not necessarily depend on the oxygen supply to their sediment surface. *Limnology and Oceanography*, *48*(2), 929–933.

Gächter, R., and Wehrli, B. 1998. Ten years of artificial mixing and oxygenation: No effect on the internal phosphorus loading of two eutrophic lakes. *Environmental Science and Technology*, *32*(23), 3659–3665.

Gächter, R., Meyer, J. S., and Mares, A. 1988. Contribution of bacteria to release and fixation of phosphorus in lake sediments. *Limnology and Oceanography*, *33*(6part2), 1542–1558.

Gächter, R., Steingruber, S. M., Reinhardt, M., and Wehrli, B. 2004. Nutrient transfer from soil to surface waters: Differences between nitrate and phosphate. *Aquatic Sciences*, *66*(1), 117–122.

Golterman, H. L. 1995. The role of the iron hydroxide-phosphate-sulphide system in the phosphate exchange between sediments and overlying water. *Hydrobiologia*, *297*, 43–54.

Golterman, H. L. 2001. Phosphate release from anoxic sediments or 'What did Mortimer really write?' *Hydrobiologia*, *450*(1–3), 99–106.

Gudasz, C., Bastviken, D., Steger, K., Premke, K., Sobek, S., and Tranvik, L. J. 2010. Temperature-controlled organic carbon mineralization in lake sediments. *Nature*, *466*(7305), 478–481.

Gunnars, A., Blomqvist, S., Johansson, P., and Andersson, C. 2002. Formation of Fe (III) oxyhydroxide colloids in freshwater and brackish seawater, with incorporation of phosphate and calcium. *Geochimica et Cosmochimica Acta*, *66*(5), 745–758.

Hecky, R. E., Campbell, P., and Hendzel, L. L. 1993. The stoichiometry of carbon, nitrogen, and phosphorus in particulate matter of lakes and oceans. *Limnology and Oceanography*, *38*(4), 709–724.

Hecky, R. E., Smith, R. E., Barton, D. R., Guildford, S. J., Taylor, W. D., Charlton, M. N., and Howell, T. 2004. The nearshore phosphorus shunt: A consequence of ecosystem engineering by dreissenids in the Laurentian Great Lakes. *Canadian Journal of Fisheries and Aquatic Sciences*, *61*(7), 1285–1293.

Higgins, S. N., and Zanden, M. V. 2010. What a difference a species makes: A meta-analysis of dreissenid mussel impacts on freshwater ecosystems. *Ecological Monographs*, *80*(2), 179–196.

Higgins, S. N., Malkin, S. Y., Todd Howell, E., Guildford, S. J., Campbell, L., Hiriart-Baer, V., and Hecky, R. E. 2008. An ecological review of Cladophora glomerata (Chlorophyta) in the Laurentian great lakes. *Journal of Phycology*, *44*(4), 839–854.

Holdren, G. C., and Armstrong, D. E. 1980. Factors affecting phosphorus release from intact lake sediment cores. *Environmental Science and Technology*, *14*(1), 79–87.

Hupfer, M., and Lewandowski, J. 2008. Oxygen controls the phosphorus release from lake sediments—A long-lasting paradigm in limnology. *International Review of Hydrobiology*, *93*(4–5), 415–432.

Huser, B. J., and Pilgrim, K. M. 2014. A simple model for predicting aluminum bound phosphorus formation and internal loading reduction in lakes after aluminum addition to lake sediment. *Water Research*, *53*, 378–385.

Huser, B. J., Egemose, S., Harper, H., Hupfer, M., Jensen, H., Pilgrim, K. M., Reitzel, K., Rydin, M., and Futter, M. 2015. Longevity and effectiveness of aluminum addition to reduce sediment phosphorus release and restore lake water quality. *Water Research*. doi:10.1016/j.watres.2015.06.051.

Hyacinthe, C., and Van Cappellen, P. 2004. An authigenic iron phosphate phase in estuarine sediments: Composition, formation and chemical reactivity. *Marine Chemistry*, *91*(1), 227–251.

Jensen, H. S., and Andersen, F. O. 1992. Importance of temperature, nitrate, and pH for phosphate release from aerobic sediments of four shallow, eutrophic lakes. *Limnology and Oceanography*, *37*(3), 577–589.

Jensen, H. S., Kristensen, P., Jeppesen, E., and Skytthe, A. 1992. Iron: Phosphorus ratio in surface sediment as an indicator of phosphate release from aerobic sediments in shallow lakes. In *Sediment/Water Interactions*, pp. 731–743. Springer, Netherlands.

Jeppesen, E., Søndergaard, M., Meerhoff, M., Lauridsen, T. L., and Jensen, J. P. 2007. Shallow lake restoration by nutrient loading reduction—Some recent findings and challenges ahead. *Hydrobiologia*, *584*(1), 239–252.

Jilbert, T., and Slomp, C. P. 2013. Iron and manganese shuttles control the formation of authigenic phosphorus minerals in the euxinic basins of the Baltic Sea. *Geochimica et Cosmochimica Acta*, *107*, 155–169.

Jørgensen, B. B. 2000. Bacteria and marine biogeochemistry. In *Marine Geochemistry*, pp. 173–207. Springer Berlin, Heidelberg.

Katsev, S., and Crowe, S. A. 2015. Organic carbon burial efficiencies in sediments: The power law of mineralization revisited. *Geology*, *43*(7), 607–610. G36626–1.

Katsev, S., and Dittrich, M. 2013. Modeling of decadal scale phosphorus retention in lake sediment under varying redox conditions. *Ecological Modelling*, *251*, 246–259.

Katsev, S., Chaillou, G., Sundby, B., and Mucci, A. 2007. Effects of progressive oxygen depletion on sediment diagenesis and fluxes: A model for the lower St. Lawrence River Estuary. *Limnology and Oceanography*, *52*(6), 2555–2568.

Katsev, S., Rancourt, D. G., and L'Heureux, I. 2004. dSED: A database tool for modeling sediment early diagenesis. *Computers and Geosciences*, *30*(9), 959–967.

Katsev, S., Sundby, B., and Mucci, A. 2006b. Modeling vertical excursions of the redox boundary in sediments: Application to deep basins of the Arctic Ocean. *Limnology and Oceanography*, *51*(4), 1581–1593.

Katsev, S., Tsandev, I., L'Heureux, I., and Rancourt, D. G. 2006a. Factors controlling long-term phosphorus efflux from lake sediments: Exploratory reactive-transport modeling. *Chemical Geology*, *234*(1), 127–147.

Katsev, S., Verburg, P., Lliros, M, and Minor, E. (in press) Tropical meromictic lakes. In Gulati, R., A. Degermendzhy, and E. Zadereev (Eds), *Ecology of Meromictic Lakes*. Springer, Netherlands.

Krom, M. D., and Berner, R. A. 1980. Adsorption of phosphate in anoxic marine sediments. *Limnology and Oceanography*, *25*(5), 797–806.

Kuo, S., and Lotse, E. G. 1974. Kinetics of phosphate adsorption and desorption by lake sediments. *Soil Science Society of America Journal*, *38*(1), 50–54.

Li, J. 2014. Sediment diagenesis in large lakes Superior and Malawi, geochemical cycles and budgets and comparisons to marine sediments. Doctoral dissertation, University of Minnesota, Minneapolis, MN.

Li, J., and Katsev, S. 2014. Nitrogen cycling in deeply oxygenated sediments: Results in Lake Superior and implications for marine sediments *Limnology and Oceanography*, *59*, 465–481.

Li, J., Crowe, S. A., Miklesh, D., Kistner, M., Canfield, D. E., and Katsev, S. 2012. Carbon mineralization and oxygen dynamics in sediments with deep oxygen penetration, Lake Superior. *Limnology and Oceanography*, *57*(6), 1634–1650.

Maerki, M., Müller, B., and Wehrli, B. 2006. Microscale mineralization pathways in surface sediments: A chemical sensor study in Lake Baikal. *Limnology and Oceanography*, *51*(3), 1342–1354.

Matisoff, G. 1995. Effects of bioturbation on solute and particle transport in sediments. In Allen, H. E. (Ed.), *Metal Contaminated Aquatic Sediments*, CRC Press.

Matisoff, G., and Wang, X. 1998. Solute transport in sediments by freshwater infaunal bioirrigators. *Limnology and Oceanography*, *43*(7), 1487–1499.

Matisoff, G., Fisher, J. B., and Matis, S. 1985. Effects of benthic macroinvertebrates on the exchange of solutes between sediments and freshwater. *Hydrobiologia*, *122*(1), 19–33.

Meile, C., and Cappellen, P. V. 2003. Global estimates of enhanced solute transport in marine sediments. *Limnology and Oceanography*, *48*(2), 777–786.

Middelburg, J. J. 1989. A simple rate model for organic matter decomposition in marine sediments. *Geochimica et Cosmochimica Acta*, *53*(7), 1577–1581.

Moore, P. A., Reddy, K. R., and Fisher, M. M. 1998. Phosphorus flux between sediment and overlying water in Lake Okeechobee, Florida: Spatial and temporal variations. *Journal of Environmental Quality*, *27*(6), 1428–1439.

Morse, J. W., Millero, F. J., Cornwell, J. C., and Rickard, D. 1987. The chemistry of the hydrogen sulfide and iron sulfide systems in natural waters. *Earth-Science Reviews*, *24*, 1–42.

Mortimer, C. H. 1942. The exchange of dissolved substances between mud and water in lakes. *The Journal of Ecology*, *30*(1), 147–201.

Mortimer, C. H. 1941. The exchange of dissolved substances between mud and water in lakes. *Journal of Ecology*, *29*(2), 280–329.

National Research Council. 1992. *Restoration of Aquatic Ecosystems*. National Academies Press, Washington, DC.

Nurnberg, G. K., Tarvainen, M., Ventelä, A. M., and Sarvala, J. 2012. Internal phosphorus load estimation during biomanipulation in a large polymictic and mesotrophic lake. *Inland Waters*, *2*(3), 147–162.

Ozersky, T., Evans, D. O., and Ginn, B. K. 2015. Invasive mussels modify the cycling, storage and distribution of nutrients and carbon in a large lake. *Freshwater Biology*, *60*(4), 827–843.

Paraska, D. W., Hipsey, M. R., and Salmon, S. U. 2014. Sediment diagenesis models: Review of approaches, challenges and opportunities. *Environmental Modelling and Software*, *61*, 297–325.

Paterson, M. J., Schindler, D. W., Hecky, R. E., Findlay, D. L., and Rondeau, K. J. 2011. Comment: Lake 227 shows clearly that controlling inputs of nitrogen will not reduce or prevent eutrophication of lakes. *Limnology and Oceanography*, *56*(4), 1545–1547.

Poulton, S. W., and Canfield, D. E. 2005. Development of a sequential extraction procedure for iron: Implications for iron partitioning in continentally derived particulates. *Chemical Geology*, *214*(3), 209–221.

Reitzel, K., Ahlgren, J., DeBrabandere, H., Waldebäck, M., Gogoll, A., Tranvik, L., and Rydin, E. 2007. Degradation rates of organic phosphorus in lake sediment. *Biogeochemistry*, *82*(1), 15–28.

Rickard, D., and Morse, J. W. 2005. Acid volatile sulfide (AVS). *Marine Chemistry*, *97*(3–4): 141–197.

Roden, E. E., and Edmonds, J. W. 1997. Phosphate mobilization in iron-rich anaerobic sediments: Microbial Fe (III) oxide reduction versus iron-sulfide formation. *Archiv für Hydrobiologie*, *139*(3), 347–378.

Roden, E. E., and Wetzel, R. G. 2002. Kinetics of microbial Fe (III) oxide reduction in fresh-water wetland sediments. *Limnology and Oceanography*, *47*(1), 198–211.

Rothe, M., Frederichs, T., Eder, M., Kleeberg, A., and Hupfer, M. 2014. Evidence for vivianite formation and its contribution to long-term phosphorus retention in a recent lake sediment: A novel analytical approach. *Biogeosciences*, *11*(18), 5169–5180.

Ruttenberg, K. C. 1992. Development of a sequential extraction method for different forms of phosphorus in marine sediments. *Limnology and Oceanography*, *37*(7), 1460–1482.

Schindler, D. W., Hecky, R. E., Findlay, D. L., Stainton, M. P., Parker, B. R., Paterson, M. J., Beaty, K. G., Lyng, M., and Kasian, S. E. M. 2008. Eutrophication of lakes cannot be controlled by reducing nitrogen input: Results of a 37-year whole-ecosystem experiment. *Proceedings of the National Academy of Sciences*, *105*(32), 11254–11258.

Senko, J. M., Dewers, T. A., and Krumholz, L. R. 2005. Effect of oxidation rate and Fe (II) state on microbial nitrate-dependent Fe (III) mineral formation. *Applied and Environmental Microbiology*, *71*(11), 7172–7177.

Shapiro, J., Lamarra, V. A., and Lynch, M. 1975. Biomanipulation: An ecosystem approach to lake restoration. In Brezonik, P. L. and Fox, J. L. (Ed), *Water Quality Management Through Biological Control*, pp. 85–96, University of Florida.

Sharpley, A. N., Chapra, S. C., Wedepohl, R., Sims, J. T., Daniel, T. C., and Reddy, K. R. 1994. Managing agricultural phosphorus for protection of surface waters: Issues and options. *Journal of Environmental Quality*, *23*(3), 437–451.

Scheffer, M. 2004. *Ecology of Shallow Lakes*. Springer Science and Business Media, New York.

Slomp, C. P., Epping, E. H. G., Helder, W., and Van Raaphorst, W. 1996. A key role for iron-bound phosphorus in authigenic apatite formation in North Atlantic continental platform sediments. *Journal of Marine Research*, *54*, 1179–1205.

Slomp, C. P., Mort, H. P., Jilbert, T., Reed, D. C., Gustafsson, B. G., and Wolthers, M. 2013. Coupled dynamics of iron and phosphorus in sediments of an oligotrophic coastal basin and the impact of anaerobic oxidation of methane. *PLoS ONE*, *8*(4), e62386.

Smith, V. H., Joye, S. B., and Howarth, R. W. 2006. Eutrophication of freshwater and marine ecosystems. *Limnology and Oceanography*, *51*(1part2), 351–355.

Sobek, S., Durisch-Kaiser, E., Zurbrügg, R., Wongfun, N., Wessels, M., Pasche, N., and Wehrli, B. 2009. Organic carbon burial efficiency in lake sediments controlled by oxygen exposure time and sediment source. *Limnology and Oceanography*, *54*(6), 2243–2254.

Søndergaard, M., Jensen, P. J., and Jeppesen, E. 2001. Retention and internal loading of phosphorus in shallow, eutrophic lakes. *The Scientific World Journal*, *1*, 427–442.

Søndergaard, M., Jensen, J. P., and Jeppesen, E. 2003. Role of sediment and internal loading of phosphorus in shallow lakes. *Hydrobiologia*, *506*(1–3), 135–145.

Spiteri, C., Van Cappellen, P., and Regnier, P. 2008. Surface complexation effects on phosphate adsorption to ferric iron oxyhydroxides along pH and salinity gradients in estuaries and coastal aquifers. *Geochimica et Cosmochimica Acta*, *72*(14), 3431–3445.

Straub, K. L., Benz, M., Schink, B., and Widdel, F. 1996. Anaerobic, nitrate-dependent microbial oxidation of ferrous iron. *Applied and Environmental Microbiology*, *62*(4), 1458–1460.

Sundby, B., Gobeil, C., Silverberg, N., and Mucci, A. 1992. The phosphorus cycle in coastal marine sediments. *Limnology and Oceanography*, *37*(6), 1129–1145.

Thibault, P. J., Rancourt, D. G., Evans, R. J., and Dutrizac, J. E. 2009. Mineralogical confirmation of a near-P: Fe = 1: 2 limiting stoichiometric ratio in colloidal P-bearing ferrihydrite-like hydrous ferric oxide. *Geochimica et Cosmochimica Acta*, *73*(2), 364–376.

Thomsen, U., Thamdrup, B., Stahl, D. A., and Canfield, D. E. 2004. Pathways of organic carbon oxidation in a deep lacustrine sediment, Lake Michigan. *Limnology and Oceanography*, *49*(6), 2046–2057.

Torres, N. T., Och, L. M., Hauser, P. C., Furrer, G., Brandl, H., Vologina, E., Sturm, M., Bürgmann, H., and Müller, B. 2014. Early diagenetic processes generate iron and manganese oxide layers in the sediments of Lake Baikal, Siberia. *Environmental Science: Processes and Impacts*, *16*(4), 879–889.

van der Zee, C., Roberts, D. R., Rancourt, D. G., and Slomp, C. P. 2003. Nanogoethite is the dominant reactive oxyhydroxide phase in lake and marine sediments. *Geology*, *31*(11), 993–996.

Waychunas, G. A., Kim, C. S., and Banfield, J. F. 2005. Nanoparticulate iron oxide minerals in soils and sediments: Unique properties and contaminant scavenging mechanisms. *Journal of Nanoparticle Research*, *7*(4–5), 409–433.

Weber, K. A., Achenbach, L. A., and Coates, J. D. 2006. Microorganisms pumping iron: Anaerobic microbial iron oxidation and reduction. *Nature Reviews Microbiology*, *4*(10), 752–764.

Welch, E. B., and Cooke, G. D. 1995. Internal phosphorus loading in shallow lakes: Importance and control. *Lake and Reservoir Management*, *11*(3), 273–281.

Williams, J. D. H., Jaquet, J. M., and Thomas, R. L. 1976. Forms of phosphorus in the surficial sediments of Lake Erie. *Journal of the Fisheries Board of Canada*, *33*(3), 413–429.

# 7 Economic and Policy Issues of Phosphorus Management in Agroecosystems

*Otto C. Doering III*

## CONTENTS

## 7.1 INTRODUCTION

Phosphorus management is an issue for achieving the benefits from phosphorus use in agriculture, and it also is a management issue for reducing the problems caused by excess phosphorus. One characteristic of this issue is that it involves a large number of different participants who may benefit or suffer. Phosphorus is beneficial and essential for plant and animal growth. Phosphorus can also create externalities where persons benefiting from the use or the discharge of phosphorus inadvertently cause damage to others without the cost of the damage being reflected back in additional costs to the person causing the damage. There is thus no compelling market signal that is received by the one causing the problem to encourage the amelioration of this externality problem. As a result, there are an extended number of stakeholders who are wittingly or unwittingly involved in management decisions about phosphorus as an economic issue and sometimes as a policy issue. Interwoven in this web are multiple trade-offs for those involved so that there is no clear management or policy solution that meets the needs and the desires of all the various stakeholders who may benefit from phosphorus or contribute to a problem. Most of the focus of policy and economic analysis on phosphorus relates to the externalities from excess phosphorus. Thus, one objective of this chapter is to show that the nature of this problem is such that the solutions do not come easily, and there is usually no single silver bullet solution. Policies, programs, and institutions that are called upon to deal with phosphorus management concerns remain challenged by the issue and the changing nature of the problem. Another objective is to demonstrate that economic and policy trade-offs are many, complex, and uncertain with problems like phosphorus management. In addition, the choice of policy instruments is also critical to the path taken and the shaping of outcomes. A final objective is to argue that economic analysis is still useful in forcing the discipline of assessing alternatives in a structured way, creating the best estimates of the consequences of alternative solutions for public choice.

## 7.2   THE PROBLEM SETTING: A WICKED PROBLEM

Phosphorus management in agroecosystems is an example of a kind of issue that can be classified as a wicked problem. Such problems are dynamic and complex as well as having many interests involved so they become socially and politically complex. Such problems are not successfully solved by experts following the linear scientific method. As an example, climate change is a wicked problem whereas putting a man on the moon is not (Kreuter et al. 2004). There are important differences between the two types of problems (Batie 2008).

### 7.2.1   PROBLEM DIFFERENCES

For tame problems, there is a clear definition of the problem that leads to the solution. The problem does not change over time. The outcome is either a successful solution to the problem or not. That is, the stopping rule applies because the task is completed when the problem is solved.

For wicked problems, there is not a universal agreement on the problem itself. Attempts to solve the problem change the problem. The problem changes over time. There is no likely solution to the problem, only the ability to make it better or worse, so the stopping rule does not apply. There is no definitive solution, and continued efforts to deal with the problem are dependent upon stakeholders, resource availability, and political forces.

### 7.2.2   THE ROLE OF STAKEHOLDERS

For tame problems, the causes of a problem are primarily determined by experts using scientific data.

For wicked problems, many stakeholders are likely to have differing ideas about what the real problem is and what the causes are.

### 7.2.3 THE NATURE OF THE PROBLEM

For tame problems, scientific-based protocols guide the choice of solutions. The problem is associated with low uncertainty as to system components and outcomes. There are also shared values as to the desirability of the outcomes.

For wicked problems, the solutions to problems are based on judgments of multiple stakeholders. The problems are associated with high uncertainty as to system components and outcomes. There are no necessarily shared values with respect to societal goals.

Wicked problems are not just technical challenges; they are shaped and governed by social concerns and political concerns. This puts them in the realm of a much bigger systems context where traditional science is much less able to manage the process toward a positive outcome. As one moves from a tame problem to a wicked problem, there is a greater conflict of values in dealing with the wicked problem and more uncertainty about the path taken and the actual outcome of the attempt to deal with it. The wicked problem itself is something of a moving target, and any economic or policy approach to the problem must deal with that. When environmental issues are involved, social preferences become critical, and the economic analysis is likely to involve numerous complex trade-offs in assessing alternative policies that may also involve nonmarket as well as market costs and benefits. What may initially appear to be a somewhat straightforward problem multiplies into an interconnected web where there are high degrees of uncertainty (Batie 2008).

## 7.3  AN EXEMPLARY WICKED PROBLEM OF NUTRIENT MANAGEMENT

Nutrients such as phosphorus and nitrogen can cause degradation in water leading to algal blooms, fish kills, and the like. The U.S. EPA has the responsibility under the Clean Water Act of regulating discharges to ameliorate such situations. As an example, Florida has suffered from the degradation of surface waters when they receive nutrients, particularly phosphorus, above some capacity to assimilate nutrients without the degradation of water quality. The high levels of nutrient pollution in Lake Okeechobee and the everglades have been of increasing concern to the citizens, the state, and the federal government. In Florida, much of this concern stems from excess phosphorus from livestock, fertilizer, sewage treatment plants, phosphorus mining, and natural concentrations of phosphorus in some waters. Florida initiated its process of addressing its nutrient problems by using a narrative (descriptive) standard to determine which waters are impaired and thus require attention. Florida is among many states that have historically taken this approach to identify impaired waters and begun to prioritize them for remediation. The Florida criteria states that "in no case shall nutrient concentrations of a body of water be altered so as to cause an imbalance in natural populations of aquatic flora or fauna" (NRC 2012, p. 22). The application of such a standard, based on the state of the natural conditions, requires a broad assessment of the biological status of a given water body, and this needs to be done for each specific water body. These assessments can be both an expensive and a time-consuming task. Water bodies not meeting the narrative standard would be listed as impaired and then the regulatory total maximum daily loads (TMDLs) may be determined for the impaired water body potentially triggering more stringent requirements for the discharges from point sources and ultimately the allocation of more stringent allowable nutrient flows from nonpoint sources as well. The state of Florida followed this procedure targeting waters that required remediation and applied resources to those waters where resources were available. Another approach is establishing a numeric standard. The main difference between the narrative approach that Florida and other states have used and the numeric standard is that the narrative standard initially focuses on the biological conditions of the water body. If the biological conditions are unacceptable, nutrients are investigated as potential causes of the unsatisfactory biological conditions. If the nutrients are found to be the problem, then the targets for loads and concentrations are set for the nutrients. In contrast, under a numeric nutrient standard, a test for nutrient concentrations set against an established threshold makes the determination that a water body is impaired. Listing then occurs if the nutrient levels are above the numeric nutrient standard that has been set.

The EPA set numeric standards for lakes and flowing waters in Florida following a judgment from a 2009 lawsuit that argued that the narrative standard used by the state of Florida to determine which waters were impaired was insufficient to accomplish the level of water quality required under the Clean Water Act. These numeric standards were given notice in the Federal Register of December 6, 2010. It should be remembered that the goal to be accomplished was and is improving the quality of Florida's lakes and flowing waters. The state of Florida, given its tourist industry and for other reasons, has devoted more resources than many other states to the effort to improve the water quality in the state. Yet the concern by some groups over a perceived lack of progress resulted in the EPA being forced by the courts to require what was believed by these groups to be a more stringent standard on the state. What is involved here is not just a science-based decision about what approach might be more effective and/or efficient but a determination by a court of the EPA's appropriate regulatory oversight of a state under the Clean Water Act.

As part of the process of setting numeric standards, the EPA was also required to conduct an economic analysis of the potential incremental costs of implementing the numeric nutrient criteria instead of the narrative criteria already in place in Florida. A National Research Council (NRC) committee was asked to assess the EPA's economic analysis of the incremental cost of the new criteria. The resulting NRC report provides an insight into just how complex, extensive, and nonlinear the process of addressing a wicked problem can be (NRC 2012). Among other things, the consequences of the rule change affect different stakeholders differently, and the priorities, the paths, the locations, and the timelines of remediation are changed from one rule to the other. A comparison of some basic differences between the two rules illustrates the differences between the potential outcomes and the trade-offs that would occur when the narrative rule is replaced by the numeric rule over a five-stage process of regulated water quality management (NRC 2012, pp. 90 and 91).

1. Waters are listed as impaired under the narrative rule based on biologically determined impairment. Under the numeric nutrient rule, phosphorus (P) and nitrogen (N) are themselves the impairment, and the concentration levels of either one of these above the numeric criteria threshold cause the water body to be so listed.

2. Under the narrative rule, once the water body is considered impaired, then there is a determination of whether P or N is causing the impairment, and the process of remediation restricting the nutrients moves on from there. While the numeric rule determines impairment on the basis of the amount of P or N according to the criteria, there is some potential flexibility. The EPA can be petitioned to approve site-specific alternative criteria (SSAC) to replace or modify the numeric nutrient criteria for P, N, or both.

3. At this point in the process, it becomes necessary to define the level of nutrient reduction required for the TMDL limits that would be set under the Clean Water Act to reduce the nutrient loads in the water body. Under the narrative rule, the water quality conditions are modeled to relate the desired biological conditions to the nutrient loads to determine the appropriate P and/or N reduction targets to meet the desired biological conditions. Under the numeric rule, the water quality is also modeled but primarily to determine the nutrient load reductions required of the offending nutrients that will result in ambient nutrient concentrations to achieve the numeric nutrient targets set under the rule.

4. Once the TMDL has been determined to meet either the narrative or the numeric rule, then its accompanying best management action plan (BMAP) has to be developed to meet the TMDL requirement. Under the narrative rule, the BMAP balances the load reductions across sources between waste load allocation (WLA) and other load allocations (LAs). Under the numeric rule, the WLA is initially set through the process of the National Pollution Discharge Elimination System (NPDES) permitting process for point sources, and then other LAs are set for nonpoint sources. Agriculture would primarily come under the nonpoint sources, while WLAs such as sewage treatment systems would be point sources.

5. Ultimately there needs to be a determination if the water quality goal has been met. Under the narrative rule, the determination is whether the desired biological conditions described in the narrative have been attained. Over the process of attainment, the nutrient requirements may be revised over time to be consistent with meeting the required biological conditions. Under the numeric rule, the goal has been met if the nutrient concentrations meet the numeric nutrient criteria or the SSAC that was negotiated with the regulatory agency for the body of water. For the numeric nutrient rule, this does not necessarily mean that the biological conditions are no longer impaired at this point.

Arguments for the numeric rule over the narrative rule were partially based on the concern that the state was moving too slowly in improving its waters under the narrative system. The belief was that under the numeric system, more bodies of water could be listed as impaired more quickly. Also, that water bodies having existing or potential problems not listed under the narrative rule might be listed under the numeric rule. The numeric rule was also seen by some as being based on a definitive scientific standard, which allowed clear distinctions to be made one way or the other simply and inexpensively. The assumption here is that a somewhat uniform threshold of nutrient content can represent the tipping point between a common perception of impaired and unimpaired water. The other assumption, of course, is that a common perception of impairment exists.

### 7.3.1 Different Results from Different Rules

The contrasts between the two approaches are more extensive than the points listed earlier might indicate—especially as one expands into the real world of implementing measures to deal with a truly wicked problem. Starting with the listing process, the desired goal is not necessarily clear or agreed on by different groups. The narrative rule's direct focus on the water body's fauna and flora might come closer to meeting some groups' perceptions of what a healthy or an unhealthy water body should look like given their sense of what activities and functions it should support. The fact that a water body has nutrient levels above the criteria established by the numeric rule indicates only that fact until the linkage is made to conditions that the general public or some other group perceives as requiring remediation. Under the narrative rule, only after the determination of a biologically based impairment is there a determination that the impairment is a result of excess nutrients. While the nutrient level itself is the standard of exactitude for the numeric rule, this does have to be adjusted to different situations as is recognized by the SSAC process. In Florida, there are natural background levels of P in some areas that are well above a numeric nutrient criteria that might be appropriate for attainment in other areas. The numeric rule for this reason and others has to have the flexibility to be modified under an SSAC that would adjust to such things as flow levels, mixing, and seasonal variations with respect to the water body. A question might be whether the SSAC would become more prevalent than the official numeric criteria. There is also the question whether the SSAC process can be applied at reasonable cost in a timely manner. The settings of the nutrient reductions required for the two approaches differ because the seemingly well-established science-based goals are different. The factors taken into consideration in the modeling of the water conditions and the requirements to meet the goals are also different. For either approach, the actual efficacy of the BMAP and the components of that plan are critical to estimating what is required and at what locations in a watershed. There is also the perception, if not a reality, that the order of attack on the problem is different under the different rules. Many point source dischargers in Florida perceived that the reduction required of them would be greater under the numeric rule because of the regulation in place under the existing NPDES permitting system and because of less technical uncertainty about the potential levels of attainment of the given nutrient reductions by point sources as compared with nonpoint sources.

Following the nature of a wicked problem, there is no end point or stopping rule in improving the water quality in Florida. Goals are to be negotiated. The problem changes over time. The nutrient loads, their location, and their character are changing over time. Florida's growing population and continuously changing land use under development pressure complicate the problem. The water table and the flows of natural aquifers are changing as well, due to withdrawals and other factors. Finally, the number of impaired and identified water bodies that need remediation is likely to be well beyond the resources available to the state to deal with under either rule in the foreseeable future. It does not necessarily help to list more water bodies as impaired if the resources to remediate these conditions are insufficient under either standard. All of this indicates that adaptive management involving modification of goals as well as paths to achieve them is going to be critically important under either rule. There can be no specific path determined ahead of time that will remain the best path forward for the duration of attention to the problem. These necessary in-course changes will change the costs and the benefits of these efforts as well as the programs and the policies required.

### 7.3.2   ROLE FOR ECONOMIC ANALYSIS

How do economists then approach a wicked problem like the case of the nutrient pollution in Florida? In that case, the economic analysis was to determine the cost difference from this point forward of applying two different rules to the task of improving the quality of the water suffering from excess nutrients. That would be the change in costs that would occur under the numeric rule as compared to the narrative rule as action on impaired waters was taken by Florida over the five stages of regulated water quality management listed earlier. (Just to add to the complexity, implied in this is the task of maintaining the water quality in those waters not suffering from excess nutrients.) The kinds of costs one has to consider are the capital, operation, maintenance, and replacement costs incurred by both point and nonpoint dischargers to reduce the amount of nutrients going into a water body to meet either standard (NRC 2012, p. 98). There are also administrative and management costs, public or private, entailed in the nutrient reduction activities that may be underestimated or ignored. These include monitoring, assessment, program, and plan development costs (determining TMDLs, SSAC activities, BMAP establishment, and a whole suite of activities involved in acquiring permits, ensuring compliance, reporting, gaining agreement to implement BMAPs, engaging in rules and challenges to them, etc.). The estimation of economic costs (or benefits) is also dependent upon where the analytical as well as the geographical boundaries are drawn for the analysis. In the Florida analysis, it was argued by some that one needed to include the water quality opportunity cost. This would be the foregone benefits not realized because the water quality was not improved to a particular level. In attempting to estimate this opportunity cost, one would have to make some assumptions about how quickly or effectively one rule improved the water quality, and where, in contrast to the other rule. Then one would have to estimate the added value that would pertain to the improved water. Even if one limits the analysis to the nutrient load control investments, the activities, and the administrative costs, there is a great deal of uncertainty about these costs. Some of the uncertainty reflects uncertainty about the effectiveness of future technology applications and administrative actions as it becomes more apparent in what works and what does not.

It may seem at this point that an economic analysis of the two approaches would be something of a useless exercise. What looks like a difficult and problematic economic exercise with little utility may in fact be a valuable exercise, not only for comparing different approaches but also for planning for a single approach.

Following the mantra that the planning process can be much more valuable than the plan, one might crosscut the five stages of regulated water quality management with the different categories of costs and do this looking forward for different periods. One can also introduce different levels of uncertainty or risk into the different costs and performance milestones. Whether one formally introduces uncertainty or not, one can be explicit about the assumptions that have been made and the degree to which something is dependent upon that assumption and the degree of confidence in the assumptions. Potentially, the process creates a formalized blueprint of different paths and the assumptions and the drivers behind each path. At the same time, it may be helpful to break down the costs and the assessments of efficacy by sector in terms of some verified reality. As an example from Florida, in the case of municipal waste plants, the NRC committee believed that it was essential to better ground-truth the cost and performance estimates as much as possible from existing experience of plants in Florida. It would also be important to estimate the proportion of plants that might come under administrative relief for certain standards of nutrient reduction (NRC 2012, p. 80). All of this does not necessarily ensure a correct estimate of the costs and the levels of achievements in the future. However, if done carefully, an economic assessment can provide a guide that forces a disciplined planning process for the effort as a whole—everything from which groups need to be involved (as administrators and as stakeholders) to a hard look at the full costs of a technology or practice over time which includes its expected efficacy over time. Economists are unlikely to be able to give the single point number estimates citizens and planners might like for attacking a wicked problem like Florida's nutrient problem. Economists should be able to help contribute to and help shape a planning process that helps citizens understand the type and the level of resources required for various aspects of attacking a problem. The role of providing a template for considering the potential costs and the likely consequences of alternative approaches might be even more important. This would include a better understanding of the various components required for action and their relative importance along with simulations of the impacts of different assumptions about the future. The choice of policy approach, the institutions involved, the values of stakeholders, and the fact of different and competing goals all play a role in the design and the usefulness of such analysis for public decisions.

## 7.4  DIFFERENT ECONOMIC AND INSTITUTIONAL APPROACHES FOR DEALING WITH EXTERNALITIES

The problem of excess phosphorus in Florida's waters is a classic externality problem. The free market by itself does not successfully deal with the problem to the satisfaction of the groups affected by the negative impacts of excess phosphorus. While the primary institutional framework for the Florida problem earlier was a regulatory one, it also involves voluntary actions, market mechanisms, and others under what is a regulatory umbrella. If one is dealing with a complex externality, one has a choice of institutional approaches that can be used singly or in combination to approach the problem. It is also important to recognize that existing institutions or institutional boundaries may actually inhibit attempts at problem solutions.

There are four common management strategies to control externalities like excess phosphorous in the environment that have evolved over the years (EPA 2011).

1. Command-and-control approach that generally involves a regulatory body setting rules of some sort over discharges of a pollutant. This is done under legal authority to limit or regulate the amount, concentration, and possibly the timing of the discharge.
2. Incentive policies or programs (often but not always government) that may involve incentives or disincentives such as subsidies, taxes, or payments to bring about behavior or technology for a particular result in the control of pollutants.
3. Market-based instruments for pollution control which may involve market-driven trading schemes, auctions, or other mechanisms to reduce pollution, sometimes at increased efficiency and reduced cost under a regulatory cap.
4. Voluntary programs in which the ends are achieved through actions that are taken or agreements that are reached between parties as a result of outreach, education, or changes in attitudes and perceptions.

Command and control with respect to discharges into water bodies is not new in the United States. In the late 1800s, there were navigation acts such as the Rivers and Harbors Act of 1899, 33 U.S.C. 403, which were meant to protect the navigability of rivers. These required a permit for discharges of material into navigable waters that might affect the course, the location, the condition, or the capacity of such waters. While much of the focus at that time was on flotsam and jetsam that might hinder navigation; the reach of such laws gradually broadened until the passage of the expanded Clean Water Act in 1972. Under a command-and-control approach, the dischargers are required to meet mandatory standards for the discharge or for the quality of the waters affected by the discharge. In practice, the ability of the government to enforce such standards develops slowly over time—especially in cases where there has not been a practice of monitoring or where monitoring is difficult or expensive. In order to accomplish goals under such conditions, the requirements might be met by dischargers applying pollution limiting technology, such as best management practices for agriculture or best available technology in other instances. As the ability to monitor improves, more precise standards can be applied focused on the quality of the water body itself, such as the numeric nutrient requirements that were at the center of the controversy in Florida. What the application of command and control reflects is the fact that the regulatory body has the power to set a standard and require actions that must be met under the force of law. One then hopes that the standard also has meaning in terms of its outcome and the attainment of the desired level of water quality.

Conservation compliance in the United States is an example of what is, in effect, a command-and-control regulation for agriculture, especially with respect to soil erosion. The Soil Conservation Act of 1984 denied major agricultural program benefits to farmers who did not take adequate care of highly erodible land, destroyed wetlands, or broke sod. These regulations included the denial of federal price supports and crop insurance. Since phosphorus moves with soil particles, this control has had an impact on reducing the phosphorus reaching streams and waterways. As of 2011, the USDA estimated that conservation compliance mechanisms applied to just over 100 million acres of U.S. cropland that was considered highly erodible. This amounted to almost one third of commodity cropland under cultivation (Claassen 2012). Conservation compliance remains in the 2014 Farm Act. It covers the major commodity and the conservation programs and includes various insurance programs. This is important as the changes in the 2014 Farm Act involved moving many benefits to the insurance provision from the commodity provision. If the insurance provision were not included in the conservation compliance, there would be less incentive to comply. While conservation compliance does not have the direct power to force an action to prevent soil erosion or sod busting, farmers will endeavor to comply to the extent that agricultural program benefits are valuable to them (Doering and Smith 2012). When prices are low, and especially with the very favorable financial provisions for farmers under the 2014 Act, the incentive to farmers to volunteer for the commodity or the insurance programs and be economically induced to follow conservation compliance is very large. Once the voluntary program enrollment step is taken, then the compliance provision effectively approaches command and control.

Government incentives or disincentives to achieve policy ends have a long history with respect to agriculture in the United States. These might be positive incentives like subsidies or negative ones like taxes. Some of the earliest incentives started with the Agricultural Adjustment Act of 1933. Farmers got the benefit of the nonrecourse loan for their program crops but had to reduce production to do so. After the 1933 Act was declared unconstitutional by the Supreme Court, the Soil Conservation and Domestic Allotment Act of 1936 became a critical instrument for paying farmers to reduce erosion and conserve soil resources. Farmers who volunteered were paid for shifting a percentage of their acreage of soil-depleting crops to soil-conserving crops, and were paid an average of $10 per acre to do so. There were additional payments of up to $1 an acre for soil building practices such as incorporating green manure (Benedict 1953). This was a substantial amount of money at the time. To adjust for inflation, one would multiply by a factor of more than 16 to equal $2015. The amounts of money spent on these conservation programs overshadowed other USDA program funding. They were a big contributor to the Roosevelt administration's need to get cash into rural areas. In addition, since farmers were undertaking conservation to earn these dollars, the transfer was politically acceptable to a public that was beginning to see the specter of the dust bowl but was still against the government giving direct relief to farmers. The program incentives that began in the 1930s, which continue to today, set the precedent that farmers would be paid for improving their farming practices on the land. Today's U.S. agricultural conservation programs like the Environmental Quality Incentive Program and the Conservation Reserve Program (CRP), which address concerns of excess nutrient pollution, operate on the same basis of voluntary participation through the encouragement of government payments.

There are a number of different permutations of market-based pollution control approaches that can relate to controlling excess phosphorus (Canchi 2006; EPA 2011).

1. Trading: Tradable water quality permits can be utilized to bring about reductions of phosphorus in water. Pollution limits must be imposed by some regulatory agency to make the scheme operate, and some entities

Economic and Policy Issues of Phosphorus Management in Agroecosystems **143**

must be able to reduce their phosphorus discharges at lower costs than others for there to be cost savings.

2. Auctions: Environmental or conservation auctions are used in agriculture where farmers may bid to provide environmental services such as pollution reduction or environmental improvements for a public or a private entity. The Bush Tender program in Australia's Victoria state is a comprehensive example of farmers bidding in bundles of environmental amenities aimed at restoring soils and improving water quality and quantity in the Murray Darling basin (Department of Sustainability and Environment 2008). Another example is the U.S. CRP.
3. Transferable quotas: Individual transferable quotas can also be a market-based tool where the quota can be sold or transferred. This can be a pollution quota or a production quota as in the case of tobacco in the United States.
4. Indemnifying risk: Risk indemnification for specified behavior may be useful to encourage the adoption of a new technology or practice that is perceived to have higher levels of risk. The extent to which a new practice increases risk while reducing excess nutrient losses may encourage a public or a private body to indemnify that risk to encourage the adoption of the practice.
5. Easement purchases: Easements where a property owner gives up certain property rights for a payment can also be used to reduce nutrient losses. An easement arrangement may result in a portion of a farm becoming a conservation buffer for a specified period of time.

These instruments are not interchangeable. They are alternative approaches with different characteristics that may be used singly or in combination. The policy objectives, the biophysical nature of the problem, the characteristics of stakeholders or actors involved, and the specific local conditions are going to determine the usefulness of the different market-based approaches.

### 7.4.1 Market-Based Tradable Permits

Tradable permits were institutionalized in the Clean Air Act Amendments of 1990. The tradable permit concept is often suggested by economists because it can utilize the market to bring about more cost-effective pollution reduction efforts. However, the trading has to be stimulated by some regulation or regulatory cap that requires a given level of pollution reduction. Since we are dealing with excess nutrients as an externality, the market by itself does not automatically engage polluting firms in expenditure to reduce their pollution. With a tradable permit scheme, firms are usually given allotments of allowable pollution levels under the cap of some overall required reduction. These permits may then be traded between firms (or bought and retired by environmental groups in some cases). A firm may meet its requirements either by physically reducing pollution to its allotted allowable level or by purchasing enough pollution permits from other firms to make up the difference between the amount it is allowed to pollute under the cap and the pollution it actually emits above its cap.

The economic efficiency aspect of the scheme rests in which firms buy and which firms sell permits, and the freedom firms have to apply innovative technology. Firms that can inexpensively meet the pollution reduction required and then reduce pollution below their cap limit can then sell the extra permits they accumulate at a profit to firms that have higher costs for reducing their pollution. A market can then develop which prices extra permits somewhere between the costs of control of the low-cost firms and the cost of control of the high-cost firms. High-cost control firms will then purchase permits, and low-cost firms will reduce their pollution below their allowable pollution levels. By utilizing tradable permits, the pollution is controlled at a lower average cost than if both the high-cost and the low-cost firms reduced their level of pollution to the cap limit through their own control measures. This is because the low-cost firms end up undertaking a higher proportion of the pollution control effort. This kind of trading was allowed under the 1990 Clean Air Act Amendments, and a market quickly developed for airborne pollutants such as sulfur dioxide, especially from electric power plants. There were some specific characteristics that led to the success of this trading. All the pollutants traded were from point sources (think smokestacks) that were already under some form of regulation and were being monitored for air quality. The baseline levels of pollution were well established, and reductions or increases could be accurately monitored, compared, and verified. A pollution permit that was based on a well-monitored point source was thus fully fungible and quantifiable.

In spite of the cachet of anything market driven, the promise for water quality trading has been difficult to realize. There are major impediments to a trading system that might mimic the one under the Clean Air Act. What was hoped for were equally beneficial results from trading between point and nonpoint water pollution sources as had been achieved in point-to-point sulfur dioxide trading (Stephenson and Shabman 2011). One critical difference is that point and nonpoint sources are under very different regulatory strictures. The Clean Water Act strictures that apply to point sources do not apply to nonpoint sources. In an impaired watershed, the primary regulatory instrument available is the NPDES permit requirement for the point source pollution, and this results in as much reduction as possible being required of the point source. The nonpoint source will likely only be tapped for reductions when further pollution reductions cannot be obtained from the point source. This is unlike the Clean Air Act case with sulfur dioxide where all parties involved were point sources, all under the same cap, and all were being monitored. In water quality trading between point and nonpoint, only the point source is under strict command and control. From an economic efficiency standpoint, the offset gained from a trade by a point source may just serve to allow a point source to continue or expand discharges to waters not meeting standards once the point source's required technical capacity to control at the source has been maximized under the Clean Water Act (Stephenson and Shabman 2011). (At this stage, the point source may have maxed out its technical capacity and its costs.) In addition, the point sources also have regulated technology requirements that may inhibit innovation in technology operation and choice. There is not a similar balanced market incentive for both nonpoint and point sources to reduce discharges. The geography of trading for water quality is also much more limited. To be operable, trades have to take place in the same watershed or the same regulatory geography in contrast to the broader air shed geography usually allowed for sulfur dioxide trading. Given that nonpoint agricultural discharges of pollutant dominate water quality problems in most watersheds, there are not enough point source trading partners so that large gains in the overall water quality can be achieved in point/nonpoint trading. These are some of the reasons that water quality trading schemes have thus far not resulted in large volumes of trades or large reductions in pollutants relative to discharges. Some more successful trading schemes like Neuse and Tar Pamlico in North Carolina have involved some common-pool management with total load management, aggregate limit monitoring, and less direct regulator involvement about how and where limits are met (Woodward et al. 2002). Tradable permits for water quality, particularly between point and nonpoint sources, have some serious obstacles to overcome, like uniform monitoring and symmetry of regulations and incentives, in order to reach the low transactions costs and efficiency achieved with airborne pollutants under the Clean Air Act Amendments.

### 7.4.2   AUCTION CONTRACTS

An auction-based contract may well be the best approach if widespread participation is required.

When the goal is to preserve or encourage management practices aimed at a particular outcome, this can be accomplished efficiently through an auction where the entities participate at their various reserve prices, which reflect their costs of doing so plus whatever incentive is required above that for participation. The CRP operates along these lines. The government of Victoria's Bush Tender program was based on an auction where landowners bid to provide various land management services and physical land and vegetation modification to reduce the salinity in the Murray Darling basin. In order to achieve success, the government had to enlist enough blocks of land with particular hydrological and vegetative characteristics in specific places to effectively reduce salinity and enhance other environmental characteristics (Department of Sustainability and Environment 2008). The government had detailed knowledge of the landscape and its hydrology and could thus pick those bids that individually provided the most benefit toward salinity reduction. At the same time, blocks of land could be assembled from different landholders with different reserve prices that synergistically provided enhanced benefits. The government of Victoria also had substantial information about the geographical area of interest so it could assess bids on the basis of their potential contribution to the overall goals. In a less precise way, the environmental benefits index used for assessing the bids for the CRP in the United States can be used in a similar way to accomplish similar goals. In the Australian case, the goal was to enhance the native vegetation and its management over the long term where it would slow the development of salinity in soils. The auction process could elicit the protection or the enhancement of the target vegetation and the required management commitment for the appropriate time span taken against the cost the landowner would bear to provide these plus whatever incentive was needed to gain participation. This was a blind auction where one landowner did not know what other landowners were being paid or what they had bid. Australian landowners also did not have the tradition that exists in the United

States of landowners being paid regularly to adopt conservation practices or reduce pollution from their land. The result was that the desired participation and attendant management was achieved at lower cost than might have been the case in the United States with something like the CRP where landowners also have more knowledge of what the government is willing to pay. One can contrast the auction approach used here with other market-based schemes in terms of their efficacy to reach given goals. The auction approach is more suitable than tradable permits when a large number or specific blocks of individuals or parcels of land need to participate to achieve a goal. Governments could incentivize various best management practices through risk indemnification. However, the geographical location of responsive landowners might not be in the concentration and the location that was essential which could better be achieved in a targeted bidding process. Easements are a possibility but often involve a property right transfer not essential in the auction for services on the land.

### 7.4.3 VOLUNTARY APPROACHES

Given the fact that excess nutrient discharges in agriculture in the United States are not regulated in comparison with other sectors of the economy or in comparison with other advanced countries, the approach to ameliorate externalities has been to encourage voluntary activities through education, persuasion, or incentives. What is involved here is participation in programs or initiatives on a voluntary basis rather than being compelled to do so. This may include unilateral voluntary initiatives, negotiated voluntary agreements that carry obligations, and public voluntary programs where the government unilaterally determines both the rewards and the obligations of participation as well as eligibility requirements (Segerson 2013). Much of the effort in the United States to reduce externalities from agriculture has involved voluntary programs with payments from the government as incentives for participation. Historically these payments were used to enhance income as well as to gain conservation objectives, as was the case in the early soil conservation programs. Today, the mantra is that payments should at least leave farmers no worse off for the effort made under a program. The key questions then become achieving participation and obtaining the environmental results desired cost effectively. Programs and payments may be structured in various ways and involve different approaches like incentives and auctions. Different goals may be more attainable through different approaches. There is also the concern for some voluntary approaches of achieving meaningful net improvements as well as the maintenance of the effort over time. A sense of stewardship possessed by an individual may also induce an individual to take unilateral action without any compensation. Incentive payments can induce participation in activities to reduce externalities, but these can be expensive—especially if there is a tradition of payments to both increase income and induce certain behavior as in the United States. A recent study indicated that only 35% of U.S. cropland is managed with best management practices for nitrogen pollution, and that extending the adoption of such practices on the remaining farmland would require a substantial increase in the federal budget devoted to incentives for nutrient management practices (Ribaudo et al. 2011). Regulatory threats can also induce voluntary participation. The threat of a potential TMDL determination can give parties the incentive to bring about reductions on their own to avoid the regulatory oversight. For example, under the Florida everglades agricultural privilege tax, farmers face higher tax payments when water quality standards for phosphorus are not met. Consumer demand is also becoming a potential force to bring about voluntary action. Consider the impact of consumer concern with dolphin deaths in tuna fishing that resulted in the industry changing its technology to be able to offer dolphin-free tuna.

There are some other factors that can affect the efficacy of voluntary approaches and other approaches that have been considered. Performance-based standards have been found to be more cost effective than practice-based standards (Antle et al. 2003). This often relates to slippage between the application of a practice and the results obtained that are not revealed in a practice-based standard. A performance-based standard also encourages innovation to design and apply practices that are the least costly to meet the performance goal. Another approach is targeting the application of practices or investments, which has been seen by economists as an important vehicle for increasing the cost effectiveness of conservation programs. The argument is that if the resources were concentrated on the proportion of a watershed that had the greatest soil erosion or excess nutrient loss, these efforts would be more effective than spreading the resources evenly across the watershed (Fox et al. 1995). Table 7.1 illustrates this for the case of soil erosion in Stratford Avon, Canada. The same simulated soil erosion control measures were applied to agricultural land in the watershed, first blanketed across 60% of the land, and then applied only to the land with a slope equal to or greater than 4%. Net benefits with targeting were increased by 75%, driven by much lower adoption costs. It is important to note that Stratford Avon has significant differences in land characteristics across the landscape that make targeting more effective.

One of the major objections to this approach has been political. Congressional representatives see the advantage in having their constituents equally able to be considered for conservation payments of relatively equal value. The term cost-effective, implying targeting, was included in the conservation provisions of farm bills in the 1990s at the urging of economists and environmental groups and then later removed in the 2002 farm bill. The emphasis by the Natural Resources Conservation Service on the Upper Mississippi River Basin Initiative has returned some aggregate level of geographical targeting. Both performance standards and targeting present opportunities to increase cost effectiveness but face political and producer hurdles.

---

**TABLE 7.1**
**Increased Net Benefits from Targeting**

| Conservation Measure | Target Variable | Area under Conservation Measure | Net Benefits Canadian Dollars |
|---|---|---|---|
| Two years corn Two years alfalfa Fall chisel plow } | No targeting | 60% | $800 |
| | Slope ≥ 4% | 18% | $1400 |

*Source:* Fox, G., G. Umali, and T. Dickinson, *Canadian Journal of Agricultural Economics*, 43, 105–118, 1995. With permission.

---

## 7.5    IS THERE EVER A WIN-WIN?

Mostly everything covered so far involves trade-offs of one sort or another. This might be the requirement for greater expenditure to induce program participation, regulations that increase costs, or market-based instruments that do not appear to lower costs or encourage more efficient technology. There is a good example of a technology development that resulted in a true win-win situation for more efficiently utilizing phosphorus and reducing excess phosphorus losses (Stahlman and McCann 2012). Inorganic phosphorus has been added to animal feed rations to meet nutrient requirements. For nonruminant animals, like poultry and swine, this can result in high phosphorus levels in the resulting manure. The added phosphorus can be the third most costly item in a feed ration after energy and protein, so there is an economic incentive for both the blender of the feed ration and the ultimate user to adopt a new technology to increase the efficiency of this phosphorus use if one is available. Phytase frees the phosphorus bound in feed grains making it more available to the animal and leads to less phosphorus being added to the animal ration. The use of phytase also reduces the amount of phosphorus in the manure—helping to curb the externality of excess phosphorus from manure working its way into streams or lakes. Because of the immediate economic advantage of this approach, it has been adopted by the feed industry and is a standard practice for over 90% of all nonruminant feed supplied to farmers in the midwestern United States. Interestingly, farmers are mostly unaware that they are even using phytase. Economics drove the widespread adoption, but it occurred at an early point in the supply chain so that farmer adoption was virtually assured and automatic. Phytase use is a case of economically induced technological innovation whose downstream effects literally yielded important social benefits rather than negative externalities.

## 7.6    CONCLUDING THOUGHTS AND POLICY IMPLICATIONS

Problems associated with economic and policy issues of phosphorus management in agroecosystems are wicked. There is not a common agreement on the problem or how pervasive it might be. Management challenges change over time, and few envision a definitive solution where this will no longer be a management issue. Experts are having to take account of the views of others whose concerns may be shaped more by perceptions or values rather than scientific knowledge. Social or political concerns, institutional structures, and legal actions may drive what is possible in terms of the management of phosphorus, as appears to be the case in the Florida nutrient example. There is no single silver bullet. In this kind of situation, economics cannot provide solutions that would be considered definitive either in terms of identifying the most cost-effective route or the route that would be universally accepted on the basis of multiple perceptions and values. Economic and policy trade-offs are numerous, complex, and imbued with a degree of uncertainty. The different policy approaches reviewed here have different outcomes and are more appropriate for some situations rather than others. Different policy instruments have different policy outcomes and affect stakeholders differently. In more limited-scale situations, economists should be able to identify and quantify economic trade-offs involved in alternative approaches to managing phosphorus in a way to reduce the negative externalities. Even for broad applications, like the management of phosphorus in Florida, the economist should be able to help others structure one or more decision frameworks covering the potential costs and trade-offs of different paths that might be taken to reduce the negative externalities that exist. This would have to take account of the many varied management requirements for point source and nonpoint source phosphorus from the many different actors and participants. Factors that are often unrepresented in engineering analyses like institutional and enforcement costs can sometimes turn a policy choice from one path to another. However, whatever might be done in terms of economics depends first upon accurate assessments by other key players. The biophysical relationships of the problem need to be known. This includes the interaction and the response of water bodies or the environment to phosphorus and the impact of those actions suggested to better manage phosphorus. If it is primarily a water problem, the hydrology needs to be known. Any technology being applied needs to be known as well—its efficacy, maintenance, longevity, etc. Wicked problems are by their nature interdisciplinary and require integrated analysis for solutions. So look to economics for guidance along the way, not for one definitive solution to problems of phosphorus management.

Policy makers are much more comfortable with silver bullet solutions—sometimes determined on the basis of a single number from a benefit/cost analysis. There is understandable reluctance to tackle the phosphorus management dilemma because it has the characteristics of a wicked problem, and there will be important complex trade-offs involved that are often no win politically. As with Lake Erie, the problem does not garner a political willingness to deal with it at the outset. The situation needs to reach near crisis dimensions before policy makers will act. To make life more difficult, a single-policy approach ends up having limited effectiveness as that policy is driven to the point where the low hanging fruit has been harvested and future gains are increasingly more expensive and less effective. Such multidimensional wicked problems require multiple policy approaches. Voluntary approaches will have to be supplemented by such things as regulation or, in the right circumstances, market-based approaches. This makes the creation of a politically consistent and acceptable solution all the more difficult. The dynamic nature of these problems of excess nutrients also means that monitoring and meaningful metrics will be critical—both biophysically and societally—to guide the policy. Enough must be known as the nutrient management process moves forward to adaptively manage and give up what does not work or is economically not cost effective. Policymaking will have to be made hand in hand with the winding course of the success or the failure of the activities on the ground. Finally the biophysically determined time lag in reducing excess nutrient problems poses a severe policy difficulty. Whatever policy is chosen is unlikely to demonstrate outstanding success by the next election. What will be required is the willingness of policy makers to innovate and experiment, to continue to be patient, to admit to what does not work, and to then try new approaches.

# REFERENCES

Antle, J., S. Capalbo, S. Mooney, E. Elliott, and K. Paustan. 2003. Spatial heterogeneity, contract design, and the efficiency of carbon sequestration policies for agriculture. *Journal of Environmental Economics and Management* 46 (4): 231–250.

Batie, S. S. 2008. Wicked problems and applied economics. *American Journal of Agricultural Economics* 90 (5): 1176–1191.

Benedict, M. R. 1953. *Farm Policies of the United States, 1750–1950.* New York: The Twentieth Century Fund.

Canchi, D. P., P. Bala, and O. Doering. 2006. *Market-Based Policy Instruments in Natural Resource Conservation.* Report for the Resource Economics and Social Sciences Division, Natural Resources Conservation Service. Washington, DC: U.S. Department of Agriculture.

Claassen, R. 2012. *The Future of Environmental Compliance Incentives in U.S. Agriculture: The Role of Commodity, Conservation and Crop Insurance Programs.* EIB-94. Washington, DC: Economic Research Service, U.S. Department of Agriculture.

Department of Sustainability and Environment. 2008. *Bush Tender: Rethinking Investment for Native Vegetation Outcomes: The Application of Auctions for Securing Private Land Management Agreements.* Melbourne, Australia: Department of Sustainability and Environment, State of Victoria.

Doering, O., and K. Smith. 2012. *Examining the Relationship of Conservation Compliance and Farm Program Incentives.* Washington, DC: The Council on Food, Agricultural and Resource Economics.

Fox, G., G. Umali, and T. Dickinson. 1995. An economic analysis of targeting soil conservation measures with respect to off-site water quality. *Canadian Journal of Agricultural Economics* 43: 105–118.

Kreuter, M. W., C. De Rosa, E. H. Howze, and G. T. Baldwin. 2004. Understanding wicked problems: A key to advancing environmental health promotion. *Health, Education, and Behavior* 31: 441–54.

National Research Council (NRC). 2012. *Review of the EPA's Economic Analysis of Final Water Quality Standards for Nutrients for Lakes and Flowing Waters in Florida.* Washington, DC: National Academies Press.

Ribaudo, M., J. Delgado, L. Hansen, M. Livingston, R. Mosheim, and J. Williamson. 2011. *Nitrogen in Agricultural Systems: Implications for Conservation Policy.* Report No. 127. Washington, DC: U.S. Department of Agriculture, Economic Research Service, Economic Research.

Segerson, K. 2013. When is reliance on voluntary approaches in agriculture likely to be effective? *Applied Economic Perspectives and Policy* 35 (4): 565–592.

Stahlman, M., and L. McCann. 2011. Technology characteristics, choice architecture, and farmer knowledge: The case of phytase. *Agriculture and Human Values* 29: 371–379.

Stephenson, K., and L. Shabman. 2011. Rhetoric and reality of water quality trading and the potential for market-like reform. *Journal of the American Water Resources Association* 47 (1): 15–28.

U.S. Environmental Protection Agency (EPA). 2011. *Reactive Nitrogen in the United States: An Analysis of Inputs, Flows, Consequences and Management Actions.* Washington, DC: EPA Science Advisory Board.

Woodward, R., R. Kaiser, and A.-M. Wicks. 2002. The structure and practice of water quality trading markets. *Journal of the American Water Resources Association* 38 (4): 967–979.

# 8 Phosphorus Fertilization and Management in Soils of Sub-Saharan Africa

*Andrew J. Margenot, Bal R. Singh,*
*Idupulapati M. Rao, and Rolf Sommer*

## CONTENTS

## 8.1   INTRODUCTION

Phosphorus (P) is an essential plant nutrient that is required for all major developmental processes and reproduction in plants (Marschner 1995). P is involved in plant energy relations and in the structure of nucleic acids. In soils, P is available to plants in the form of hydrated orthophosphate in the soil solution. Purple or bronze leaves are common P deficiency symptoms (Figure 8.1). In plants, reductions in leaf expansion (Rao and Terry 1989) and also number of leaves (Lynch, Läuchli, and Epstein 1991) are the most obvious symptoms of P deficiency. Compared with shoot growth, root growth is less inhibited under P deficiency, leading to a typical decrease in shoot/root ratio (Fredeen, Rao, and Terry 1989). P is also a major constituent of fertilizers required to improve and sustain crop yields. Low P soils support 70% of all terrestrial biomass and also account for over half of the total agricultural land (Lynch 2011). Soils containing insufficient amounts of plant-available P not only produce economically unacceptable yields but also decrease the efficiency of other inputs, particularly nitrogen (N). Thus, there is an urgent need to seek strategies by which P fertilizers can be used more effectively in those farming systems where P is currently deficient and where their use is economically feasible.

In sub-Saharan Africa (SSA), soil P availability is declining because of soil degradation. The low availability of soil P is a major constraint to agroecosystem productivity in SSA due to interrelated material (e.g., soil properties, crop germplasm), knowledge (e.g., input management), and socioeconomic (e.g., market access, input costs, best practices) factors. Material investments are necessary to overcome low P reserves resulting from weathering processes, strong cultivation intensity, and/or limited P additions. These material investments must be complemented by knowledge investments, the lack of which constrain and undermine resource use efficiency and farmers' capacity and willingness to make future investments. Such knowledge investments include management practices like application rates and methods, which in turn are supported by soil surveys and field trials to determine soil- and crop-specific P management. Both material and knowledge investments in P management face unique but interrelated challenges in SSA.

(a)

(b)

**FIGURE 8.1** **(See color insert.)** (a) Phosphorus-deficient 20-day maize exhibiting characteristic purple veining and stunting consistent with low available soil P (1.9 mg kg$^{-1}$ soil, Olsen test). (Courtesy of International Plant Nutrition Institute, Peachtree Corners, Georgia, http://media.ipni.net/.) (b) Phosphorus omission trial demonstrating severe stunting in the check (no P added; left) as compared to a 33 kg P ha$^{-1}$ treatment (right). (Courtesy of Dr. Thomas Morris.)

Low inherent nutrient reserves, especially for P in highly weathered soils that are prevalent in SSA, combine with soil degradation to contribute to food insecurity. For example, in East Africa, the majority of agricultural soils are estimated to experience P limitations, e.g., >50% in Tanzania (Okalebo et al. 2007) and 80% in Kenya (Jama and Van Straaten 2006). P deficiency limits the efficiency of other resources like N and water, which may often limit agricultural productivity in resource-poor regions of SSA. However, the projected increase in N inputs with stagnant P inputs is expected to increasingly unbalance the input N:P ratio in SSA agroecosystems, with negative effects on future crop yields (van der Velde et al. 2014).

In addition to replenishment of nutrients as a sound management practice, net addition of P is necessary—sometimes termed *investments in soil P capital* (Buresh, Smithson, and Hellums 1997; Sanchez et al. 1997; Sanchez 2002). Increasing cultivation intensity and nutrient export of long-cultivated and/or weathered soils with already low P availability is largely not being compensated by P inputs. For each ton of maize dry matter harvested, 2–4 kg of P is removed (Simpson, Okalebo, and Lubulwa 1996). The result is mining of soil P, which in SSA is estimated at 3 kg ha$^{-1}$ year$^{-1}$ (Stoorvogel, Smaling, and Janssen 1993; Mwangi 1996). While mining of soil P is generally lower than that of other nutrients like N, there is no opportunity for fixing P from an atmospheric pool. Indeed, the declines in crop productivity with continued cropping in regions like East Africa are largely attributable to negative P balances (Mnkeni, Semoka, and Buganga 1991; Bekunda, Bationo, and Ssali 1997). The strategic management of P including, external inputs, is needed to improve and sustain yields in SSA.

Nutrient balances alone cannot adequately describe P deficiencies, because a high proportion of soil P can be fixed in biologically unavailable forms (Smaling, Nandwa, and Janssen 1997). Such phosphate fixation capacities are another constraint and tend to co-occur in weathered soils with generally low P stocks. Acidity for such soils is also common, meaning that farmers often face P deficiencies in tandem with Al toxicity (Roy and McClellan 1986). High P saturation capacities of weathered soils compromise input-only approaches to P deficiencies.

The presence of regionally and locally available resources such as phosphate rock (PR), manure, and green manure crops provides lower cost and ultimately more realistic options for improving soil P availability. Recent geological surveys have identified P resources in East Africa, including PR and lime in southern Tanzania (Kalvig et al. 2012). These offer local inputs for regional use in a relatively unprocessed form that may remove the dependency on expensive processed and/or imported inputs. Yet P management in SSA has yet to address the master variable of soil fertility: pH. Given the strong pH control on P availability and fixation, and the common co-manifestation of Al toxicity in acid soils, we emphasize the potential of pH management as an indirect way of managing soil P in SSA. In contrast to other regions with similar soil constraints such as the Brazilian *cerrados* or the Colombian *llanos*, liming in SSA is relatively underutilized despite its proven track record to markedly increase the productivity of acid, weathered soils (Sánchez and Salinas 1981). Liming offers resource-strapped SSA farmers greater mileage on of limited P inputs and overall improvements in soil fertility from acidity amelioration (Margenot and Sommer 2014).

Strategic P management in SSA also requires investment in knowledge resources, such as mapping of soils and establishing P response trials to determine appropriate soil P test levels. These are the less commonly addressed challenges to P management specific to SSA. The unavailability and/or the affordability of analytical methods for (1) assessing the quality of low-cost, local alternatives to concentrated and soluble but high-cost P inputs like triple super phosphate (TSP) and (2) diagnosing P deficiency and fixation potential limits research and outreach in regions such as East Africa (Nziguheba 2007). Field data increasingly support the notion that one-size-fits-all approaches are not useful and are potentially problematic for the management of nutrients like P in a region as diverse in soils and climates as SSA. For example, a recent study dismisses the idea of a best P management when considering the net financial returns to farmers, instead highlighting the interactions of P input type (PR vs. TSP), other nutrient inputs, and the high variability of the agronomic response and the financial returns to the same P management strategies (Lamers, Bruentrup, and Buerkert 2015a,b).

Finally, knowledge capital applies to soil management. The high variability of soil conditions in SSA suggests that investments in soil P capital will be most successful if management recommendations and practices are regionally tailored to agroecological zones (Bationo et al. 2006). Broad generalizations are not helpful for the site-specific management that is needed. The diversity of soil conditions, compounded by management, encompasses different kinds of constraints and opportunities for increasing available soil P.

The objective of this review is to assess the developments in P management in SSA, with specific reference to the soil, crop, and economic conditions unique to the subcontinent. To this end, we provide a brief summary of the basic principles of P chemistry and plant uptake and how these principles play out under the confluence of soil conditions, resource availability, and managements. Furthermore, we strive to identify promising and underexplored approaches to P management.

## 8.2   PHOSPHORUS CHEMISTRY AND CYCLING IN SOILS

Phosphorus is the tenth most abundant element on earth yet the second most limiting nutrient across biomes (Walker and Syers 1976; Canfield, Erik, and Bo 2005). It is essential to life because of the structural (DNA, RNA, phospholipids) and functional (ATP, NADPH) roles it performs in cells. Inherent chemical properties make P unique in its biogeochemical cycling, particularly in comparison to the other major nutrients for ecosystem productivity. In contrast to other elements such as carbon (C), N, and sulfur (S), the challenges to monitoring biogeochemical transformations of P result from its unique chemistry: P has no gas phase, only one stable isotope ($^{31}$P), and exists almost exclusively in one oxidation state (+5) (Blake, O'Neil, and Surkov 2005). Due to its largely linear biogeochemical cycle, P is highly limited in weathered soils, which constitute a significant (29%) portion of soils globally (Sanchez and Logan 1992; Vitousek et al. 2010). Generally, plant-available inorganic P is scarce relative to organic P ($P_{org}$, 20–90% of the total soil P), through which the existing P in terrestrial systems is cycled (Harrison 1987; Celi and Barberis 2005; Jones and Oburger 2011; Turner and Engelbrecht 2011).

In soils, contrasting but coupled abiotic and biotic processes regulate P cycling and availability, including sorption to minerals and mineralization of organic P (Oberson et al. 2011). The mineral weathering of phosphates occurs at low rates that are generally insufficient to meet crop demand, exacerbated by P export via harvesting and losses (chiefly erosive). The result is that agroecosystems face simultaneous strong demands for P and limitations on its availability, a dynamic that is exacerbated in input-limited systems of SSA.

This section provides a brief overview of the soil conditions governing P availability. For further reading on P cycling and geochemistry, other reviews are available. As a starting point, the reader is referred to a review of P geochemistry during pedogenesis (Walker and Syers 1976) and a discussion of the drivers of P limitation in terrestrial systems (Vitousek et al. 2010). Other reviews are also recommended to understand the different aspects of organic P cycling (Turner, Frossard, and Baldwin 2005) and the role of biological drivers of P cycling for its management in agroecosystems (Oberson et al. 2011).

### 8.2.1 FACTORS CONTROLLING P AVAILABILITY IN SOILS

The chemistry of phosphate in soil involves interactions with mineral and organic soil components, which in turn govern its availability for crop uptake and mediate soil P dynamics in response to management practices like P inputs. The strong influence of edaphic properties, most notably interactions of pH and mineralogy, mean that soil P cycling and its availability to crops can strongly vary by soil type.

Soil P chemistry is dominated by pH-dependent binding processes and consequent precipitation of P (Sharpley 1995). Plants uptake phosphate $\left(PO_4^{3-}\right)$ from the soil solution in the form of orthophosphate: $HPO_4^{2-}$ or $H_2PO_4^-$ depending on the soil solution pH (Tisdale, Nelson, and Beaton 1985). As an anion, phosphate can bind to positively charged binding sites on the surfaces of soil minerals and soil organic matter (SOM). Strong binding of phosphate by cations and consequent precipitation are strong controls on P availability: $Fe^{2+}$ and $Al^{3+}$ at low pH and $Ca^{2+}$ at high pH (Figure 8.2).

Adsorption to mineral surfaces is influenced by pH, because soil solution proton concentration affects the surface charge of minerals and SOM and thus the number and strength of P-binding sites. For variable charge minerals, commonly present in weathered minerals (e.g., kaolinite), soil pH also strongly influences P sorption.

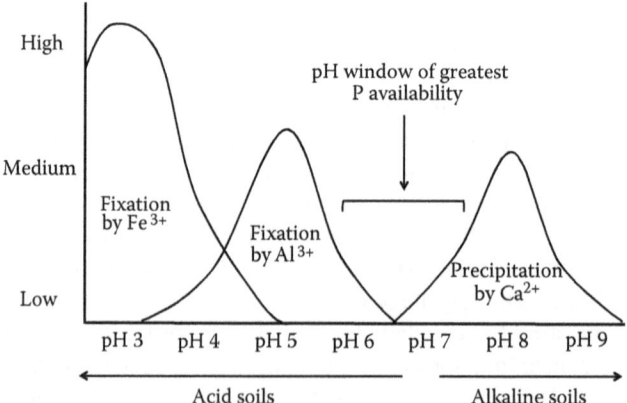

**FIGURE 8.2**  Influence of soil pH on soil phosphorus availability by binding of cations.

For example, the maximum adsorption of phosphate to kaolinite was observed at pH values of 4–5 and rapidly decreased above pH 6 (Chen, Butler, and Stumm 1973). Phosphate sorption to iron (hydr)oxide goethite is maximized at pH 5–6 and decreases to 60% of this value at pH 9 (Hawke, Carpenter, and Hunter 1989). Surface area effects are critical because they increase potential binding sites of plant-available inorganic P. For this reason, amorphous Fe and Al (hydr)oxides possessing greater surface area than crystalline forms have greater P retention potential. Likewise, SOM exhibits pH-dependent binding sites for phosphate and other anions, which would be expected to substantially influence P sorption at high SOM levels and low pH values. However, P sorption to SOM is thermodynamically weaker than its bonds with Al, Fe, and Ca.

## 8.2.2 ORGANIC P

In unmanaged and many managed systems, P in biomass and thus largely in organic forms is the chief input to soils. Organic P has a unique role in meeting crop P demand in tropical regions because of (1) the high potential turnover of organic P into plant-available inorganic P (mineralization) under warmer climates and (2) the lower fixation of organic P forms by mineral surfaces relative to inorganic P in weathered soils that are prevalent in such regions.

The mineralization of organic P can contribute up to 25 kg P ha$^{-1}$ year$^{-1}$ in mineral soils and as much as 160 kg P ha$^{-1}$ t year$^{-1}$ in organic soils (Sharpley 1995). Thus, enzymes performing mineralization of organic P, known as phosphatases, can significantly influence P availability (Vance, Uhde-Stone, and Allan 2003; Turner and Engelbrecht 2011). Phosphatases are able to mineralize P which is subsequently competed for by geochemical and biological sinks (Esberg et al. 2010). Weathered soils are generally P-limited and consequently display increased enzymatic pressure on P cycling (Stursova and Sinsabaugh 2008; Sinsabaugh, Hill, and Follstad Shah 2009; Sinsabaugh and Follstad Shah 2012; Waring, Weintraub, and Sinsabaugh 2014). Organic P tends to accumulate in soils largely as inositol P forms (Turner et al. 2002), the recalcitrance of which may reflect stabilization by mineral weathering products such as iron and aluminum oxides (Turner, Richardson, and Mullaney 2007; Turner and Blackwell 2013).

## 8.2.3 SOIL CONDITIONS RELEVANT TO P MANAGEMENT IN SSA

In general, P deficiency is one of the largest constraints to crop production in SSA agroecosystems owing to (1) low native soil P and (2) high P fixation capacity. This reflects pedogenic factors: a high degree of weathering and/or a low concentration of P in the parent material (van der Waals and Laker 2008; Sugihara et al. 2012). Ultisols and Oxisols therefore represent as much as 70% of P-deficient soils globally (Fairhurst et al. 1999). These two soil orders are estimated to constitute 20.5% of the total land area in Africa (Table 8.1). Soil acidity has a strong control on P availability with P availability maximized between pH 6 and 8 (Figure 8.2), yet less than 22% of African soils exhibit pH values in this range. Specifically, 14.7% of soils exhibit pH < 5.5, which favors P fixation by Al and Fe, and nearly half of all African soils are above pH 8.5, which favors P fixation by Ca (Table 8.2). However, the majority of these less weathered (Entisols, Inceptisols) and alkaline soils tend to occur in North Africa. As a result, the majority of soils in SSA, in particular cultivated areas, are generally weathered and low in pH (Figure 8.3).

In addition to pH and mineralogy, soil texture and morphology can influence P availability. For example, low-permeability horizons in the profile can prevent the downward movement of P inputs (Allen et al. 2006). Although P is generally considered less mobile than N, P translocation can be significant even in fine-textured soils. For example, on a clay soil, up to 45% of P applied at 50 kg P ha$^{-1}$ year$^{-1}$ for 4 years

**TABLE 8.1**
**Prevalence of Soil Orders in Africa, USDA Classification**

| Soil Order | Area (10³ km²) | Percentage of Total Land Area |
|---|---|---|
| Andisols | 49 | 0.2 |
| Histosols | 15 | 0.1 |
| Spodosols | 31 | 0.1 |
| Oxisols | 4389 | 14.3 |
| Vertisols | 990 | 3.2 |
| Aridisols | 8076 | 26.4 |
| Ultisols | 1906 | 6.2 |
| Mollisols | 70 | 0.2 |
| Alfisols | 3200 | 10.4 |
| Inceptisols | 2378 | 7.8 |
| Entisols | 7506 | 24.5 |
| Nonsoil surface | 2063 | 6.7 |

*Source:* Eswaran, H., R. Almaraz, E van den Berg, P. Reich. *Geoderma*, 1.77:1–18, 1997.

**TABLE 8.2**
**Estimated Distribution of Soil pH across Africa**

| pH | Area (10³ km²) | Perecentage of Total Land Area |
|---|---|---|
| <3.5 | 31 | 0.1 |
| 3.5–4.2 | 1193 | 3.9 |
| 4.2–5.5 | 3278 | 10.7 |
| 5.5–6.5 | 4306 | 14 |
| 6.5–8.5 | 6997 | 22.8 |
| >8.5 | 14,845 | 48.4 |

*Source:* Eswaran, H., R. Almaraz, E van den Berg, P. Reich. *Geoderma*, 1.77:1–18, 1997.

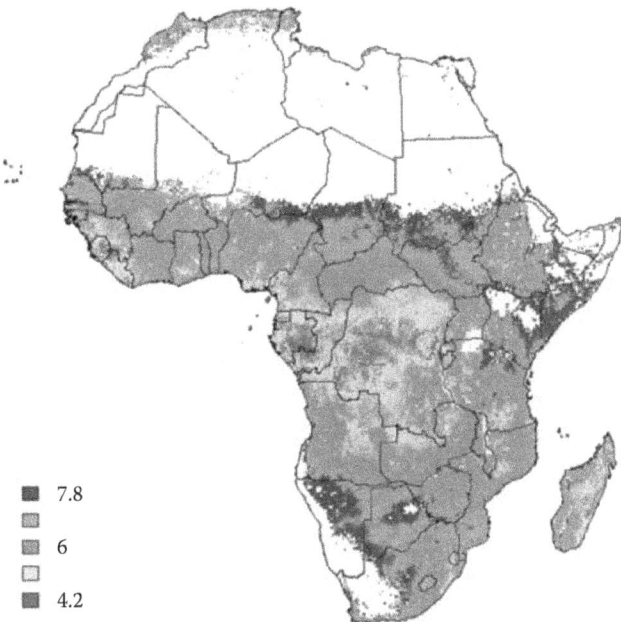

7.8

6

4.2

**FIGURE 8.3**    **(See color insert.)** Working map of predicted pH of surface soil (0–5 cm) developed by AfSIS. Note the prevalence of acidic (pH < 7) soils in SSA. (Courtesy of AfSIS, Wageningen, Netherlands, http://soilgrids.org.)

was not accounted for in 0–15 cm layer, and up to 69% for a sandy soil. The lower mineral oxide content in the sandy soil likely resulted in greater leaching (Sugihara et al. 2012). Such results suggest the need to literally dig deeper and consider soil depths beyond typical surface sampling depths to establish meaningful soil P budgets. Although the vertical distributions of total and labile P are generally surficial compared to other nutrients (Jobbágy and Jackson 2001), soils in the tropics, including weathered soil orders, can exhibit variable vertical P trends (Dieter, Elsenbeer, and Turner 2010).

### 8.2.4    Knowledge Constraints on Soils Affect P Management

The residual effect of P inputs and the effect on soil P by other parameters sensitive to land use history require improved soil maps and soil testing. Better recommendations can be supported by high-resolution soil maps, in conjunction with trials that calibrate P application rates to match specific locations with local recommendations. Here, the importance of soil maps is discussed; soil testing for P is addressed in Section 8.5.

As for many parts of the world, soil maps for SSA are often too low resolution to be of use for land management (Sanchez et al. 2009). Soils maps are developed at different scales for different purposes. For example, in USDA soil taxonomy, order 5 is 1:5,000,000 and order 1 is 1:1000. A map at 1:1,000,000 is limited to showing soil suborder groups, which can be useful for showing regional soil constraints to agriculture (Soil Survey Staff 2015). Order 1:190,000 is commonly used for county-wide projections in the United States for agricultural mapping. The small size and heterogeneity that defines SSA smallholdings (Vanlauwe et al. 2014) as well as their strong management-induced soil fertility gradients (Tittonell et al. 2013) suggests the need for high-resolution soil maps, with emphasis on soil properties useful to and/ or influenced by management, such as cation exchange capacity (CEC), pH, and C content (Sanchez et al. 2009). Thus, at the scale of smallholder agriculture (<2 ha) in East Africa, conventional soil maps at order 1 are appropriate.

Nevertheless, legacy maps are often the only soil maps available for many regions of SSA. For example, soil mapping done in 1980 (Sombroek, Braun, and van der Pouw 1982) is used as a basis for soil characterizations in Kenya, despite the 1:1,000,000 scale used. As Sombroek, Braun, and van der Pouw (1982) pointed out, this soil map was exploratory and meant to provide general information on the land use potential. However, their maps continue to be the basis for many soil descriptions, e.g., the Acrisol–Ferralsol association in western Kenya. Such low-resolution maps and broad soil classification are not helpful for practical soil fertility management.

To this end, rapid advancements are being made on techniques and approaches to construct high-resolution soil maps for SSA. The African Soil Information Service (AfSIS) develops digital soil maps for 17.5 million $km^2$ across 42 nations on the continent. With a resolution of 1000–250 m, these soil maps offer improvement over legacy maps and provide information on currently unmapped regions. AfSIS and its collaborators employ remote sensing (satellite-based) measurements and legacy soil profiles ($n = 12,000$) for a total of 28,000 sampling points, in tandem with regression techniques (Vågen et al. 2010; Leenaars 2013). A related effort is the infrared-based prediction of soil properties to provide maps based on soil C, pH, or clay content (Figure 8.3) (Sanchez et al. 2009). Like remote sensing, these approaches are regression based. A trade-off of their rapidity is the limitation to maps of surface soils.

Though digital maps can provide accurate predictive maps of certain soil param-eters useful for basic fertility baselines (e.g., clay content, CEC, pH) (Minasny et al. 2009; Terhoeven-Urselmans et al. 2010), the prediction of soil P is more difficult. Infrared spectroscopy is well suited to predicting soil properties relevant to P man-agement (pH, mineralogy, P fixation capacity), but less so to P pools, in particular labile P (Bogrekci and Lee 2005; Hu 2013; Soriano-Disla et al. 2013). Interactions of P with soil components means that its availability can be decoupled from C, in contrast to N (Turner, Frossard, and Baldwin 2005) and is often fundamentally different from other nutrient cations like potassium (K), creating a challenge for its spatial prediction relative to other nutrients. For example, across 385 smallholdings in Niger involved in yam cultivation and encompassing a diversity of management practices, the spatial modeling of N and K could successfully provide site-specific recommendations for these nutrients (Jemo et al. 2014). In contrast, soil P was poorly spatially modelled, and as a result, the management recommendations for P were limited to regional scale, in contrast to N and K.

## 8.3  SSA VERSUS GLOBAL P FERTILIZER USE AND TRENDS

The application of concentrated fertilizers to soils in Africa has been very low in comparison to other regions of the world (Figure 8.4). Fertilizer use, in the continent, especially P and K, has been stagnant in the last three decades but has markedly increased in other regions of the developing world, such as East and South Asia and Latin America. For example, by the turn of the century, fertilizer use in Africa was 8 kg ha$^{-1}$, compared with 96 kg ha$^{-1}$ in East and Southeast Asia and 101 kg ha$^{-1}$ in South Asia (Morris 2007). The consumption of P fertilizers in all SSA countries except Ethiopia has generally decreased or stagnated in the past decade and, in some countries such as in Uganda and Mozambique, remains very low (Table 8.3).

The limited use of fertilizers is determined by a variety of reasons. These include high purchasing costs, especially after market reforms removed subsidies in African nations, inefficient marketing systems, and restricted markets for outputs that constrain investment opportunities (Bekunda, Sanginga, and Woomer 2010; Sommer et al. 2013). Additional constraints on the profitability of fertilizer use include low access to agricultural technologies like seeds and irrigation. Transportation costs can also be a significant obstacle to fertilizer accessibility. As a result, in nations like Nigeria, the reduction of transportation costs is estimated to have a greater effect on fertilizer profitability than fertilizer subsidies (Liverpool-Tasie et al. 2015).

The trend of low fertilizer use in SSA has been partially reversed with the introduction of targeted subsidies by African governments such as Malawi and Kenya (Sommer et al. 2013). As a result, the prospects of the increased use of mineral fertilizers in such nations are promising. In 2013, it was estimated that African governments allotted nearly US$ 1 billion annually to fertilizer subsidies (Jayne and Rashid 2013). If the average application rates of inorganic fertilizer in SSA rose to 50 kg ha$^{-1}$, from 10 kg ha$^{-1}$ currently, there would be a substantial impact on agricultural yields (Larson and Frisvold 1996). However, recent work highlights the significant role of the illicit use of funds in smart subsidy programs in overestimating fertilizer use in case study nations of Kenya, Malawi, and Zimbabwe (Jayne et al. 2013). Additionally, conventional benefit-cost analyses tend to overestimate

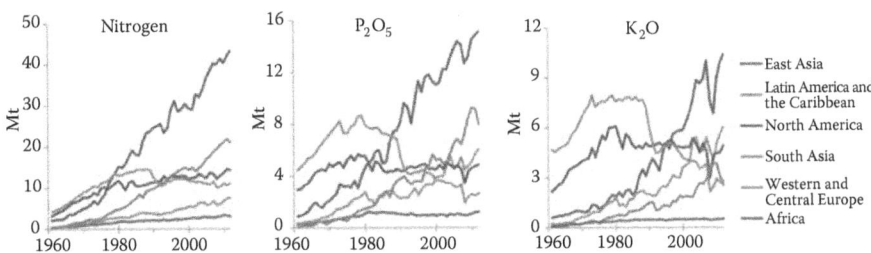

**FIGURE 8.4**  (**See color insert.**) Total annual nitrogen, phosphate (P$_2$O$_5$ equivalent), and potassium (K$_2$O equivalent) fertilizer consumption (Mt) from 1961–2015 for various regions of the world. (Courtesy of International Fertilizer Industry Association, Paris, France, http://ifadata.fertilizer.org/ucSearch.aspx, 2015.)

**TABLE 8.3**
**Phosphorus ($P_2O_5$ Equivalent) Fertilizer Consumption ($10^3$ t year$^{-1}$)**
**in Selected SSA Countries**

| Country | Year | | | | |
|---|---|---|---|---|---|
| | 2002 | 2004 | 2006 | 2008 | 2010 |
| Tanzania | 5.28 | 9.73 | 11.55 | 9.26 | 8.06 |
| Zambia | 13.47 | 11.41 | 8.55 | 8.89 | 8.43 |
| Zimbabwe | 48.09 | 20.41 | 45.30 | 35.18 | 48.08 |
| Ethiopia | 69.98 | 45.60 | 64.48 | 122.25 | 162.06 |
| Uganda | 2.13 | 3.30 | 2.17 | 5.49 | 2.81 |
| Mozambique | 2.00 | 3.47 | 2.84 | 10.36 | 2.59 |

*Source:* Food and Agriculture Organization of the United Nations—Statistics Division, Rome, Italy, http://faostat.fao.org/.

the benefit of such subsidy programs by not taking into account the effects on local fertilizer markets (Jayne et al. 2013).

## 8.4   TYPES OF P INPUTS

### 8.4.1   INORGANIC P SOURCES

Inorganic P inputs consist of PR mined from geological deposits or products derived from PR. The acidulation of PR provides a more concentrated and soluble P input, such as TSP. Total and highest-quality PR deposits are globally concentrated in North Africa (Morocco), although there are local deposits of lower but appreciable quality and quantity throughout SSA. PR is a lower-cost alternative to more soluble inputs like TSP, which are more expensive. Unawareness of the proper use of high- and low-cost inputs can discourage their use. The dissolution of PR translates to slower crop response to lower-cost P inputs, typically at least 3 years at application rates affordable to smallholders, which can be long enough to discourage their continued use (Bationo, Mughogho, and Mokwunye 1986). Other problems with the use of PR may include bulkiness, low market availability, and high content of heavy metals (Vanlauwe and Giller 2006).

The strong residual effect of P inputs, especially PR, highlights both the longer-term importance of P management and concurrent soil testing to understand soil P fertility changes. The strong residual effect of P is the basis for the concept of *investments in soil P capital*, as changes in pH and P stocks improve P fertility as a function of time. The profitability of repeated inorganic and organic additions of P has been shown to gradually increase over the years (4+) (Lamers, Bruentrup, and Buerkert 2015b).

The majority of studies on P inputs in SSA are short term (<4 years) and thus do not sufficiently address medium- to long-term effects of certain P management strategies. This can potentially create bias in P fertilization recommendations.

Given the propensity of P to accumulate in soils, this has important implications for effective and economical P use. For example, in western Kenya, the national extension services (Kenya Agricultural and Livestock Research Organisation, KALRO, formerly abbreviated as KARI) promotes an application rate of 40 kg P ha$^{-1}$ year$^{-1}$ based on 20 kg P ha$^{-1}$ per maize crop, which is in close agreement of the estimated optimal returns at 38 kg P ha$^{-1}$ year$^{-1}$ (Kihara and Njoroge 2013). These rates are likely to satisfy crop requirements within several years, necessitating lower maintenance rates (Woomer 2007). Preliminary results from a long-term trial by the International Center for Tropical Agriculture (CIAT) on an Oxisol in western Kenya showed that P additions in the form of PR and TSP at these rates for 11 years increased the labile P levels by 2.5-fold greater than maize minimums (Margenot et al. 2014).

### 8.4.1.1    Phosphate Rock: The Basis of Concentrated P Inputs

PR refers to a broad class of phosphate-rich minerals that typically contain calcium and fluoride, and exhibit high diversity in mineralogy, geographical distribution, and value for agricultural management of P fertility (Notholt 1994). The majority of global PR deposits are sedimentary (80–85%), followed by igneous (17%), which are more suitable for P fertilizers than the remaining metamorphic deposits (Sabiha et al. 2009; Notholt 1994). PR can be directly applied to soils or used as the feedstock for concentrated and soluble P fertilizers, by treatment with sulfuric acid or phosphoric acid to produce SSP and TSP. Over 80% of mined PR is used for the manufacture of water-soluble P fertilizers (Pur'Homme 2010). The net effect of PR on soil P and crop performance depends on the interaction of PR quality (e.g., P content, solubility) and soil conditions (e.g., pH, mineralogy, and soil Ca and P concentrations).

### 8.4.1.2    PR Chemistry and Quality

The quality of PR is important for its use as a direct amendment and is the result of its chemical composition and surface area. The chemical composition dictates the potential P input (concentration) and the availability over time (solubility) (Chien, Hammond, and Leon 1987). Sedimentary PRs generally consist of apatite with isomorphic substitution for phosphate by carbonate and fluoride (i.e., carbonate apatites, fluoroapatites). Important for their direct application as P fertilizers, sedimentary PRs express high net surface area due to microcrystallinity and internal surface area, which facilitates greater solubilization in soils relative to coarsely crystalline igneous and metamorphic deposits, despite higher P content of igneous PR (Khasawneh and Doll 1979). Sedimentary calcium apatites are economically viable as fertilizers; in addition to providing P, they also are a significant source of Ca, which in soils of high exchangeable acidity serves as a liming agent.

PR is effectively water insoluble (Bolan and Hedley 1989, 1990), and its solubilization in soils is made possible by soil conditions like pH and sinks for dissolution products, chiefly Ca and P. The congruent dissolution equation of PR idealized as fluoroapatite (Equation 8.1) highlights the importance of soil pH and Ca and P sinks as the three drivers of this process (Chien 1977a; Robinson and Syers 1990; Rajan, Watkinson, and Sinclair 1996). The percentage of Ca saturation, the P sorption capacity, and the Ca-exchange capacity can be used to predict the extent of the PR dissolution in soils (Mackay et al. 1986).

$$Ca_{10}(PO_4)_6F_2 + 12H_2O \rightarrow 10Ca^{2+} + 6H_2PO_4^- + 2F^- + 12OH^- \qquad (8.1)$$

A sufficiently high Ca content can inhibit PR dissolution despite a low pH, because $CaCO_3$ in the apatite lattice preferentially dissolves relative to $Ca_3(PO_4)_2$ (Robinson and Syers 1990). Evidence suggests the Ca sink is the most significant control on PR dissolution (Robinson and Syers 1990). At ideal conditions of low pH (4.5) and strong Ca and P sinks, the dissolution of PR can be nearly complete (95%) within 44 days (Robinson and Syers 1990). In a review of SSA field experiments comparing PR and soluble P fertilizers like TSP, Szilas et al. (2007) identifies pH and soil P availability as the best predictors of yield response to PR but noted that less acidic soils can still benefit from PR application if there is a strong P sink (i.e., low soil P).

Given the low base saturation and high exchangeable acidity of acid, weathered soils, it is not unexpected that studies show that these soils strongly benefit from PR application. Acid, weathered soils display the greatest PR dissolution, which consequently increases crop available soil P (Kanabo and Gilkes 1988). In fact, the high Ca content of PR makes it a high-quality Ca amendment (Khasawneh and Doll 1979), which is critical for weathered soils with inherent low Ca status further exacerbated by cultivation (Sale and Mokwunye 1993; Vitousek et al. 2010). Ca deficiency is only now being recognized as a limiting factor in many SSA soils. In regions like western Kenya, Ca deficiency may match P as a constraint on yields (Kihara and Njoroge 2013).

PR thus offers a two-for-one benefit from the P management perspective: in addition to providing P, it can ameliorate soil acidity, which typically accompanies and enforces P deficiencies (see Section 8.4.6). The extent of the increase in crop available soil P with PR reflects the decrease in P fixation via a liming effect and net P addition. The liming effect of PR results from proton consumption and addition of base cation of $Ca^{2+}$, and particularly for sedimentary PR, $CO_3^-$. PRs can have calcium carbonate equivalency (CCE) values of >50% (Sikora 2002). For example, the northern Tanzania Minjingu PR (MPR) can have up to 68% CCE, making it a low-grade liming agent (Nekesa et al. 2005). The liming effects of PR also reflect its particle size (Tisdale, Nelson, and Beaton 1985), known as the fineness factor (FF) (Peters 1996).

As new PR deposits are explored in SSA, their analysis should be standardized to ensure the comparability of studies. Laboratory measurements are helpful to accurately determine the PR liming potential based on specific PR and soil properties. In general, the phosphate content is a more sensitive proxy for predicting the liming potential of PR than directly quantifying the carbonate content because of its greater relative molar quantity.

On the other hand, the liming potential of PR makes it less ideal for circum-neutral or alkaline soils to address P deficiencies and potentially decreases its efficacy in initially acid soils with repeated application over time as the liming effect increases pH. The decreases in exchangeable acidity tend to be greater than the pH increases because of the high Ca loading from PR. Since exchangeable $Al^{3+}$ markedly decreases at pH > 5.5 in weathered soils, and because short-term (1–3 years) pH increases following PR applications at agronomic rates are typically +0.5 pH units (Pearson 1975), decreases in $Al^{3+}$ from PR can be appreciable in soils with pH ~5.5 (Chien and Friesen 2000).

These chemical principles have been corroborated by field experiments in SSA. A compilation of crop yield response to MPR versus soluble P fertilizers like TSP finds (1) a lag in P availability for PR due to dissolution, which entails (2) 74% relative yield with MPR compared to soluble P fertilizers in the first year, but (3) an increase in MPR relative yield to 94% in year 2, and 104% in year 3 and thereon (Szilas, Semoka, and Borggaard 2007). The overyield effect of PR relative to soluble P fertilizers after multiple seasons is attributable to the addition of $Ca^{2+}$ and the decrease in exchangeable acidity from PR.

### 8.4.1.3   PR Deposits in SSA

SSA constitutes 22% of earth's land area but contains only 2% of known PR deposits (Sheldon and Davidson 1987). These deposits are of variable reactivity and size (Roy and McClellan 1986), and there are likely additional undiscovered PR deposits due to the inadequate exploration of deep soils in the subcontinent (Sheldon and Davidson 1987). Due to their extent and the quality of deposits, East Africa has two highly used PR deposits: Minjingu (Tanzania) and Busumbu (Uganda). Additional deposits include Tilemsi (Mali), Matam (Senegal), Dorowa (Zimbabwe), and Sukulu (Uganda) (Van Kauwenbergh 1991). An excellent overview of PR resources in SSA is provided by van Straaten (2002).

Of all PR sources in SSA, only those from Tilemsi (Mali), Matam (Senegal), and Minjingu (Tanzania) are known to have a solubility in 2% citric acid exceeding the 10% threshold for agronomically significant dissolution (Vanlauwe and Giller 2006). MPR is the most commonly studied PR in SSA and is a relatively large (10 million tons) and high-quality (5.6% neutral ammonium citrate solubility) deposit (Van Kauwenbergh 1991; van Straaten 2002; Jama and Kiwia 2009). The increasing abundance of agronomic trials examining the effectiveness of various PRs makes the compilation of these data a useful resource because field trial data are needed to complement lab analyses of chemical composition.

### 8.4.2   Strategies to Improve P Solubility of PR

The low solubility of PR means that its ability to meet crop P demand can be limited, particularly at low application rates and/or in initial seasons. Various chemical, physical, and biological approaches can be used to improve solubility.

### 8.4.2.1   Biological Approaches

Composting with PR is one of the pillars of integrated soil fertility management (Sanginga and Woomer 2009), but the empirical data from SSA reviewed by Vanlauwe and Giller (2006) suggest no benefit to P availability in the short term. The increases in available P from co-composting PR reflect P mineralized from the organic matter (OM) rather than increased dissolution of PR. Using $^{32}$P-labeled synthetic (i.e., francolite) and mined PR (labeled by neutron bombing) allowed P from PR to be distinguished from P mineralized from manure. Isotope labels demonstrated no significant increase in PR dissolution during composting (Mahimairaja, Bolan, and Hedley 1994). This could be for two reasons: First, acidification during composting and even nitrification from added ammonium is of insufficient magnitude to significantly increase PR dissolution (Mahimairaja, Bolan, and Hedley 1994; Mowo 2000). Compost products express near-neutral pH not favorable for increasing PR dissolution (Vanlauwe and Giller 2006). Second, organic materials used in composting, in particular manure, have high Ca concentrations which thermodynamically disfavor PR dissolution (Equation 8.1) (Robinson and Syers 1991). From a P availability standpoint, PR application is most effective when directly added to soils, especially those with lower pH values. Despite this, the inclusion of PR in composting is still proposed as a means to enhance the PR improvement of soil P at the extension level, such as in Mali and Burkina Faso (Vanlauwe and Giller 2006). Although there are logistic advantages to combining inputs in a single application, these practices do not improve PR as a P input and may be a misuse of PR if the target soil exhibits low pH, as this favors for dissolution of PR applied directly.

Additional biological approaches to improving PR solubility involve phosphate-solubilizing microorganisms (PSM) and arbuscular mycorrhizal fungi (AMF) to scavenge P (Maheshwari 2011). Fomenting mycorrhizal associations can improve P uptake of low native soil P and added PR alike. In acid, weathered soils, AMF alone can also improve the P uptake from soluble P inputs such as $Ca(H_2PO_4)_2$ (Cozzolino, Di Meo, and Piccolo 2013), which is consistent with evidence that AMF improve P acquisition regardless of the source in weathered soils (Alloush, Zeto, and Clark 2000). AMF associations are most beneficial at low soil P levels and under conditions of low soil disturbance such as reduced tillage (Grant et al. 2005). Inoculation with PSB and PSM can improve the availability of added PR added (Miyasaka and Habte 2001). Generally, inoculation with PSM improves the availability of PR, although weaker effects on plant growth occur on more strongly P-fixing soils (Osorio and Habte 2015). AMF can desorb P from mineral surfaces by secreting organic chelators such as oxalic acid, although the efficacy of this strategy depends on the mineral type (Osorio and Habte 2013a). The low desorption of P by AMF from allophane suggests that such mechanisms are least effective in volcanic soils, followed by weathered soils with iron oxides and kaolinite, and most effective in less weathered soils dominated by 2:1 phyllosilicate minerals such as montmorillonite.

However, AMF contributions to plant P uptake are typically observed at high P application rates (Bâ and Guissou 1996; Alloush, Zeto, and Clark 2000; Satter et al. 2006), which may not be feasible for smallholders. For example, inoculation of *Leucaena leucocephala* with AMF (*Glomus fistulosum*) significantly increased its P uptake at PR application rates equal to or greater than 300 mg P kg$^{-1}$ soil in an acid, weathered soil (Typic Haploustox, pH 4.9) (Osorio and Habte 2013a). Assuming a bulk density of 1.10 g cm$^{-3}$ and an incorporation depth of 0.25 m, this translates to a minimum application rate of 825 kg P ha$^{-1}$.

Coupling PSM with arbuscular mycorrhizae can improve the plant response to PR, highlighting the importance of soil–microbe interactions for PR dissolution. Many PSM are rhizosphere-associated and can increase the plant P supply by secreting organic acids to solubilize plant-unavailable P pools, including fixed P (Fe- and Al-oxide bound) and Ca–P in parent materials or PR. An excellent overview of biological strategies to improve the P availability in weathered tropical soils is provided by Oberson et al. (2006).

Using PR can alter the populations of native soil PSM, with potential conse- quences for the PR efficacy over long-term management. The application of P at $40 \, \text{kg}^{-1} \, \text{ha}^{-1} \, \text{year}^{-1}$ as MPR increased fungal diversity and the PSM population by up to +90%, whereas TSP application significantly reduced overall bacterial diversity and the PSM populations of PSB by up to −69%, with comparable yield increases (Waigwa, Othieno, and Okalebo 2003).

### 8.4.2.2 Physical and Chemical Treatments

Grinding to increase surface area and partial acidulation to improve the solubility can improve the efficacy of PR in the short term (initial season), albeit at the cost of a lower residual effect (Chien et al. 1990). In general, finer texture entails greater surface area and thus dissolution rates for a given PR. The benefits of partial acidulation on solubilizing PR depend on its particular composition (Chien et al. 1990). For example, the acidulation (50% with $H_2SO_4$) of Tahoui PR (Niger) did not improve its solubility because of its Al and Fe contents, in contrast to Parc PR (Niger), which exhibited com- parable efficacy as SSP (91%) in terms of millet yield in the first and second seasons following application. Acidulation may not be necessary for PR with high reactiv- ity. The partial acidulation of MPR did not improve clover biomass on a Vertisol in Ethiopia (pH 5.5), but it did for low-reactivity Chilembwe PR (CPR; Zambia) (Haque, Lupwayi, and Ssali 1999). This was attributed to twofold greater soluble P content of MPR and high surface area typical of sedimentary deposits, in contrast to the igneous- origin CPR of lower P solubility and surface area (see Section 8.4.1.2).

Reduction–oxidation (redox) dynamics during wet–dry cycles can be exploited as a mechanism to manage P, particularly in lowland rice. For example, the reduc- tive dissolution of $Fe^{3+}$ under anaerobic conditions of soil submergence led to the release of Fe-bound P (Amery and Smolders 2012). The magnitude of the effect of redox cycles on freeing sorbed P, and on potential P loss, is influenced by not only the iron oxide content, but also the soil texture and the duration of these wet–dry cycles. Under the anaerobic conditions that occur in flooded soils, OM decomposi- tion can further drive Fe reduction. Applying this concept, OM additions to flooded, weathered soils markedly increased the available P relative to flooding alone (Amery and Smolders 2012). Coarser-textured soils are more prone to leaching of desorbed P but, on the other hand, are less likely to be waterlogged as compared to fine-textured soils (Scalenghe et al. 2014). The reduction of $Fe^{3+}$ via frequent wet–dry cycles used in the system of rice intensification (SRI) has been used to explain its high success on acid, mineral oxide-rich soils such as in Madagascar (Dobermann 2004), which could include effects on P availability. Furthermore, the diffusion of phosphate in submerged soils toward the crop roots is increased by a decrease in the diffusion impedance factor (Jungk and Claassen 1997), as well as potentially mineralizable organic P.

As a result, PR can be used in submerged acid soils as in lowland rice agroecosys- tems in West Africa (Nakamura et al. 2013). Field trials have found that PR is able to improve rice yield, consistent with previous studies of the solubility of PR under submerged acid (pH 4.8) soils (Chien 1977b). As in nonsubmerged soils, the interplay of the PR dissolution rate and the sorption of newly available P account for a net change in available P. Consequently, highly P-fixing soils cultivated under submerged conditions may not experience increases in available P despite the dissolution of PR (Yampracha et al. 2006). Co-additions of OM may aid in the dissolution of PR in submerged soils via the chelation of soluble cations by organic acids. For example, the co-addition of legume biomass (mucana, cowpea) with Ogun PR (Nigeria) in acid soils (pH 5.2–6.6) significantly improved water-soluble P (Olajumoke Adesanwo, Adetunji, and Diatta 2012), although as discussed in Section 8.4.2.1, this may reflect P mineralization from residues.

Combining PR with soil-acidifying inputs (e.g., ammonium and urea) can improve the dissolution of PR via proton-producing processes such as nitrification (Akande et al. 2004; Vanlauwe et al. 2006). PR can also be mixed with zeolites, aluminosilicate minerals that serve as a sink for $Ca^{2+}$ to drive its dissolution. The coapplication of PR with zeolite can produce comparable soil available P increases and maize biomass with 25% less PR application (Aainaa et al. 2014).

## 8.4.3  SOIL PROPERTIES: INTERACTIONS AND TRADE-OFFS FOR INORGANIC P INPUTS

Understanding the residual effects of P inputs is important for P management, because the effects of such inputs carry over into the second, third, and even fourth cropping seasons (Delve et al. 2009). The residual effect of P reflects strong interactions with soil properties such as P fixation capacity and texture (see Section 8.2.1). The high solubility of super phosphates may be a disadvantage in soils of coarser texture in which P leaching can limit the residual effect (Ojo, Akinrinde, and Akoroda 2010) and may partly explain why yields increase over time with PR relative to super phosphates. Slower dissolution of PR means that available soil P and crop response can lag behind for PR for 2–3 years following application compared to more soluble inputs (Msolla, Semoka, and Borggaard 2005). The modeling of residual P explained the crop response to continued P inputs over 14 years, which more strongly influenced maize yield than N application by years 3–4 (Janssen 2011).

The residual effect of PR offers a way for resource-poor smallholders to manage limited cash and/or unstable market prices, since it can extend P availability across years when P inputs are not possible. An application of 60 kg P ha$^{-1}$ as MPR led to an initial increase in pH, available P, and yield that declined to preapplication levels by the fourth growing season (Ndung'u et al. 2006). Soils with low P and associated soil conditions disfavoring its availability such as high exchangeable acidity may therefore need prolonged P investments before less frequent and/or lower maintenance applications can be used.

The high concentrations and solubility of P in these input types means that on P-deficient soils, application rates as low as 4.4 kg P ha$^{-1}$ can be profitable to farmers (Njui and Musandu 1999). At higher P application rates, N can become co-limiting and decrease PUE (see Section 8.6.2). For example, Njui and Musandu (1999) found that applications beyond 26.2 P kg ha$^{-1}$ on an P-deficient soil in western Kenya (pH 4.5, Mehlich 1.7 mg P kg$^{-1}$) required 40 kg N ha$^{-1}$ to produce a P response. In a recent assessment of maize response to P fertilizers in western Kenya across 25 studies and 126 fields, Kihara and Njoroge (2013) identified 50 kg ha$^{-1}$ year$^{-1}$ as the upper limit for yield response, and 38 kg ha$^{-1}$ year$^{-1}$ as an optimal rate. Approximately 50% lower yields were observed for unfertilized soils relative to P-fertilized treatments.

Interactions of P inputs and soil properties engender management trade-offs across time and space. Spatially, P inputs can be applied via placement or broadcasting. The placement of P, also known as microdosing, bottle cap method, or Coca-Cola technique (Tabo et al. 2007; Twomlow et al. 2011), is a better option when less P fertilizer is available. Manual placement on a per-plant basis produced higher yields at rates of <50 kg P ha$^{-1}$ as TSP (van der Eijk, Janssen, and Oenema 2006). However, since plant location typically changes with each season, the coincidence of subsequent plantings on previous placements is not guaranteed, meaning that residual P effects are typically lower for placement versus broadcast. For this reason, the placement of lower P applications across growing seasons can be more effective than a single large placement in the first season.

Broadcasting is effective at high application rates, and depending on P input and soil type, there may or may not be advantages to applying P in a single higher application versus repeated lower applications across seasons. Using less-soluble inputs like PR makes concentrating application in a single season more optimal for crop response (Lamers, Bruentrup, and Buerkert 2015b). In contrast, highly soluble inputs like TSP generally show fewer differences between single high applications and repeated lower applications across seasons. Greater soil contact by broadcasting promotes PR dissolution, but high solubility of super phosphates can lead to P fixation. This may explain why single high doses of PR have been found to provide the greatest economic return relative to other P input types and application strategies (Chien et al. 1990; Lamers, Bruentrup, and Buerkert 2015b). Microdosing with PR may be less effective than with super phosphates, presenting a trade-off for PR between the amount applied and the application area. For example, broadcasting PR at 13 kg P ha$^{-1}$ showed comparable economic returns on millet in the Sahel relative to microdosing in mounded hills at 4 kg P ha$^{-1}$ (Lamers, Bruentrup, and Buerkert 2015b).

Input distribution across time is less significant at high application rates because sufficient crop available P will be solubilized from a larger PR dose. Over the course of 5 years, the application of a single 250 kg P ha$^{-1}$ in the first year or 50 kg P ha$^{-1}$ annually in the form of PR or TSP provided comparable maize yield and economic returns (Jama and Kiwia 2009). Because of its stronger residual effect, high PR application in the first year can be tapered to a low maintenance rate (25% of initial application) within 2–3 years, or alternatively, supplemented with super phosphates (Bonzi et al. 2011).

Practical consideration should also be given to the temporal distribution of P inputs. Lower and frequent P inputs are favored under conditions of competing demands on cash, labor, and time requirements, and avoid triggering strong weed competition observed under high applications (van der Eijk, Janssen, and Oenema 2006).

## 8.4.4 Organic P Resources

High quality OM amendments can provide comparable or superior improvements in soil P availability relative to inorganic P inputs. The benefits of organic inputs for soil available P include slow release of mineralizable organic P, increased microbial P cycling, and indirectly, improvement of non-P properties such as soil tilth and micronutrient additions. A general disadvantage of organic P inputs is low P content, both plant-available inorganic P and total P, and the unpredictability of organic P mineralization. Organic P inputs are therefore generally more complicated when considering short-term P management. Additionally, the availability of OM inputs in many regions of SSA is typically insufficient to meet crop P demands.

### 8.4.4.1 Mechanisms by Which Organic Inputs Improve P Availability

There are five potential mechanisms by which organic inputs can improve soil P availability (Guppy et al. 2005a,b; Iyamuremye and Dick 1996; Oberson et al. 2006, 2011): (1) inducing ligand exchange (i.e., organic sorbates competing with P for mineral binding sites); (2) buffering soil acidity; (3) providing a large sink for exchangeable P; (4) providing a potential P source via mineralization; and (5) stimulating biological cycling of P, chiefly via microbial drivers.

Guppy et al. (2005b) argue that at realistic application rates, OM inputs such as manure increase available soil P due to the addition of P in OM rather than organomineral interactions such as ligand exchange. For example, mesocosm studies typically use low-weight organic acids as model OM in concentrations of 3–5 orders of magnitude greater than in field experiments. Although OM may effectively compete with sorption sites, it can also increase the sorption of P to its chelated metal cations. Similarly, increases in available P during co-composting of PR with OM largely reflect the mineralization of organic P (Mahimairaja, Bolan, and Hedley 1994).

Nonetheless, there are improvements to P availability that can be achieved through the use of OM inputs like residues. The application of plant biomass from species like tithionia (*Tithonia diversifolia*) can produce increases in soil available P and maize yields comparable to inorganic P inputs (Jama et al. 2000; Nziguheba et al. 2000) while uniquely reducing P fixation (Nziguheba et al. 1998). Ameliorating P fixation with OM is thought to reflect decreases in binding sites and/or exchangeable acidity as a result of OM complexation of exchangeable $Al^{3+}$ and consumption of $H^+$ via decomposition processes and OM buffering.

Biological cycling of P can play a significant role in provisioning plant-available P, in particular for weathered soils with high potential for fixation. Organic amendments can affect biological P cycling through microbial biomass P (MBP) and enzymes that hydrolyze organic P (phosphatases).

Microbial biomass can serve as a labile reservoir of plant-available P, which in P-fixing soils is an important mechanism for avoiding the geochemical capture of P by a microbial competition for soil solution P (Liu et al. 2008). Depending on management effects and microbial biomass size, inorganic P pulses to soils can be rapidly immobilized into microbial biomass (50% within 3 days) in cultivated soils (Oehl et al. 2001), with the turnover of MBP estimated to be 80% at 9 days in forested Oxisols (Achat et al. 2010). The subsequent turnover of microbial biomass allows scavenged P to become transiently available to plants (Oberson and Joner 2005). On the other hand, a high turnover rate of MBP could also entail competition with plants for soil solution P (Achat et al. 2010). Furthermore, the dynamic nature of the soil microbial biomass and its sensitivity to management also make it difficult to predict. Lack of or excessive P fertilization can reduce the size of the microbial biomass, while organic amendments and inorganic–organic mixtures can increase the microbial biomass (Malik, Marschner, and Khan 2012). Evidence suggests that low-input systems can maximize the benefits of microbial P cycling by combinations of organic and inorganic P fertilizers rather than during separate application (Ayaga, Todd, and Brookes 2006). In this study, MBP positively correlated with maize grain yield in P-fixing soils, and there were greater relative increases in MBP on soils with greater P fixation. Similarly, increases in MBP followed additions of OM and inorganic P in strongly P-fixing but not weakly P-fixing soils (Koutika et al. 2013). Over 32 weeks, there was greater variability in MBP, from 22.5 $\mu$g P g$^{-1}$ soil in week 1 following the addition of manure (10 g kg$^{-1}$) and KH$_2$PO$_4$ (18.4 mg P kg$^{-1}$) to 4.8 $\mu$g P g$^{-1}$ soil in week 2, then increasing to 15 $\mu$g P g$^{-1}$ soil in week 16. Since microbial biomass and MBP increase with soil OM (Achat et al. 2010), OM additions can increase the MBP response to inorganic P inputs in P-fixing soils (Gichangi, Mnkeni, and Brookes 2010; Malik, Marschner, and Khan 2012).

Inputs of OM may additionally provide enhanced biological cycling of P via phosphatases in the soil, in particular extracellular phosphatases. An excellent overview of soil phosphatases and their role in biological P cycling to increase its crop availability is provided by Nannipieri et al. (2011). The changes in soil phosphatase activity may be a benefit of co-composting PR with OM, despite evidence that this method does not improve PR dissolution. The addition of compost produced from a mixture of coffee pulp, manure, gypsum, and PR to an Oxisol increased potential activities of soil enzymes with increased compost additions (10–80 g kg$^{-1}$ soil) after 28 days, including both acid and alkaline phosphomonesterases, in tandem with increased microbial respiration (Oliveira and Ferreira 2014).

### 8.4.4.2  Manure

Manure is a valuable resource for smallholders in SSA and can be used to meet P demands and improve soil conditions influencing P availability. However, the quantity and the quality of manures in SSA generally limit its ability to meet crop P needs. Low manure production and small herd size in much of SSA limit utility of P applications in the form of manure (Kibunja et al. 2012). The P content of manure can also vary across scales in SSA. For example, the total P content of manures varied from 2500 to 800 mg P kg$^{-1}$ in East Africa and 5700 to 600 mg P kg$^{-1}$ in West Africa (Probert et al. 1992). In western Burkina Faso, the variability in the P content of manure produced on-site across 98 farms entailed application rates of 2.2–5.1 t ha$^{-1}$ (Blanchard et al. 2014). Total P ranged from 880 ± 440 mg kg$^{-1}$ in low quality manure and 2200 ± 1320 mg kg$^{-1}$ in high quality manure, classified as *low* and *high* quality by farmers. Composts were comparable to low-quality manure in P content, with no differences between compost quality classifications. In contrast to manure, compost showed much greater variability in P content relative to N, likely reflecting stronger effects of feedstock on P, and convergence of C:N during composting. The species and diet of livestock and the manure storage method can significantly affect its quality. This means that organic P amendments based on manure can vary seasonally by livestock diet (Romney, Thorne, and Thomas 1994). Additionally, the bulkiness of organic inputs like manure often means that these are less often used on outfields (Dembélé et al. 2000), a problem that extends to PR (Vanlauwe and Giller 2006).

The lack of manure for appreciable effects on crop yields reflects the nutrient management situation of many smallholders across SSA. For smallholders in East and South Africa, manure production is limited by ownership of cattle and grazing assets (Zingore et al. 2011; Tittonell and Giller 2013). Overall, there is an insufficient cattle population in SSA to meet crop nutrient demand via manure alone (Tittonell and Giller 2013), and nutrient export via harvesting is typically orders of magnitude greater than nutrients that are returned in manure (Bationo and Mokwunye 1991). Furthermore, the high bulk and low P content of such inputs necessitate high application rates for appreciable crop response. For example, applying manure (1800 mg P kg$^{-1}$) at 17 t ha$^{-1}$ represented an annual P application rate of 30.6 kg P ha$^{-1}$ (Zingore et al. 2008). After three seasons at such rates, homefields and outfields on fine and coarse texture soils showed unexpected decreases in Olsen P (e.g., fine-texture outfield 6.6 mg kg$^{-1}$ soil) relative to the unfertilized control (7.2 mg kg$^{-1}$ soil). This could reflect the drop in pH from 5.1 to 4.9, resulting in increased P fixation, as well as greater P export of increased harvested yield. Nitrification during manure decomposition may also cause pH decreases (Murwira and Kirchmann 1993). As with inorganic P inputs, when manure is limited, placement application improves its ability to increase P for individual plants (Rusinamhodzi et al. 2013).

### 8.4.4.3   Plant-Based OM: Residues and Green Manures

Residues and green manures offer farmers a means to scavenge P in marginal or low-fertility lands and conduct P transfers among holdings. However, for this reason, they do not represent a net P input (Jama et al. 2000). Throughout SSA, agroforestry species managed as a source of residues to improve soil P fertility include tithonia (*Tithonia diversifolia*), tephrosia (*Tephrosia vogelii*), and crotalaria (*Crotalaria grahmiana*) (Gachengo et al. 1998; George et al. 2002a). Of these, tithonia has received much attention, particularly in East Africa. Originating in Mesoamerica, tithonia is widespread throughout the tropics and is recognized as a green manure of high potential for P management.

The P availability of green manures depends on residue quality. Relative to other residues, the high P content (3.7 g P kg$^{-1}$ dry mass) and low C:P (110) of tithonia accounts for its unique ability to provide an appreciable and sustained supply of crop-available P (Mustonen, Oelbermann, and Kass 2011). The residue C:P determines the complementary or the competitive role of the microbial biomass. Under the conditions of limited soil P but nonlimited C, soil microbes can immobilize 20–50% of the soil P during the decomposition of residues (Walbridge 1991). The high P content and the quality of residues like tithonia prevent microbial competition for mineralized P, allowing the accumulation of P in microbial biomass without decreasing plant-available soil solution P (Nziguheba et al. 1998; Kwabiah et al. 2001; Pypers et al. 2005). On a low-P soil in southern Kenya (pH 5.9, 5 mg kg$^{-1}$ soil resin-extractable P), tithonia not only replenished resin-extractable P but also met microbial P demands, because the high P content of tithonia facilitated low microbial biomass P:C ratio (Kwabiah et al. 2003).

Due to its local availability and lower cost, tithonia generally offers greater returns compared to other P amendments. When combined with inorganic P, maize yields increased with the increase in proportion of tithonia and showed a nonadditive increase in yield (+0.6 t ha$^{-1}$) when tithonia provided 36% or more of the total P input (at a rate of 15.5 kg ha$^{-1}$) (Nziguheba et al. 2002).

Improved fallows entail the intentional planting of uncropped fields with rapidly growing species that capitalize on the intercrop time gap to mobilize the nutrients in biomass that become available in a subsequent growing season following the incorporation of fallow biomass (Sanchez 1999). Although originally designed to improve N management, several fallow species have strong potential for improving soil P status. In Uganda, soil P availability tracked the dry matter production of fallow species (Mubiru and Coyne 2009). Improved fallows produced greater biomass and thus biomass P relative to native fallow (3.2 kg biomass P ha$^{-1}$), with concurrent greater release of P in the ensuing cropping season. Greatest soil P increases were observed for canavalia (*Canavalia ensiformis*, 10.6 kg biomass P ha$^{-1}$) across sites and growing seasons, followed by mucuna (*Mucuna pruriens*, 6.5 kg biomass P biomass ha$^{-1}$) and lablab (*Lablab vulgarisi*, 5.6 kg biomass P ha$^{-1}$). Canavalia is a multipurpose forage legume adapted to drought stress (Douxchamps et al. 2013) and thus poses additional benefits to improving PUE (see Sections 8.4.4.5 and 8.6.2).

Green manures are typically grown on the same farm or area in which they are used, and therefore may not represent a net P input. The use of biomass transfer is limited for a certain time because continued nutrient mining will eventually limit the biomass P. Nutrient mining below critical thresholds for residue production is more likely on marginal soils used for this purpose and therefore already on a thin nutrient balance. The risk is undermining the continued use of off-field biomass. Biomass transfers can lead to P depletion in as quickly as one growing season (George et al. 2002a). In this resulting state of degradation (low P, soil acidity), the poor growth of a tithonia fallow in degraded fields in Rwanda prevented its continued use as green manure source (Bucagu, Vanlauwe, and Giller 2013).

The repeated use of green manures like tithonia grown on marginal lands can also compromise their quality as P amendments. A survey in western Kenya found the potential of tithonia biomass transfer to improve soil P availability to depend on the source of tithonia (George et al. 2001). Specifically, tithonia from nutrient-depleted soils (unfertilized agricultural fields) were a less effective source of P and K, via biomass transfer, than tithonia from unmanaged margins and hedges. Specifically, there was higher leaf P concentration for tithonia from nonmanaged hedges (3.2 g P kg$^{-1}$) than that from unfertilized fields (2.2 g P kg$^{-1}$) (George et al. 2001). As a result, the net mineralization of P from tithonia depended on the source, with 90% of the leaves from hedges, but by only 14% from unfertilized fields meeting the critical P concentration of 2.5 g kg$^{-1}$ for mineralization.

#### 8.4.4.4    Overlooked P Sources: Waste Streams and Ash

Waste streams are a significant gap in the global P cycle and account for the majority of P losses from agroecosystems. An estimated two-thirds of P harvested and exported in biomass is not returned (Karunanithi et al. 2015). For example, bones from livestock slaughter in Ethiopia were estimated to represent 17,000–36,000 t P year$^{-1}$, which could meet 28–58% of this nation's agricultural P demand with a value of US\$ 50–104 million (Simons et al. 2014). The rapid urbanization in SSA presents a significant gap and an opportunity to improve local and regional P cycling. Urban centers are a sink for P from farmers' fields and a potential source for water pollution and human health hazards. This in part reflects the isolation of agricultural, wastewater, and sanitation sectors in SSA (Timmer and Visker 1998). Waste management in urban areas represents an opportunity to simultaneously improve public and ecosystem health and P deficiencies in surrounding agroecosystems.

The majority of P in urban (and rural) waste streams is contained in food waste and human excreta, chiefly urine. For example, the amount of P lost from human waste streams (excrement and food waste) in Nairobi's district of Kibera (population: 235,000) is estimated at 0.47 kg P person$^{-1}$ year$^{-1}$, amounting to 9 t P month$^{-1}$ (Kelderman et al. 2009). Of this, an estimated 65% is lost as raw sewage effluent or surface runoff, representing an annual P loss of 70.2 t. In general, constraints to the recycling of P from waste streams stem from the potential hygiene risks and infrastructure challenges, such as the high cost of transport from unprocessed waste and wastewater treatment and infrastructure for receiving human waste and storing and delivering recycled P products (Cordell, Drangert, and White 2009). Even so, there are some current opportunities to collect urban P deposits, such as human waste concentrated in pit latrines. In the Bwaise slum in Kampala, Uganda, pit latrines were estimated to retain 99% of P in the waste stream (Nyenje 2014).

Social and cultural challenges to reusing waste streams for agricultural production can stem from the perception of waste, which vary across cultures and may be complicated. For example, in southern Ghana, household interviews and group discussion identified a general negative attitude to interacting with fresh excreta, and despite the acknowledgment of its potential as a fertilizer, participants were unwilling to use it on their own crops or consume crops grown with excreta (Nimoh et al. 2014). Yet in northern Ghana (Tamale and Bolgatanga), farmers recognize the economic and agronomic benefits of using sewage sludge despite social ridicule and potential health hazards from its improper handling (Cofie, Kranjac-Berisavljevic, and Drechsel 2005). As a result, 90% of "night soil" in Tamale is used for agricultural production, and its increased use was concurrent with declining fertilizer imports in Ghana (Owusu-Bennoah and Visker 1994). In contrast, across much of Uganda, human waste is composted and applied to a variety of crops, including root and agroforestry species.

Ash residues from home stoves and agricultural waste are common in SSA (Karekezi and Turyareeba 1995), and its use as a dual P and liming amendment for acid soils in SSA is receiving increased attention (Materechera and Mkhabela 2002; Bougnom et al. 2011; Materechera 2012). Ash may be more useful for its liming effect, attributable to high base cation content and presence of Ca and Mg (hydr)oxides and carbonates (Lerner and Utzinger 1986; Pitman 2006). The combination of rapidly reacting cation (hydr)oxides and less reactive carbonates produces an immediate and sustained liming effect across the growing season following application (Ulery, Graham, and Amrhein 1993; Materechera 2012). Depending on the feedstock and the burn temperature and duration, ash can be a moderate-quality liming agent (26–59% CCE) (Ohno and Susan Erich 1990).

More so than manure, the P content of ash can greatly vary among feedstocks, e.g., 10.5% for cereals and 1% for straw (Schiemenz et al. 2011) and is not significantly influenced by combustion temperature, in contrast to other nutrients like N (Sarabèr, Cuperus, and Pels 2011). Like manure and other waste-based amendments, the ability of ash to supply sufficient P is limited by amendment availability. For example, P in ashes from cacao residues in Cote D'Ivoire represented only 2.1% of the cacao crop P needs (Sarabèr, Cuperus, and Pels 2011). Beneficial increases in soil pH and P availability are often observed at high ash application rates, such as the addition of 2 t ha$^{-1}$ for maize in South Africa (Materechera 2012).

When large amounts of vegetative biomass are combusted as in slash-and-burn agriculture, ash can significantly contribute to crop P need. These effects depend on the temperature and the duration of the burn (Galang, Markewitz, and Morris 2010), as well as the amount and the type of biomass burned. Slash-and-burn may also increase P by pyromineralization of litter and organic surface horizons rich in organic P (Giardina et al. 2000). With the increasing temperature of the burn, the negative effects on P cycling include (1) increases in P sorption capacity of surface soil (0–10 cm) due to increased surface area of mineral oxides and free and/or amorphous $Fe^{3+}$ and $Al^{3+}$ (Kwari and Batey 1991; Romanya, Khanna, and Raison 1994; Yusiharni and Gilkes 2012), (2) death of microbial biomass and loss of extracellular phosphatase activity (Saa et al. 1993), and (3) increased risk of soil erosion with concurrent P loss. As a result, increasing burning temperature generally leads to net decreases in soil available P (Yusiharni and Gilkes 2012). In contrast to other nutrients like N, K, and Ca, P levels do not return to pre-slash-and-burn levels but experience a net decrease following abandonment (Kleinman, Pimentel, and Bryant 1995). Since burning can induce mineral changes that enhance P sorption in surface soils over the medium term, once pyromineralized P is consumed, the burning can exacerbate P limitations and increase the need for P inputs (Ketterings, van Noordwijk, and Bigham 2002). P loss can also occur via volatilization at temperatures attainable during burning (280°C) (Raison, Khanna, and Woods 1985), although its airborne export is more commonly via aerosolized particulates or wind-blown ash. At field scales, this can account for substantial P losses of, for example, 11 kg P ha$^{-1}$ (55% of ash P) (Giardina, Sanford, and Døckersmith 2000) or 8–11 kg P ha$^{-1}$ (Sommer et al. 2004), and at the global scale accounts for a major P flux from terrestrial to atmospheric pools (1.8 Tg P year$^{-1}$) (Wang et al. 2015).

### 8.4.4.5    Crop Rotations: Legume Interactions with P Inputs

In unmanaged tropical forest ecosystems, leguminous species can improve biomass P accumulation because N fixation allows greater investment in N-expensive phosphatases to scavenge P (Wang, Houlton, and Field 2007; Houlton et al. 2008). In managed agroecosystems, integrating legume intercrops with P inputs can provide greater mileage on such inputs while simultaneously increasing N fixation (Giller 2002). This is a strategy of double dipping a single P input targeted to the legume rotation, because residual P and residual fixed N contribute to the nutrient demands of the following cereal crop. The inclusion of a legume such as mucuna or soybean reduced the need for P application for continuous maize from once every 2 years to once every 3 years (Kihara et al. 2010).

Legume-P interactions for improving maize yield and overall productivity are the basis of specific fertilization packages in East Africa, such as managing beneficial interactions in legume intercrops (MBILI) (Tungani, Mukhwana, and Woomer 2002). MBILI was modified to include PR as low-cost P input in the PR evaluation project package (PRE-PAC) for P-deficient soils in western Kenya (Rutto et al. 2011). The PREP-PAC package provides sufficient inputs for 25 $m^2$ plots at 100 kg P ha$^{-1}$ as MPR, 40 kg N ha$^{-1}$, 125 g food grain legume seed (biological N-fixation component), the legume inoculant (*Rhizobium*), a biodegradable adhesive, and lime seed pelleting to favor inoculation (Nekesa et al. 1999).

### 8.4.5 COMBINING INORGANIC AND ORGANIC AMENDMENTS

Combining organic inputs like tithonia with inorganic P sources can improve overall PUE and crop response through a number of mechanisms. Favorable interactions for soil P by combining inorganic and organic inputs may be mediated by microbial biomass responses, both total biomass and species abundance (Malik, Marschner, and Khan 2012). Generally, higher-quality residues (high P concentration, low C:P) determine whether these interactions are favorable for soil available P. Favorable interactions are generally found at higher P input rates (50–250 kg P ha$^{-1}$) but are less clear at more realistic lower rates (e.g., 15 kg P ha$^{-1}$) (Nziguheba et al. 1998).

A specific benefit of combining organic and inorganic P inputs is lifting non-P nutrient limitations to increase the P response. This may explain higher yields with mixtures at the same P rate (Mukuralinda et al. 2009). In Rwandan coffee systems, P inputs explained more variation (74%) in coffee yield in Rwanda over three seasons than N (56%) and K (64%) inputs, and tephrosia mulch significantly increased the coffee yield relative to NPK, although this may reflect greater P content of tephrosia rather than an interaction effect (Bucagu, Vanlauwe, and Giller 2013). The addition of inorganic P may also improve parameters favorable for mineralization of organic P in the OM component, e.g., C:P (Nziguheba et al. 2000). Sanchez et al. (1997) found that the combination of tithonia biomass with 250 kg P ha$^{-1}$ as MPR increased maize yields fivefold in western Kenya partially due to 60 kg K ha$^{-1}$ supplied with the plant material.

### 8.4.6 INDIRECT P MANAGEMENT BY LIMING TO AMELIORATE FIXATION

#### 8.4.6.1 Liming Decreases Sorption of Native and Added P in Soils

Soil acidity is a major constraint to agricultural production on weathered soils, which if managed properly, makes these soil types one of the most potentially productive (Sánchez and Salinas 1981). Lime can be used to improve the availability of native and/or added P by increasing soil pH because mineral oxide binding of P decreases as the pH increases from 4–7 (Haynes 1982; Havlin et al. 2013). Liming therefore has tremendous potential to improve agricultural production on weathered soils like Oxisols (Sánchez and Salinas 1981; Fageria and Baligar 2008). The amount of lime necessary to improve crop production depends on crop type (species and cultivar), the quality of liming material (measured relative to $CaCO_3$ as the percentage of CCE), and the current and target soil pH. Liming rates can be determined by soil pH, base saturation or its converse measurement of exchangeable acidity, or Al saturation. An excellent discussion of lime, P, soil, and plant interactions is provided by Haynes (1982).

Unlike regions with similar soil conditions, e.g., Brazil, liming of acid soils is not prevalent in SSA. This may reflect low awareness and/or knowledge of lime's potential to improve yields, as well as market availability, transport constraints, and economic costs (Okalebo et al. 2009). Research and extension efforts have not sufficiently evaluated and promoted lime as part of P management and an overall soil quality improvement practice in acid soils. Liming can also be useful to correct for soil acidification from inputs such as urea and diammonium phosphate (DAP) (Okalebo et al. 2009).

### 8.4.6.2 Considerations on the Use of Lime in Acid, P-Deficient Soils

Improvements to soil P fertility with lime can occur in the short term (<1 month), but such changes are generally observed at high applications rates (5–20 t ha$^{-1}$) (Gichangi and Mnkeni 2009). Rapid increases in available P following the reaction of added lime to desorb fixed P have been termed the *P spring effect* (Mike and Nanthi 2003). Additionally, plants are able to take up more P (along with other nutrients) because of decreased $Al^{3+}$ toxicity. This may entail an increase in rhizosphere mining of organic P since some crop species are able to induce significant mineralization of rhizosphere organic P via phosphatase exudation (George et al. 2002b).

Liming effects on the biological cycling of P are not well characterized but are expected to be favorable for soil P availability, since microbial activity is generally sensitive to low pH. This may account for observed increases in soil organic P and C:$P_o$ with increasing soil acidity (Turner and Blackwell 2013). Liming also affects soil enzyme activities because enzymes have pH optima. Although acid phosphomonoesterase activity decreased with pH increases induced by liming, alkaline phosphomonesterase and potentially phosphodiesterase were significantly increased, and correlated with microbial biomass C (Acosta-Martínez and Tabatabai 2000; Ekenler and Tabatabai 2003). Consequently, mineralization of organic P occurs with liming and associated pH increase (Haynes 1982), potentially contributing to the P spring effect of lime.

The quality of liming material influences its potential for improving soil P availability. Lime quality can be measured by CCE. Like PR, particle size is a major control on reaction rate, with finer textures offering a more rapid liming effect (Tisdale, Nelson, and Beaton 1985). The particle size efficiency of lime increases with its fineness factor (FF) (Huang, Fisher, and Argo 2007). As a result of these physical properties, there can be lags in lime effects on yield, typically in the first season following application (Tabu et al. 2007). CaO (quicklime or burned lime) will react more rapidly than less-soluble $CaCO_3$ (limestone or unburned lime). At lower application rates (1.4 t ha⁻¹), there were greater differences in short-term potato yield between burned and unburned lime sources, but there were no differences at higher rates (4.8 t ha⁻¹) (Nduwumuremyi et al. 2013). Differences in CCE of two local lime sources and their processing by burning resulted in differences in potato yields (Nduwumuremyi et al. 2013). Exchangeable $Al^{3+}$ was decreased below the toxicity threshold of 5% (Abbott 1987) with high application (4.2 t ha⁻¹) of agricultural burned and unburned lime (86% CCE) from 58.4% exchangeable $Al^{3+}$ (Nduwumuremyi et al. 2013). This corresponded to an increase in pH from 4.8 to 5.7 and a 44% increase in Bray-II P to 5.6 mg kg⁻¹ soil; potato yields increased by >50% (21.9 t ha⁻¹). The yield benefits of lime applications plateau with increasing application rates and/or time as pH increases. This may reflect non-P nutrient limitations. For example, doubling lime applications from 1.4 to 2.9 t ha⁻¹ did not double yields (Nduwumuremyi et al. 2013).

Liming to increase pH benefits soil fertility beyond P supply because the majority of nutrients are less available at low pH (<5) (Havlin et al. 2013), including secondary nutrients and micronutrients. Lime can also provide trace nutrients, depending on the source. The provisioning of base cations such as Ca and Mg by lime has strong effects on the overall fertility of weathered soils (Fageria and Baligar 2008), as well as micronutrients such as Mo and B (Clark 1984), which are limiting in such soils (Vitousek et al. 2010). A meta-analysis of maize yields and P constraints ($n = 2800$) in western Kenya found that the maximum yield obtained in research stations stagnated at 7 t ha⁻¹, well below the 10 t ha⁻¹ potential (Smaling and Janssen 1993), likely as a result of secondary nutrient (Ca, Mg) and micronutrient deficiencies (Kihara and Njoroge 2013).

### 8.4.6.3   Liming for P Management in SSA

Given the expense of concentrated P inputs and the decrease in PUE from geochemical fixation, lime is a way to increase the mileage of P inputs in acid soils. Lime generally decreases P requirements for a given yield threshold because less added P is fixed. The uptake of P and consequent biomass growth of coffee (*Coffea arabica*) was greater with lime additions (0.69 g $CaCO_3$ kg⁻¹ soil) to acid soils from Rwanda, although soil P was still considered deficient (Bucagu, Vanlauwe, and Giller 2013). Similarly, coffee yields increased with lime additions, strongly correlating with pH (Cyamweshi et al. 2014). In northwest Cameroon, the addition of lime and OM (mucuna residues) on strongly P-sorbing soils reduced P fertilizer requirement by 45–83% for maize, bean, and potato (Yamoah et al. 1996).

In East Africa, liming has been shown to have great potential to alleviate P constraints, as well as offer co-benefits to crop production from amelioration of other nutrient deficiencies and $Al^{3+}$ phytotoxicity. For example, in western Kenya, the majority of soils are weathered and exhibit pH < 5.5 (Smithson et al. 2003). Extension recommendations are 40 kg P ha⁻¹ year⁻¹, in agreement with optimal rates estimated at 38 kg P ha⁻¹ (Kihara and Njoroge 2013). However, a lower rate of 26 kg P ha⁻¹ can increase available P above the critical threshold for maize of 10 mg Olsen P kg⁻¹ soil and secure high maize yields (6 t ha⁻¹ vs. unfertilized farmer control of 0.5 t ha⁻¹) when co-applied with lime (2 t ha⁻¹), as well as N (75 kg ha⁻¹) (Okalebo et al. 2009). A single addition of 2 t lime ha$^{-1}$ increased the pH from 5.8 to 6.5, above the critical pH threshold of P fixation by Al (Figure 8.2).

The negative effects of liming include a potential for increased weed competition and decreased availability of micronutrients such as Zn, as was observed with liming that induced an increase in pH > 5 for Ultisols in Nigeria (Friesen, Juo, and Miller 1980). Since lime increases soil Ca and pH, its use precludes subsequent PR application (Tabu et al. 2007). Lime is better paired with soluble P inputs, whereas PR presents a low-cost alternative to both inputs. In hand-tilled systems, lime incorporation is limited to surface horizons, to which crop root growth may be constrained. The lack of deep lime incorporation can be an issue because P sorption generally increases with depth (Eze and Loganathan 1990).

## 8.5  ASSESSMENT OF P AVAILABILITY IN SOIL

There are three reasons for measuring soils for available and total P: (1) determine the appropriate application rate for specific soil conditions, (2) adjust subsequent applications rates to account for residual effect of previous inputs, and (3) assess past and predict future management effects on soil P stocks, from available to organic to fixed P.

The comparison of different nutrient management strategies in western Kenya demonstrates the importance of soil testing to determine the need for P inputs over time. In western Kenya, nutrient management interventions can perform poorly (low maize yield) despite soil P increases, suggesting a critical threshold of 15 mg Bray-I P kg$^{-1}$ soil (Woomer 2007). This threshold is comparable to the Olsen-P threshold of 10 mg P kg$^{-1}$ soil proposed by Okalebo et al. (2002) assuming an Olsen:Bray-I conversion factor of 2. As a result, fertilization packages in western Kenya should consider multiseasonal effects of repeated P applications. For example, MBILI promoted in Kenya entails an initial 20 kg P ha$^{-1}$ application as DAP (Tungani, Mukhwana, and Woomer 2002). Other projects, such as the PREP (Nekesa et al. 1999), involve a one-time P recapitalization application of 100 kg P ha$^{-1}$ as MPR, or in the case of the fertilizer use recommendation project, an initial P application of 100 kg P ha$^{-1}$ as DAP followed by maintenance applications of 20 kg P ha$^{-1}$ season$^{-1}$ (Kenya Agricultural Research Institute 1994).

High P rates are currently recommended for maize in western Kenya by the national extension services (Kenya Agricultural Research Institute [KARI]): 40 kg P ha$^{-1}$ per year$^{-1}$ based on 20 kg P ha$^{-1}$ per crop. This is in close agreement with estimated optimal returns at 38 kg P ha$^{-1}$ year$^{-1}$ (Kihara and Njoroge 2013), but the trials used for this estimation were of short term (<10 years). As Woomer et al. (2007) suggested, these rates are often necessary to satisfy the crop requirements initially but in subsequent years are likely too high, necessitating lower maintenance rates. Corroborating Woomer's hypothesis, P additions at these rates for 11 years increased labile P over 200% above minimum thresholds for maize on an Oxisol in western Kenya (Margenot et al. 2014). Even on short-term (<2 years) scales, the recommended application rates are too high for lifting P availability because of other non-P limitations, including micronutrients (Woomer 2007; Kihara and Njoroge 2013).

### 8.5.1  SOIL PHOSPHORUS TESTS

Soil P tests are commonly used to refer to a soil-appropriate measure of plant-available P. The objective of soil P tests is to develop a relationship between the method of soil P analysis, typically through an extraction, and crop P uptake or more typically crop yield. The soil P test is therefore most meaningful in providing P management guidelines when it is calibrated through P fertilization response trials. Given the strong effect of pH as the master variable on soil P availability, three types of tests are used based on the pH of the measured soils: the Olsen test for alkaline soils using 0.5 M $NaHCO_3$ (Watanabe and Olsen 1965), Bray test for acid soils using 0.03 M $NH_4F$ and 0.025 M HCl (Bray and Kurtz 1945), and Mehlich test for acid soils using 0.2 N $CH_3COOH$ + 0.25 N $NH_4NO_3$ + 0.013 N $HNO_3$ + 0.015 N $NH_4F$ + 0.001 M ethylenediaminetetraacetic acid in Mehlich-III (Mehlich 1984). Additional tests include chelation by organic acids such as oxalate, acetate, or lactate, and salt extractions, such as 0.02 M KCl and 0.01 M $CaCl_2$ (Menon, Chien, and Hammond 1989). Following extraction, the inorganic P in the solution is quantified by colorimetry or inductively coupled plasma atomic emission spectrometry or optimal emission spectrometry.

The strong effect of pH on soil P speciation and thus its removal by a specific extractant makes paramount the matching of a test method to soil pH. Although soil P test values are generally correlated, pH extremes and the presence of carbonates affect the correlations for the three more commonly employed soil tests, Bray, Mehlich, and Olsen. These tests have the potential to be highly standardized among many different labs (Wolf and Baker 1985). The standardization of soil tests and other soil P analyses is important for providing recommendations, partly because altering the test parameters (e.g., shaking time) can change results depending on soil properties. Such modifications explained the discrepancies among Bray test results in Rwanda (Drechsel, Mutwewingabo, and Hagedorn 1996). Important for this standardization is the use of a control or a lab standard to check for errors or reagent staling.

Soil P test recommendations are specific to crop type (species, potentially cultivar) and soil-climate conditions. Recommendations are for crop response, typically yield, but also crop uptake (e.g., foliar P, total biomass, grain yield and/or nonharvestable yield) P content. Like soil mapping to understand broad edaphic constraints on agricultural use, the establishment of soil test guidelines for P management requires specificity. For example, trials on the response of maize to P fertilization in a single district in Tanzania (Morogoro) found that even among P-deficient sites, optimum P application rates ranged from 16 to 50 kg P ha$^{-1}$ to achieve the critical threshold of 10.5 mg Olsen P kg$^{-1}$ soil (Ussiri et al. 1998). Despite the low pH (<7) of these soils, Olsen P gave stronger predictions of maize yield as compared to other commonly used soil test methods, including Bray-I and Mehlich-III which are considered to be more suitable predictors of plant-available P in acid soils. This again highlights the need for field trials and lab tests to establish site-specific recommendations.

### 8.5.2  SINK-BASED METHODS

Sink-based soil tests measure P available from the soil to a passive sink, such as a plant root idealized by a positively charged surface such as iron- and/or aluminum-impregnated or anion exchange surface in the form of membranes or resins. The determination of P availability using anion exchange resin has been shown to better simulate P removal from soil surrounding the rhizosphere environment than other extractants (Cooperband and Logan 1994). Analogous to the roots, as P is removed from the soil solution by adsorption to a sink, P not in the soil solution (e.g., weakly sorbed to mineral or OM) replenishes the soil solution P by equilibrium processes (Qian and Schoenau 2002). Because they measure exchangeable P, sink methods are considered to be a more realistic indication of soil P available for crop uptake. In contrast, extracts represent instantaneous snapshots of P in the soil solution during extraction—P availability at a particular moment, versus available P (Beckett and White 1964). For this reason, sink-based methods typically involve longer assay times (18–24 hours) than extractions (e.g., 5 min for Bray, 30 min for Olsen). Loading sink surfaces with counterions such as carbonate further mimic ion exchange processes in the root uptake of P (Sibbesen 1978; Qian and Schoenau 2002). The use of anion exchange membranes (AEMs) allows more accurate determination of microbial biomass by difference between nonfumigated and fumigated samples in strongly P-fixing soils such as Andosols and Oxisols by competing with soil-binding sites for lysed microbial P (Kouno, Tuchiya, and Ando 1995; Oberson et al. 1997).

### 8.5.3   P Sorption Measurements

Complementary to soil P tests, sorption of P by a soil can be measured to determine the potential P fixation, or at the other extreme of soil P saturation, estimate the risk of P loss. P sorption can be measured in two ways: exposing soil to a high P spike (single-point sorption) or determining a sorption isotherm by exposing a series of soil samples to different concentrations of P to calculate the maximum P sorption ($Q_{max}$) (Essington 2004). P sorption tests can be useful to (1) estimate P additions for maintaining a specific P concentration in soil solution (Fox and Kamprath 1970) or conversely, (2) assess P input efficiency given the proportion of the added P that would become plant unavailable (Nwoke et al. 2003). For example, single-point sorption tests have been used to estimate P fertilization requirements for sugarcane production on P-fixing soils in South Africa (Gichangi, Mnkeni, and Muchaonyerwa 2008). In Guinea, sorption isotherms were used to estimate P application to highly fixing soils based on calculated P fixation for the upper plow horizon (Nwoke et al. 2003).

### 8.5.4   Sequential Extraction: Hedley Fractionation

Sequential chemical extraction can be used to divide soil P into different inorganic ($P_i$) and organic ($P_o$) fractions, although the translation of these fractions to plant availability is not straightforward and risks chemical reductionism. The Hedley method employs a sequence of chemical extractions that define a series of $P_i$ and $P_o$ fractions by increasing the chemical lability (Hedley, Stewart, and Chauhan 1982): resin exchange membrane or AEM-$P_i$, NaHCO$_3$-$P_i$ and -$P_o$, NaOH-$P_i$ and -$P_o$, and HCl-$P_i$ and -$P_o$. These are broad, operational measures of soil P defined by the chemical form of P (e.g., Ca-P represented by $HCl-P_i$) (Cross and Schlesinger 1995; Condron and Newman 2011). An excellent review on the ability of the Hedley fractionation to provide information on land use and management effects on P cycling is provided by Negassa and Leinweber (2009).

Hedley fractionations have provided insight on management effects on organic P, specifically (1) highlighting the gradual depletion of organic P with continued cultivation and (2) corroborating the significance of P cycling through organic forms in P-limited weathered soils. Decreases in soil available P over time are thought to largely reflect the mineralization of organic P, as has been found across chronosequences and shifting cultivation systems in Nigeria and Tanzania (Adepetu and Corey 1976; Tiessen, Salcedo, and Sampaio 1992; Sugihara et al. 2012). This is consistent with greater decomposition and SOM turnover in tropical zones (Olson 1963; Ayanaba and Jenkinson 1990; Feller and Beare 1997; Six et al. 2002), which would entail greater mineralization of organic P. The application of $^{33}P$ labeling demonstrated the potential of the Hedley fractionation to identify differences in P dynamics among managements in a weathered soil (Oxisol) (Buehler et al. 2002). Soils under positive P balance from fertilization accumulated newly added P across all $P_i$ fractions (resin $P_i$, $NaHCO_3-P_i$, $NaOH-P_i$, $HCl-P_i$), but not $P_o$ fractions. In contrast, soils with low or no fertilization showed preferential accumulation of P in $NaOH-P_o$ and $HCl-P_o$ fractions, indicating that (1) the differences in P cycling in agroecosystems reflect P management history, chiefly P balance, and (2) the organic cycling of P is important under conditions of P limitation.

## 8.6  IMPROVING P EFFICIENCY OF CROPS FOR LOW P SOILS

Plants possess a number of adaptive mechanisms to cope with soil P deficiency, leading to changes at morphological, physiological, biochemical, and molecular levels (Zhang, Liao, and Lucas 2014). Developing P-efficient plants by modifying their adaptive strategies represents a sustainable approach that does not compromise environmental quality and limited P resources. From an agronomic point of view, and in an operational sense, the genotypic differences in P efficiency of crop plants need to be defined as the differences in growth or in yield of crops when grown in a P-deficient soil (Marschner 1995; George et al. 2011). A P-efficient plant is able to produce a higher yield in a low P soil compared to a standard genotype (Graham 1984). Higher P acquisition efficiency (PAE) from soils and improved internal PUE are two basic strategies that plants utilize to adapt to soils with low plant-available P (Vance, Uhde-Stone, and Allan 2003; Lynch 2011; Richardson and Simpson 2011; Veneklaas et al. 2012; López-Arredondo et al. 2014; Zhang, Liao, and Lucas 2014).

The differences in P efficiency can be quantified using several measures (Hammond et al. 2009). Agronomic PUE is the increase in yield per unit of added P fertilizer (g DM $g^{-1}$ $P_f$). This is equivalent to the product of the increase in plant P content per unit of added P fertilizer (g P $g^{-1}$ $P_f$), often referred to as plant P uptake efficiency (PUpE), and the increase in yield per unit increase in plant P content (g DM $g^{-1}$ P), or P utilization efficiency. There are four other measures of PUE: (1) P efficiency ratio, which is yield divided by the amount of P in the plant (g DM $g^{-1}$ P) or the reciprocal of tissue P concentration if the entire plant is harvested; (2) physiological PUE, which is yield divided by tissue P concentration at a given P concentration in the rooting medium ($g^2$ DM $g^{-1}$ P); (3) the critical value required for 90% yield, which is the amount, or the concentration, of P in the rooting medium required for a given percentage of maximum yield (g P), expressed as the $K_m$ value required for half-maximal yield; and (4) critical tissue P concentration, which is tissue P concentration required for a given percentage of maximal yield.

The genetic variation in plant P efficiency is well documented, and numerous quantitative trait loci (QTL) encoding traits for crop P efficiency have been identified in a variety of crops including rice, maize, common bean, and soybean (for reviews see López-Arrendondo et al. 2014 and Zhang, Liao, and Lucas 2014). Although conventional breeding showed significant progress in developing P-efficient cultivars, particularly for soybean in China (Wang, Yan, and Liao 2010), success with marker-assisted breeding has been limited due to significant environmental effects on P efficiency traits. As a result, most identified QTL have made small contributions to overall P efficiency. However, in rice the use of QTL Pup1 (Phosphorus uptake 1) in marker-assisted breeding has led to the development of rice lines showing a dramatic increase in PUpE when grown in P-deficient soils (Chin et al. 2010). The increase in rice yield on P-deficient soil was demonstrated by the overexpression of PSTOL1— the gene responsible for Pup1 QTL—indicating high potential for further genetic enhancement of P efficiency in rice (Gamuyao et al. 2012). Although transgenic approaches also yielded significant experimental results in improving P efficiency in different crops (López-Arredondo et al. 2014; Zhang, Liao, and Lucas 2014), there has not yet been a transgenic plant line produced for improving P efficiency that has been released for commercial use.

### 8.6.1  P Acquisition Efficiency

An effective management strategy for soils with low P content and/or P fixation is to enhance the plant's efficiency in acquiring soil P (Lynch 2011). Improved P acquisition by crop plants can be achieved by using three approaches (Ramaekers et al. 2010). First, traditional plant breeding for enhanced P acquisition is a feasible strategy as shown by a range of inheritance studies and the breeding of improved crops with greater P acquisition and better tolerance to low P soils (Ramaekers et al. 2010; Lynch 2011; Gabasawa and Yusuf 2013; Beebe et al. 2014; Kugblenu et al. 2014; Mendes et al. 2014; Leiser et al. 2015). Second, genetic engineering can be used to introduce genes that improve P acquisition and growth of crop plants (López-Arredondo et al. 2014; Zhang, Liao, and Lucas 2014). A third strategy focuses on the use of agricultural practices to enhance plant growth under P-deficient conditions through inoculation with plant growth-promoting rhizobacteria and mycorrhizae (Ramaekers et al. 2010; Richardson and Simpson 2011).

Although enhancing the PAE of crops can improve agroecosystem productivity in the short term, a longer-term consequence is that the increased PAE will lower the total soil P content over time. The rate of this decrease will depend on the initial total P content of the soil, the cropping intensity, and the crop P requirement (Ramaekers et al. 2010). To avoid ending up with overall low P soil contents, it is advisable to add small amounts of soluble P inputs or less-soluble slow-release forms such as rock phosphate. This strategy avoids the rapid loss of added P through soil leaching in coarse-textured soils that are unable to retain P. In P-fixing soils, this strategy additionally reduces the fixation of applied P not taken up by the plant. Combining this strategy with an improved plant efficiency to acquire P will ensure a higher recovery of applied P, lowering P fertilizer requirements.

One of the key mechanisms to increase plant access to P is greater topsoil exploration resulting from root architectural, morphological, and anatomical traits (Lynch 2011; Richardson and Simpson 2011). The ideotype of topsoil foraging has been proposed for improving the PAE by roots (Lynch 2011; White et al. 2013; Lynch and Wojciechowski 2015). It is possible to breed for this ideotype to develop crops for low P soils (Lynch 2011, 2013; Lynch and Wojciechowski 2015). Enhancing topsoil foraging is essential to improve PAE of crop plants because only 20% of the topsoil is explored by the roots during crop growth and development. Efficient genotypes of common bean and maize have shallow roots in the topsoil, and a shallower root growth angle of the axial or seminal roots increases topsoil foraging and thereby contributing to greater values of PAE.

In addition to root architectural traits, root morphological traits such as root length, diameter, surface area, volume, presence of root hairs, and length of root hairs contribute to inter- and intraspecific variation in PAE. The formation of root cortical aerenchyma, which convert living cortical tissue to air space through programmed cell death, improves the PAE by reducing the metabolic cost of soil exploration (Lynch and Wojciechowski 2015). A cost-benefit analysis of root traits indicated that root hairs have the greatest potential for improving PAE relative to their cost of production (Brown et al. 2013). Greater gains in PAE can be achieved through increased length and longevity of root hairs, as compared to increasing their density. The genetic variation in root hair length can be exploited to develop crop cultivars with improved PAE due to their ability to expand the effective P depletion zone around the root axis (Lynch and Wojciechowski 2015). Dimorphic root architecture of axial roots with a greater range of growth angles could also be considered for improving the PAE of crop plants grown in low P soils. In common bean (*Phaseolus vulgaris* L.), the QTL for root architecture are associated with QTL for PAE (Liao et al. 2004), allowing for breeding favorable traits like drought resistance and tolerance to Al toxicity (Yang, Rao, and Horst 2013).

Increased production and secretion of organic acids and enzymes such as phosphatases and ribonucleases in the rhizosphere can increase PAE by the mobilization of P in the rhizosphere (Zhang, Liao, and Lucas 2014). Purple acid phosphatases are a class of acid phosphatases implicated in plant response to P starvation and encompass a diversity of enzymes with intracellular and extracellular (via secretion) activities to scavenge and recycle P (Tian and Liao 2015). Secreted organic acids solubilize P by chelating Al, Fe, and Ca via carboxylate moieties from insoluble Al-P, Fe-P and Ca-P species, respectively, thereby rendering P soluble. Soil organic P is also not directly available to plants unless hydrolyzed or mineralized into $P_i$ by released root phosphatases. The association between plant roots and AMF is another means of improving PAE in several crops (Ramaekers et al. 2010; Richardson and Simpson 2011). The general understanding is that a major mechanism through which

AMF-associated plants acquire higher growth and drought resistance might be increased P nutrition, although additional processes independent of P nutrition may also be important (Suriyagoda et al. 2014). Further research is needed to exploit the beneficial effects of AMF association in relation to P supply for improving crop adaptation to low P soils.

## 8.6.2 P USE EFFICIENCY

Plant adaptation to P-limited soils can also be affected by genotypic differences in PUE. A potential strategy to reduce dependency on P fertilizer is to enhance the plant's internal PUE. Plants with enhanced PUE show higher growth for the same amount of P taken up. A comparison of the amount of P taken up by various crops to produce 1 t of yield indicated that common bean and soybean stand out as the most P-demanding crops per unit of economic yield (Rao, Friesen, and Osaki 1999). Modern crop varieties use P more efficiently than older varieties mainly as a result of improvements in harvest index (HI), which is related to plant structural and C allocation traits. Beebe et al. (2008) reported that the selection for drought resistance in common beans through improved partitioning of photosynthates to grain also improved yield under low P soil conditions, indicating the importance of improved HI on improving the P efficiency of this species. Recent work on low-input maize systems indicated that across several maize landraces ($n = 20$), PUE in conditions of low soil P were increased by greater internal P utilization and a high HI (Bayuelo-Jiménez and Ochoa-Cadavid 2014).

Higher values of PUE imply effective partitioning and remobilizing P among organs and tissues, and perhaps more importantly, a capacity to accumulate dry matter as a result of efficient use of P in metabolic processes (Veneklaas et al. 2012). Genotypic differences in PAE confound PUE rankings because genotypes with higher PAE suffer a lower degree of P stress, resulting in lower PUE (Rose et al. 2011). Improved PUE can be achieved by plants that have overall low P concentrations and by optimal distribution and redistribution in the plant allowing maximum growth and biomass allocation to harvestable plant parts (Wang, Yan, and Liao 2010). Significant decreases in plant P pools may be possible through reductions of superfluous ribosomal RNA and replacement of phospholipids by sulfolipids and galactolipids. Improvements in P distribution within the plant may be possible by increased remobilization from tissues with decreased P need such as senescing leaves and reduced partitioning of P to developing grains. These changes are expected to prolong and enhance the productive use of P in photosynthesis and have nutritional and environmental benefits (Veneklaas et al. 2012).

Su et al. (2009) identified six QTL controlling PUE in Chinese winter wheat pot and field trials. Moreover, positive linkages were observed between the QTL for PAE and PUE at two loci, suggesting the possibility of improving PAE and PUE simultaneously. Field research on the genetic architecture of PUE in maize cultivated in a low P soil showed that approximately 80% of QTL mapped for PAE co-localized with those for PUE, indicating that the efficiency in acquiring P is the main determinant of PUE (Mendes et al. 2014). Measurements of PAE, PUE, and grain yield in the same environments risk autocorrelations that may mask underlying genotypic relations. Recently, Leiser et al. (2015) tested the value of PAE and PUE traits for a selection of sorghum for performance in P-limited environments and found that PAE and PUE traits independent of HI were of similar importance for grain yield under low P conditions in statistically independent trials.

A comprehensive understanding of plant adaptive responses is required to simultaneously improve PAE and PUE, together with agronomic approaches that can collectively help meet the sustainability challenge of P delivery to crops. Targeted research is needed to quantify the magnitude of PAE and PUE gains that may be obtained through different mechanisms and their variation associated with genetic and environmental factors. Recent approaches such as genome-wide expression (transcription) QTL analyses and genome-wide association studies using next-generation sequencing could help to identify loci related to and controlling plant PAE and PUE (Veneklaas et al. 2012; Zhang, Liao, and Lucas 2014). Greater focus on plant-based P management ("feed the crop, not the soil") offers a means to markedly decrease the need for P inputs to agroecosystems by averting low-efficiency P additions to soils and instead use low, crop-targeted P inputs (Withers et al. 2014).

### 8.6.3 Precision Agriculture

Precision agriculture holds a strong potential to improve the agronomic efficiency of P inputs, but the high cost of its technological applications makes it currently untenable for many agroecosystems across SSA. One of the goals of precision agriculture is to improve the efficiency of inputs by targeting inputs to temporal and spatial variabilities in soil nutrient concentration and uptake by crops. This is predicated on quantifying the in-field variability of soil properties, typically by spectroscopic and/ or remote-sensing technologies, in order to appropriately match inputs (Mulla 2013). The intensive use of technology in precision agriculture (Stafford 2000) engenders high capital costs, and as a result, its application is largely nonexistent across SSA (Seelan et al. 2003). However, limited case studies have demonstrated a strong potential of spatial mapping of variable, if low soil fertility in small-scale fields in SSA, such as millet production in the West Sahel (Florax, Voortman, and Brouwer 2002). On the other hand, microdosing P fertilizer (see Section 8.4.3) arguably represents a low-technology manifestation of precision agriculture because it can improve the agronomic efficiency of added P by spatial and temporal targeting of inputs to individual crop plants. A second constraint to precision agriculture for P management is that methods such as near-infrared spectroscopy used to characterize the spatial variability of soil nutrients have limited ability to quantify predict available soil P (see Section 8.2.4) (Chang et al. 2001).

## 8.7 MODELING CROP RESPONSE TO P

Field research on crop response to P inputs is time and labor intensive. Such research also has its limitations where the scaling of results beyond the particular soil, field, region, and/or agroecosystem is concerned. On the other hand, the theoretical understanding of P processes in soils, especially under tropical conditions where the organic fraction of P plays a crucial role, has steadily improved over the past decades. It is therefore not surprising that scientists attempt to describe soil P processes, P crop uptake, and the influence of environment (e.g., soils, climate, and management) on these with the aid of computer simulation models. Besides scientific curiosity, the goal has been to employ models to determine P dynamics in a predictive fashion (spatially as well as temporarily) beyond the system under study.

Initial work to describe the P movement in soils and plant uptake by computer models began in the late 1960s with the work of Nye and Marriott (1969). The Barber and Cushman model followed in the late 1970s/early 1980s (Barber 1995), the concept of which was incorporated into the model of nutrient uptake (Claassen and Steingrobe 1999), as well as the phosmod model (Greenwood, Karpinets, and Stone 2001) and the model by Mollier et al. (2001, 2008). Further details about these models can be found in the study by Ryan et al. (2012). These models all have one thing in common: they never made it into wider use. None of these models has been frequently applied the model developers themselves.

Mollier et al. (2008) modeled maize P uptake well under nonlimiting conditions in southwestern France, but the results under P-deficient conditions were less promising. In agroecosystems where P fertilizer is available and abundantly applied, there is low need for detailed crop models to evaluate alternative P management (Probert and Keating 2000), even though recognized environmental damage of decades of high P inputs merit such modeling. The opposite is true for P-limited soils and for regions where P fertilizer is not applied in sufficient quantities—often two co-occurring scenarios in many regions of SSA.

In comparison to the wealth of research publications dealing with N dynamics, crop N uptake, and approaches to optimize N management, P model development and application has been notably lagging behind. Not surprisingly, there are few additional widely known models that are capable of simulating soil P dynamics and crop responses.

The first one is the erosion-productivity impact calculator (Jones et al. 1984) and the second, the CENTURY model (Parton 1996). A number of studies have been published on the application of CENTURY, yet the majority address OM turnover only. CENTURY's P routine includes five inorganic P ($P_i$) pools (labile, sorbed, strongly sorbed, occluded, and parent) and six organic P ($P_o$) pools (passive, slow, active, microbial, soil litter, surface litter) (Figure 8.5). The CENTURY model was used by Gijsman et al. (1996) to simulate the C, N, and P dynamics of a highly weathered Oxisol under savanna grassland in Colombia. The authors were unable to do so satisfactorily and consequently pointed out that the various P pools would need revision to better reflect tropical conditions. It also turned out problematic to evaluate the results of a model that builds on conceptual pools rather than soil P fractions that can actually be measured.

As far as more widely used crop models that are still maintained and updated on a regular basis are concerned, a P module was first integrated into Crop Environment Resources Synthesis (CERES) and Crop Growth (CROPGRO) models within the decision support system for agrotechnology transfer (DSSAT) software in early 2000 (Daroub et al. 2003). This P module included datasets from Tanzania but did not involve substantial further testing. Using maize datasets from Ghana, Dzotsi et al. (2010) modified the P module in DSSAT. Specifically, the authors separated the

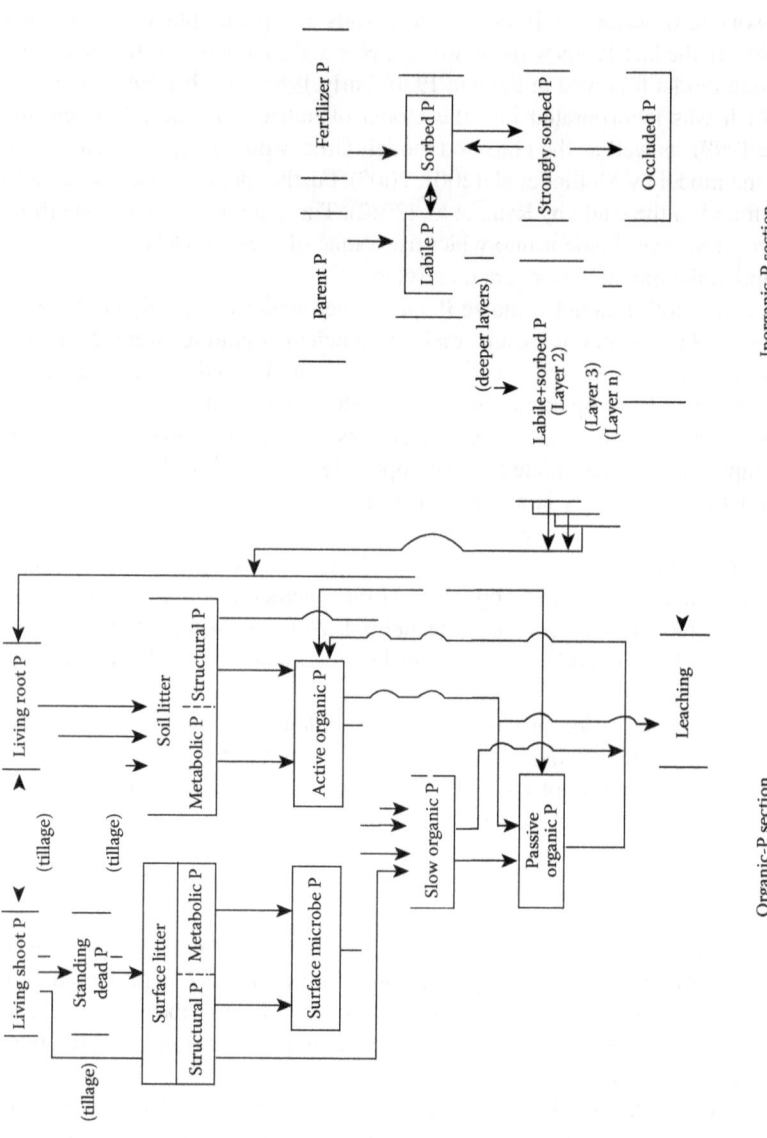

**FIGURE 8.5** The conceptual layout of the P module of CENTURY. (From Gijsman, A. J., A. Oberson, H. Tiessen, and D. K. Friesen, *Agronomy Journal*, 88, 894–903, 1996. With permission.)

inorganic P pool into two spatially distinct pools, a pool close to the root zone (i.e., available for plant uptake) and a second pool beyond plant root reach. To facilitate ease of use, the authors developed a set of regression equations to relate equilibrium P concentrations and P transformation rates to soil properties (e.g., texture, water content, organic matter). This version is currently applicable to maize, sorghum, and, as of recently, peanut (Naab et al. 2015). However, the bulk of studies on the application of DSSAT in SSA focus on trials where P is nonlimiting due to fertilization. A few studies employing DSSAT use the built-in rudimentary soil fertility limitation factor, and whereas other studies simply ignore P limitations, even if these are present. Thus, the module remains to be thoroughly tested with datasets for representative agroecosystems in SSA to improve model predictions, as well as to mainstream P simulations to the same level as N.

Besides DSSAT, the agricultural production systems simulator (APSIM) is the second best known and widely applied crop model. A P module was introduced into APSIM in early 2000 (Keating et al. 2003; Probert 2004) and was tested on contrasting soil types (Delve et al. 2009). As is the case for DSSAT, there have been only a limited number of publications building on the P module in APSIM. Past and ongoing participatory modeling work in SSA (Dimes, Twomlow, and Carberry 2003), ignore P limitations, despite the goal of advising farmers on resource use and optimization; certainly a problematic approach.

Ongoing research by the CIAT deals with a more sophisticated way of quantifying P pools. Using soils across a variety of land uses in Malawi and Tanzania, CIAT scientists are working to relate P fractions obtained by sequential extraction (Tiessen, Stewart, and Moir 1983) to land use history, and to predict fractions using mid-infrared spectroscopy (MIRS). The idea behind this is to explain land-use driven changes in P fractions as well as develop a rapid and sophisticated method to initialize DSSAT P pools by MIRS, bypassing tedious and expensive lab fractionations (see Section 8.5.4) or error-prone application of default regression equations to derive these pools.

## 8.8 CONCLUSIONS AND RESEARCH PERSPECTIVES

As a result of the interactions of soil properties, cropping demands and resources, and socioeconomic conditions unique to the subcontinent, P is a key limiting nutrient for agroecosystem productivity in SSA. Net P inputs to recapitalize soils in SSA are ultimately necessary to lift P limitations. To this end, there are substantial opportunities and low-hanging fruit that offer high mileage on small investments to improve P management. SSA has not only limited but also underexplored P resources, including PR and lime. As current inputs rates in SSA are the lowest of any global region, even minor increases in net P inputs will lessen P limitations in SSA. Biological mechanisms can contribute to crop P uptake but are generally significantly in low-input systems with low soil P. Liming holds a significant potential to decrease P fixation, improve PUE of costly inputs, and ameliorate additional non-P nutrient and toxicity limitations on PUE. Despite the proven efficacy of liming to improve the soil P status in similar soil-climate conditions such as South America and East Africa, there is limited work and even less implementation of pH management to address P deficiency.

Additional research on the affordability and the accessibility of much needed inputs is a critical but oft-overlooked component of P management in SSA. Temporal and spatial specificities of P management hold strong if underutilized potential to maximize the effect of limited P inputs. There is certainly a role that can be played by computer modeling of crop P responses, but a wider-scale application of available tools is yet to come. Breeding efforts to reduce the demand of crops for P show promise, and there are advancements in identifying the genetic basis of traits involved in P uptake and utilization by crops. PAE and PUE can be improved at numerous points along the soil–microbe–plant pathway(s) of P transformation. Increased productivity will require that organic matter inputs be supplemented with inorganic P inputs for crops to provide net P inputs to soils with negative P balances. The combined use of mineral fertilizers (rapid nutrient release) with organic fertilizers (slow nutrient release) is capable of synchronizing P supply with demand, thereby maximizing PUE and minimizing environmental impacts of P loss.

Although input-based strategies are inevitable to improve soil P status and general soil fertility in SSA, there exist considerable economic, social, and even political roadblocks, as well as knowledge constraints. The complexity of these, and their traditional separation from agronomic approaches to soil fertility management, constrains the implementation of well-established knowledge.

## ACKNOWLEDGMENTS

This work was partially supported by a U.S. Borlaug Fellowship in Global Food Security and the Consultative Group for International Agricultural Research Program on Grain Legumes.

## REFERENCES

Aainaa, H. N., O. H. Ahmed, S. Kasim, and N. M. Ab Majid. 2014. Reducing Egypt rock phosphate use in *Zea mays* cultivation on an acid soil using clinoptilolite zeolite. *Sustainable Agriculture Research* 4 (1):56.

Abbott, T. S. 1987. *BCRI Soil Testing: Methods and Interpretation*. Biological and Chemical Research Institute, NSW Agriculture and Fisheries, Orange, New South Wales.

Achat, D. L., C. Morel, M. R. Bakker, L. Augusto, S. Pellerin, A. Gallet-Budynek, and M. Gonzalez. 2010. Assessing turnover of microbial biomass phosphorus: Combination of an isotopic dilution method with a mass balance model. *Soil Biology and Biochemistry* 42 (12):2231–2240.

Acosta-Martínez, V., and M. A. Tabatabai. 2000. Enzyme activities in a limed agricultural soil. *Biology and Fertility of Soils* 31 (1):85–91.

Adepetu, J. A., and R. B. Corey. 1976. Organic phosphorus as a predictor of plant-available phosphorus in soils of southern Nigeria. *Soil Science* 122 (3):159–164.

Adesanwo, O. O., M. T. Adetunji, and S. Diatta. 2012. Effect of legume incorporation on solubilization of Ogun phosphate rock on slightly acidic soils in SW Nigeria. *Journal of Plant Nutrition and Soil Science* 175 (3):377–384.

Akande, M. O., F. I. Oluwatoyinbo, J. A. Adediran, K. W. Buari, and I. O. Yusuf. 2004. Soil amendments affect the release of P from rock phosphate and the development and yield of okra. *Journal of Vegetable Crop Production* 9 (2):3–9.

Allen, S. C., V. D. Nair, D. A. Graetz, S. Jose, and P. K. Ramachandran Nair. 2006. Phosphorus loss from organic versus inorganic fertilizers used in alleycropping on a Florida Ultisol. *Agriculture, Ecosystems & Environment* 117 (4):290–298.

Alloush, G. A., S. K. Zeto, and R. B. Clark. 2000. Phosphorus source, organic matter, and arbuscular mycorrhiza effects on growth and mineral acquisition of chickpea grown in acidic soil. *Journal of Plant Nutrition* 23 (9):1351–1369.

Amery, F., and E. Smolders. 2012. Unlocking fixed soil phosphorus upon waterlogging can be promoted by increasing soil cation exchange capacity. *European Journal of Soil Science* 63 (6):831–838.

Ayaga, G., A. Todd, and P. C. Brookes. 2006. Enhanced biological cycling of phosphorus increases its availability to crops in low-input sub-Saharan farming systems. *Soil Biology and Biochemistry* 38 (1):81–90.

Ayanaba, A., and D. S. Jenkinson. 1990. Decomposition of carbon-14 labeled ryegrass and maize under tropical conditions. *Soil Science Society of America Journal* 54 (1):112–115.

Bâ, A. M., and T. Guissou. 1996. Rock phosphate and vesicular-arbuscular mycorrhiza effects on growth and nutrient uptake of Faidherbia albida (Del.) seedlings in an alkaline sandy soil. *Agroforestry Systems* 34 (2):129–137.

Barber, S. A. 1995. *Soil Nutrient Bioavailability: A Mechanistic Approach*. John Wiley & Sons, New York.

Bationo, A., and A. U. Mokwunye. 1991. Role of manures and crop residue in alleviating soil fertility constraints to crop production: With special reference to the Sahelian and Sudanian zones of West Africa. In *Alleviating Soil Fertility Constraints to Increased Crop Production in West Africa*, edited by Mokwunye, A. U. Springer, Amsterdam.

Bationo, A., S. K. Mughogho, and U. Mokwunye. 1986. Agronomic evaluation of phosphate fertilizers in tropical Africa. In *Management of Nitrogen and Phosphorus Fertilizers in Sub-Saharan Africa*. Springer, Amsterdam, the Netherlands.

Bationo, S., E. Hartemink Alfred, O. Lungo, M. Naimi, P. Okoth, E. M. A. Smaling, and L. Thiombiano. 2006. African soils: Their productivity and profitability of fertilizer use. Paper read at *African Fertilizer Summit, Abuja, Nigeria*.

Bayuelo-Jiménez, J. S., and I. Ochoa-Cadavid. 2014. Phosphorus acquisition and internal utilization efficiency among maize landraces from the central Mexican highlands. *Field Crops Research* 156 (0):123–134.

Beckett, P. H. T., and R. E. White. 1964. Studies on the phosphate potentials of soils. *Plant and Soil* 21 (3):253–282.

Beebe, S. E., I. M. Rao, C. Cajiao, and M. Grajales. 2008. Selection for drought resistance in common bean also improves yield in phosphorus limited and favorable environments. *Crop Science* 48 (2):582–592.

Beebe, S. E., I. M. Rao, M. J. Devi, and J. Polania. 2014. Common beans, biodiversity, and multiple stresses: Challenges of drought resistance in tropical soils. *Crop and Pasture Science* 65 (7):667–675.

Bekunda, M. A., A. Bationo, and H. Ssali. 1997. Soil fertility management in Africa: A review of selected research trials. In *Replenishing Soil Fertility in Africa*, edited by Buresh, R. J., P. A. Sanchez, and F. Calhoun. Soil Science Society of America and American Society of Agronomy, Madison, WI.

Bekunda, M., N. Sanginga, and P. L. Woomer. 2010. Restoring soil fertility in Sub-Sahara Africa. In *Advances in Agronomy*, edited by Donald, L. S. Academic Press, Cambridge, MA.

Blake, R. E., J. R. O'Neil, and A. V. Surkov. 2005. Biogeochemical cycling of phosphorus: Insights from oxygen isotope effects of phosphoenzymes. *American Journal of Science* 305 (6–8):596–620.

Blanchard, M., K. Coulibaly, S. Bognini, P. Dugue, and É. Vall. 2014. Diversity in the quality of organic manure produced on farms in West Africa: What impact on recommendations for the use of manure? *Biotechnologie, Agronomie, Société et Environnement* 18 (4):512–523.

Bogrekci, I., and W. S. Lee. 2005. Spectral soil signatures and sensing phosphorus. *Biosystems Engineering* 92 (4):527–533.

Bolan, N. S., and M. J. Hedley. 1989. Dissolution of phosphate rocks in soils: 1. Evaluation of extraction methods for the measurement of phosphate rock dissolution. *Fertilizer Research* 19 (2):65–75.

Bolan, N. S., and M. J. Hedley. 1990. Dissolution of phosphate rocks in soils: 2. Effect of pH on the dissolution and plant availability of phosphate rock in soil with pH dependent charge. *Fertilizer Research* 24 (3):125–134.

Bonzi, M., F. Lompo, N. Ouandaogo, and P. M. Sédogo. 2011. Promoting uses of indigenous phosphate rock for soil fertility recapitalisation in the Sahel: State of the knowledge on the review of the rock phosphates of Burkina Faso. In *Innovations as Key to the Green Revolution in Africa*, edited by Bationo, A., B. Waswa, J. M. Okeyo, F. Maina, and J. M. Kihara. Springer, Amsterdam.

Bougnom, B. P., B. A. Knapp, F.-X. Etoa, and H. Insam. 2011. Possible use of wood ash and compost for improving acid tropical soils. In *Recycling of Biomass Ashes*, edited by Insam, H., and B. A. Knapp. Springer, Berlin Heidelberg.

Bray, R. H., and L. T. Kurtz. 1945. Determination of total, organic, and available forms of phosphorus in soils. *Soil Science* 59 (1):39–46.

Brown, L. K., T. S. George, L. X. Dupuy, and P. J. White. 2013. A conceptual model of root hair ideotypes for future agricultural environments: What combination of traits should be targeted to cope with limited P availability? *Annals of Botany* 112 (2):317–330.

Bucagu, C., B. Vanlauwe, and K. E. Giller. 2013. Managing Tephrosia mulch and fertilizer to enhance coffee productivity on smallholder farms in the Eastern African Highlands. *European Journal of Agronomy* 48 (0):19–29.

Buehler, S., A. Oberson, I. M. Rao, D. K. Friesen, and E. Frossard. 2002. Sequential phosphorus extraction of a P-labeled Oxisol under contrasting agricultural systems. *Soil Science Society of America Journal* 66 (3):868–877.

Buresh, R. J., P. C. Smithson, and D. T. Hellums. 1997. Building soil phosphorus capital in Africa. *Replenishing Soil Fertility in Africa*. In *Replenishing Soil Fertility in Africa*, edited by Buresh, R. J., P. A. Sanchez, and F. Calhoun. Soil Science Society of America, Madison, WI.

Canfield, D. E., E. Erik, and T. Bo. 2005. The phosphorus cycle. In *Advances in Marine Biology*, edited by Donald, E. K., E. Canfield, and T. Bo. Academic Press, Cambridge, MA.

Celi, L., and E. Barberis. 2005. Abiotic stabilization of organic phosphorus in the environment. In *Organic Phosphorus in the Environment*, edited by Turner, B. L., E. Frossard, and D. S. Baldwin. CABI Publishing, Wallingford, UK. 113–132.

Chang, C.-W., D. A. Laird, M. J. Mausbach, and C. R. Hurburgh. 2001. Near-infrared reflectance Spectroscopy–Principal components regression analyses of soil properties. *Soil Science Society of America Journal* 65 (2):480–490.

Chen, Y.-S. R., J. N. Butler, and W. Stumm. 1973. Adsorption of phosphate on alumina and kaolinite from dilute aqueous solutions. *Journal of Colloid and Interface Science* 43 (2):421–436.

Chien, S. H. 1977a. Dissolution of phosphate rocks in a flooded acid soil. *Soil Science Society of America Journal* 41 (6):1106–1109.

Chien, S. H. 1977b. Thermodynamic considerations on the solubility of phosphate rock. *Soil Science* 123 (2):117–121.

Chien, S. H., and D. K. Friesen. 2000. Phosphate fertilisers and management for sustainable crop production in tropical acid soils. In *Management and Conservation of Tropical Acid Soils for Sustainable Crop Production*. International Atomic Energy Agency, Vienna, Austria.

Chien, S. H., L. L. Hammond, and L. A. Leon. 1987. Long-term reactions of phosphate rocks with an Oxisol in Colombia. *Soil Science* 144 (4):257–265.

Chien, S. H., J. Henao, C. B. Christianson, A. Bationo, and A. U. Mokwunye. 1990. Agronomic evaluation of two unacidulated and partially acidulated phosphate rocks indigenous to Niger. *Soil Science Society of America Journal* 54 (6):1772–1777.

Chin, J. H., X. Lu, S. M. Haefele, R. Gamuyao, A. Ismail, M. Wissuwa, and S. Heuer. 2010. Development and application of gene-based markers for the major rice QTL Phosphorus uptake 1. *Theoretical and Applied Genetics* 120 (6):1073–1086.

Claassen, N., and B. Steingrobe. 1999. Mechanistic simulation models for a better understanding of nutrient uptake from soil. In *Mineral Nutrition of Crops: Fundamental Mechanisms and Implications*, edited by Rengel, Z. Haworth Press Inc, New York.

Clark, R. B. 1984. Physiological aspects of calcium, magnesium, and molybdenum deficiencies in plants. In *Soil Acidity and Liming*, edited by Adams, F. American Society of Agronomy and Crop Science Society of America, Madison, WI.

Cofie, O. O., G. Kranjac-Berisavljevic, and P. Drechsel. 2005. The use of human waste for peri-urban agriculture in Northern Ghana. *Renewable Agriculture and Food Systems* 20 (02):73–80.

Condron, L. M., and S. Newman. 2011. Revisiting the fundamentals of phosphorus fractionation of sediments and soils. *Journal of Soils and Sediments* 11 (5):830–840.

Cooperband, L. R., and T. J. Logan. 1994. Measuring in situ changes in labile soil phosphorus with anion-exchange membranes. *Soil Science Society of America Journal* 58 (1):105–114.

Cordell, D., J.-O. Drangert, and S. White. 2009. The story of phosphorus: Global food security and food for thought. *Global Environmental Change* 19 (2):292–305.

Cozzolino, V., V. Di Meo, and A. Piccolo. 2013. Impact of arbuscular mycorrhizal fungi applications on maize production and soil phosphorus availability. *Journal of Geochemical Exploration* 129 (0):40–44.

Cross, A. F., and W. H. Schlesinger. 1995. A literature review and evaluation of the Hedley fractionation: Applications to the biogeochemical cycle of soil phosphorus in natural ecosystems. *Geoderma* 64 (3–4):197–214.

Cyamweshi, R. A., N. L. Nabahungu, A. Mukashema, V. Ruganzu, M. C. Gatarayiha, A. Nduwumuremyi, and J. J. Mbonigaba. 2014. Enhancing nutrient availability and coffee yield on acid soils of the central plateau of southern Rwanda. *Global Journal of Agricultural Research* 2 (2):44–55.

Daroub, S. H., A. Gerakis, J. T. Ritchie, D. K. Friesen, and J. Ryan. 2003. Development of a soil-plant phosphorus simulation model for calcareous and weathered tropical soils. *Agricultural Systems* 76 (3):1157–1181.

Delve, R. J., M. E. Probert, J. G. Cobo, J. Ricaurte, M. Rivera, E. Barrios, and I. M. Rao. 2009. Simulating phosphorus responses in annual crops using APSIM: Model evaluation on contrasting soil types. *Nutrient Cycling in Agroecosystems* 84 (3):293–306.

Dembélé, I., D. Koné, A. Soumaré, D. Coulibaly, Y. Koné, B. Ly, and L. Kater. 2000. Fallows and field systems in dryland Mali. In *Nutrients on the Move: Soil Fertility Dynamics in African Farming Systems*, edited by Hilhorst, T., and F. M. Muchena. International Institute for Environment and Development, London.

Dieter, D., H. Elsenbeer, and B. L. Turner. 2010. Phosphorus fractionation in lowland tropical rainforest soils in central Panama. *CATENA* 82 (2):118–125.

Dimes, J., S. Twomlow, and P. S. Carberry. 2003. Application of new tools: Exploring the synergies between simulation models and participatory research in smallholder farming systems In *Global Theme 3: Water, Soil and Agrodiversity Management for Ecosystem Resilience*. International Crops Research Institute for the Semi-Arid Tropic, Bulawayo, Zimbabwe.

Dobermann, A. 2004. A critical assessment of the system of rice intensification (SRI). *Agricultural Systems* 79 (3):261–281.

Douxchamps, S., I. M. Rao, M. Peters, R. Van Der Hoek, A. Schmidt, S. Martens, J. Polania et al. 2013. Farm-scale tradeoffs between legume use as forage versus green manure: The case of *Canavalia brasiliensis*. *Agroecology and Sustainable Food Systems* 38 (1):25–45.

Drechsel, P., B. Mutwewingabo, and F. Hagedorn. 1996. Effect of modifications of the P-Bray no. 1 method on soil test results. *Zeitschrift für Pflanzenernährung und Bodenkunde* 159 (4):409–410.

Dzotsi, K. A., J. W. Jones, S. G. K. Adiku, J. B. Naab, U. Singh, C. H. Porter, and A. J. Gijsman. 2010. Modeling soil and plant phosphorus within DSSAT. *Ecological Modelling* 221 (23):2839–2849.

Ekenler, M., and M. A. Tabatabai. 2003. Responses of phosphatases and arylsulfatase in soils to liming and tillage systems. *Journal of Plant Nutrition and Soil Science* 166 (3):281–290.

Esberg, C., B. du Toit, R. Olsson, U. Ilstedt, and R. Giesler. 2010. Microbial responses to P addition in six South African forest soils. *Plant and Soil* 329 (1–2):209–225.

Essington, M. E. 2004. *Soil and Water Chemistry: An Integrative Approach*. Taylor & Francis, Abingdon-on-Thames.

Eswaran, H., R. Almaraz, E. van den Berg, and P. Reich. 1997. An assessment of the soil resources of Africa in relation to productivity. *Geoderma* 1.77:1–18.

Eze, I. C., and P. Loganathan. 1990. Effects of pH on phosphate sorption of some paleudults of southern Nigeria. *Soil Science* 150 (3):613–621.

Fageria, N. K., and V. C. Baligar. 2008. Ameliorating soil acidity of tropical Oxisols by liming for sustainable crop production. In *Advances in Agronomy*, edited by Donald, L. S. Academic Press, Cambridge, MA.

Fairhurst, T., R. Lefroy, E. Mutert, and N. Batjes. 1999. The importance, distribution and causes of phosphorus deficiency as a constraint to crop production in the tropics. *Agroforestry Forum* 9:2–8.

Feller, C., and M. H. Beare. 1997. Physical control of soil organic matter dynamics in the tropics. *Geoderma* 79 (1):69–116.

Florax, R. J. G. M., R. L. Voortman, and J. Brouwer. 2002. Spatial dimensions of precision agriculture: A spatial econometric analysis of millet yield on Sahelian coversands. *Agricultural Economics* 27 (3):425–443.

Fox, R. L, and E. J. Kamprath. 1970. Phosphate sorption isotherms for evaluating the phosphate requirements of soils. *Soil Science Society of America Journal* 34 (6):902–907.

Fredeen, A. L., I. Madhusudana Rao, and N. Terry. 1989. Influence of phosphorus nutrition on growth and carbon partitioning in *Glycine max*. *Plant Physiology* 89 (1):225–230.

Friesen, D. K., A. S. R. Juo, and M. H. Miller. 1980. Liming and lime-phosphorus-zinc interactions in two Nigerian Ultisols: I. Interactions in the soil. *Soil Science Society of America Journal* 44 (6):1221–1226.

Gabasawa, A. I., and A. A. Yusuf. 2013. Genotypic variations in phosphorus use efficiency and yield of some groundnut cultivars grown on an Alfisol at Samaru, Nigeria. *Journal of Soil Science and Environmental Management* 4 (3):54–61.

Gachengo, C. N., C. A. Palm, B. Jama, and C. Othieno. 1998. Tithonia and senna green manures and inorganic fertilizers as phosphorus sources for maize in Western Kenya. *Agroforestry Systems* 44 (1):21–35.

Galang, M. A., D. Markewitz, and L. A. Morris. 2010. Soil phosphorus transformations under forest burning and laboratory heat treatments. *Geoderma* 155 (3–4):401–408.

Gamuyao, R., J. H. Chin, J. Pariasca-Tanaka, P. Pesaresi, S. Catausan, C. Dalid, I. Slamet-Loedin, E. M. Tecson-Mendoza, M. Wissuwa, and S. Heuer. 2012. The protein kinase Pstol1 from traditional rice confers tolerance of phosphorus deficiency. *Nature* 488 (7412):535–539.

George, T. S., A.-M. Fransson, J. P. Hammond, and P. J. White. 2011. Phosphorus nutrition: Rhizosphere processes, plant response and adaptations. In *Phosphorus in Action.* Springer, Amsterdam.

George, T. S., P. J. Gregory, J. S. Robinson, R. J. Buresh, and B. A. Jama. 2001. *Tithonia diversifolia*: Variations in leaf nutrient concentration and implications for biomass transfer. *Agroforestry Systems* 52 (3):199–205.

George, T. S., P. J. Gregory, J. S. Robinson, R. J. Buresh, and B. Jama. 2002a. Utilisation of soil organic P by agroforestry and crop species in the field, western Kenya. *Plant and Soil* 246 (1):53–63.

George, T. S., P. J. Gregory, M. Wood, D. Read, and R. J. Buresh. 2002b. Phosphatase activity and organic acids in the rhizosphere of potential agroforestry species and maize. *Soil Biology and Biochemistry* 34 (10):1487–1494.

Giardina, C. P., R. L. Sanford, and I. C. Døckersmith. 2000. Changes in soil phosphorus and nitrogen during slash-and-burn clearing of a dry tropical forest. *Soil Science Society of America Journal* 64 (1):399–405.

Giardina, C. P., R. L. Sanford, I. C. Døckersmith, and V. J. Jaramillo. 2000. The effects of slash burning on ecosystem nutrients during the land preparation phase of shifting cultivation. *Plant and Soil* 220 (1–2):247–260.

Gichangi, E. M., and P. N. S. Mnkeni. 2009. Effects of goat manure and lime addition on phosphate sorption by two soils from the Transkei eegion, South Africa. *Communications in Soil Science and Plant Analysis* 40 (21–22):3335–3347.

Gichangi, E. M., P. N. S. Mnkeni, and P. C. Brookes. 2010. Goat manure application improves phosphate fertilizer effectiveness through enhanced biological cycling of phosphorus. *Soil Science & Plant Nutrition* 56 (6):853–860.

Gichangi, E. M., P. N. S. Mnkeni, and P. Muchaonyerwa. 2008. Phosphate sorption characteristics and external P requirements of selected South African soils. *Journal of Agriculture and Rural Development in the Tropics and Subtropics (JARTS)* 109 (2):139–149.

Gijsman, A. J., A. Oberson, H. Tiessen, and D. K. Friesen. 1996. Limited applicability of the CENTURY model to highly weathered tropical soils. *Agronomy Journal* 88 (6):894–903.

Giller, K. E. 2002. Targeting management of organic resources and mineral fertilizers: Can we match scientists' fantasies with farmers' realities? In *Integrated Plant Nutrient Management in Sub-Saharan Africa*, edited by Vanlauwe, B., N. Sanginga, J. Diels, and R. Merckx. (Bevat: Balanced Nutrient Management Systems for the Moist Savanna and Humid Forest Zones of Africa.) CAB International, Wallingford.

Graham, R. D. 1984. Breeding for nutritional characteristics in cereals. In *Advances in Plant Nutrition,* edited by Tinker, P. B., and A. Lauchli. Praeger Publishers, New York.

Grant, C., S. Bittman, M. Montreal, C. Plenchette, and C. Morel. 2005. Soil and fertilizer phosphorus: Effects on plant P supply and mycorrhizal development. *Canadian Journal of Plant Science* 85 (1):3–14.

Greenwood, D. J., T. V. Karpinets, and D. A. Stone. 2001. Dynamic model for the effects of soil P and fertilizer P on crop growth, P uptake and soil P in arable cropping: Model description. *Annals of Botany* 88 (2):279–291.

Guppy, C. N., N. W. Menzies, F. P. C. Blamey, and P. W. Moody. 2005a. Do decomposing organic matter residues reduce phosphorus sorption in highly weathered soils? *Soil Science Society of America Journal* 69 (5):1405–1411.

Guppy, C. N., N. W. Menzies, P. W. Moody, and F. P. C. Blamey. 2005b. Competitive sorption reactions between phosphorus and organic matter in soil: A review. *Soil Research* 43 (2):189–202.

Hammond, J. P., M. R. Broadley, P. J. White, G. J. King, H. C. Bowen, R. Hayden, M. C. Meacham et al. 2009. Shoot yield drives phosphorus use efficiency in *Brassica oleracea* and correlates with root architecture traits. *Journal of Experimental Botany* 60 (7):1953–1968.

Haque, I., N. Z. Lupwayi, and H. Ssali. 1999. Agronomic evaluation of unacidulated and partially acidulated Minjingu and Chilembwe phosphate rocks for clover production in Ethiopia. *European Journal of Agronomy* 10 (1):37–47.

Harrison, A. F. 1987. *Soil Organic Phosphorus: A Review of World Literature*. CAB International, Wallingford.

Havlin, J., S. L. Tisdale, W. L. Nelson, and J. D. Beaton. 2013. *Soil Fertility and Fertilizers*. Pearson, New York.

Hawke, D., P. D. Carpenter, and K. A. Hunter. 1989. Competitive adsorption of phosphate on goethite in marine electrolytes. *Environmental Science & Technology* 23 (2):187–191.

Haynes, R. J. 1982. Effects of liming on phosphate availability in acid soils. *Plant and Soil* 68 (3):289–308.

Hedley, M. J., J. W. B. Stewart, and B. S. Chauhan. 1982. Changes in inorganic and organic soil phosphorus fractions induced by cultivation practices and by laboratory incubations. *Soil Science Society of America Journal* 46 (5):970–976.

Houlton, B. Z., Y.-P. Wang, P. M. Vitousek, and C. B. Field. 2008. A unifying framework for dinitrogen fixation in the terrestrial biosphere. *Nature* 454 (7202):327–330.

Hu, X.-Y. 2013. Application of visible/near-infrared spectra in modeling of soil total phosphorus. *Pedosphere* 23 (4):417–421.

Huang, J., P. R. Fisher, and W. R. Argo. 2007. Container substrate-pH response to differing limestone type and particle size. *HortScience* 42 (5):1268–1273.

International Fertilizer Industry Association. 2015. *Production and International Trade Statistics*. International Fertilizer Industry Association, Paris.

Iyamuremye, F., and R. P. Dick. 1996. Organic amendments and phosphorus sorption by soils. In *Advances in Agronomy*, edited by Donald, L. S. Academic Press, Cambridge, MA.

Jama, B., and A. Kiwia. 2009. Agronomic and financial benefits of phosphorus and nitrogen sources in western Kenya. *Experimental Agriculture* 45 (03):241–260.

Jama, B., and P. Van Straaten. 2006. Potential of East African phosphate rock deposits in integrated nutrient management strategies. *Anais da Academia Brasileira de Ciências* 78 (4):781–790.

Jama, B., C. A. Palm, R. J. Buresh, A. Niang, C. Gachengo, G. Nziguheba, and B. Amadalo. 2000. *Tithonia diversifolia* as a green manure for soil fertility improvement in western Kenya: A review. *Agroforestry Systems* 49 (2):201–221.

Janssen, B. H. 2011. Simple models and concepts as tools for the study of sustained soil productivity in long-term experiments: I. New soil organic matter and residual effect of P from fertilizers and farmyard manure in Kabete, Kenya. *Plant and Soil* 339 (1–2):3–16.

Jayne, T. S., and S. Rashid. 2013. Input subsidy programs in sub-Saharan Africa: A synthesis of recent evidence. *Agricultural Economics* 44 (6):547–562.

Jayne, T. S., D. Mather, N. Mason, and J. Ricker-Gilbert. 2013. How do fertilizer subsidy programs affect total fertilizer use in sub-Saharan Africa? Crowding out, diversion, and benefit/cost assessments. *Agricultural Economics* 44 (6):687–703.

Jemo, M., O. J. Jayeoba, T. Alabi, and A. Lopez Montes. 2014. Geostatistical mapping of soil fertility constraints for yam based cropping systems of North-central and Southeast Nigeria. *Geoderma Regional* 2–3 (0):102–109.

Jobbágy, E. G., and R. B. Jackson. 2001. The distribution of soil nutrients with depth: Global patterns and the imprint of plants. *Biogeochemistry* 53 (1):51–77.

Jones, D. L., and E. Oburger. 2011. Solubilization of phosphorus by soil microorganisms. In *Phosphorus in Action*, edited by Bünemann, E., A. Oberson, and E. Frossard. Springer, Berlin Heidelberg.

Jones, C. A., C. V. Cole, A. N. Sharpley, and J. R. Williams. 1984. A simplified soil and plant phosphorus model: I. Documentation. *Soil Science Society of America Journal* 48 (4):800–805.

Jungk, A., and N. Claassen. 1997. Ion diffusion in the soil-root system. *Advances in Agronomy* 61:53–110.

Kalvig, P., N. Fold, J. B. Jønsson, and E. Elisaimon Mshiu. 2012. Rock phosphate and lime for small-scale farming in Tanzania, East Africa. *Geological Survey of Denmark and Greenland Bulletin* 26:85–88.

Kanabo, I. A. K., and R. J. Gilkes. 1988. The effect of the level of phosphate rock application on its dissolution in soil and on bicarbonate-soluble phosphorus. *Fertilizer Research* 16 (1):67–85.

Karekezi, S., and P. Turyareeba. 1995. Woodstove dissemination in Eastern Africa—A review. *Energy for Sustainable Development* 1 (6):12–19.

Karunanithi, R., A. A. Szogi, N. Bolan, R. Naidu, P. Loganathan, P. G. Hunt, M. B. Vanotti, C. P. Saint, Y. S. Ok, and S. Krishnamoorthy. 2015. Phosphorus recovery and reuse from waste streams. In *Advances in Agronomy*. Academic Press, Cambridge, MA.

Keating, B. A., P. S. Carberry, G. L. Hammer, M. E. Probert, M. J. Robertson, D. Holzworth, N. I. Huth et al. 2003. An overview of APSIM: A model designed for farming systems simulation. *European Journal of Agronomy* 18 (3–4):267–288.

Kelderman, P., D. K. Koech, B. Gumbo, and J. O'Keeffe. 2009. Phosphorus budget in the low-income, peri-urban area of Kibera in Nairobi (Kenya). *Water Science and Technology* 60 (10):2669–2676.

Kenya Agricultural Research Institute. 1994. *Fertilizer Use Recommendations*. Kenya Agricultural Reesearch Institute, Nairobi.

Ketterings, Q. M., M. van Noordwijk, and J. M. Bigham. 2002. Soil phosphorus availability after slash-and-burn fires of different intensities in rubber agroforests in Sumatra, Indonesia. *Agriculture, Ecosystems & Environment* 92 (1):37–48.

Khasawneh, F. E., and E. C. Doll. 1979. The use of phosphate rock for direct application to soils. *Advances in Agronomy* 30:159–206.

Kibunja, C. N., F. B. Mwaura, D. N. Mugendi, P. T. Gicheru, J. W. Wamuongo, and A. Bationo. 2012. Strategies for maintenance and improvement of soil productivity under continuous maize and beans cropping system in the sub-humid highlands of Kenya: Case study of the long-term trial at Kabete. In *Lessons Learned from Long-term Soil Fertility Management Experiments in Africa*, edited by Bationo, A., B. Waswa, J. Kihara, I. Adolwa, B. Vanlauwe, and K. Saidou. Springer, Amsterdam.

Kihara, J., and S. Njoroge. 2013. Phosphorus agronomic efficiency in maize-based cropping systems: A focus on western Kenya. *Field Crops Research* 150 (0):1–8.

Kihara, J., B. Vanlauwe, B. Waswa, J. M. Kimetu, J. Chianu, and A. Bationo. 2010. Strategic phosphorus application in legume-cereal rotations increases land productivity and profitability in western Kenya. *Experimental Agriculture* 46 (01):35–52.

Kleinman, P. J. A., D. Pimentel, and R. B. Bryant. 1995. The ecological sustainability of slash-and-burn agriculture. *Agriculture, Ecosystems & Environment* 52 (2–3):235–249.

Kouno, K., Y. Tuchiya, and T. Ando. 1995. Measurement of soil microbial biomass phosphorus by an anion exchange membrane method. *Soil Biology and Biochemistry* 27 (10):1353–1357.

Koutika, L.-S., T. E. Crews, G. Ayaga, and P. C. Brookes. 2013. Microbial biomass P dynamics and sequential P fractionation in high and low P fixing Kenyan soils. *European Journal of Soil Biology* 59 (0):54–59.

Kugblenu, Y. O., F. K. Kumaga, K. Ofori, and J. J. Adu-Gyamfi. 2014. Evaluation of cowpea genotypes for phosphorus use efficiency. *Journal of Agricultural and Crop Research* 2 (10):202–210.

Kwabiah, A. B., C. A. Palm, N. C. Stoskopf, and R. P. Voroney. 2003. Response of soil microbial biomass dynamics to quality of plant materials with emphasis on P availability. *Soil Biology and Biochemistry* 35 (2):207–216.

Kwabiah, A. B., N. C. Stoskopf, R. P. Voroney, and C. A. Palm. 2001. Nitrogen and phosphorus release from decomposing leaves under sub-humid tropical conditions. *Biotropica* 33 (2):229–240.

Kwari, J. D., and T. Batey. 1991. Effect of heating on phosphate sorption and availability in some north-east Nigerian soils. *Journal of Soil Science* 42 (3):381–388.

Lamers, J. P. A., M. Bruentrup, and A. Buerkert. 2015a. Financial performance of fertilisation strategies for sustainable soil fertility management in Sudano–Sahelian West Africa 1: Profitability of annual fertilisation strategies. *Nutrient Cycling in Agroecosystems* 102:137–148.

Lamers, J. P. A., M. Bruentrup, and A. Buerkert. 2015b. Financial performance of fertilization strategies for sustainable soil fertility management in Sudano–Sahelian West Africa. 2: Profitability of long-term capital investments in rock phosphate. *Nutrient Cycling in Agroecosystems* 102:149–165.

Larson, B. A., and G. B. Frisvold. 1996. Fertilizers to support agricultural development in sub-Saharan Africa: What is needed and why. *Food Policy* 21 (6):509–525.

Leenaars, J. G. B. 2013. *Africa Soil Profiles Database: Version 1.1 A Compilation of Geo-Referenced and Standardized Legacy Soil Profile Data for Sub Saharan Africa*. Africa Soil Information Service (AFSIS) project. ISRIC—World Soil Information, Wageningen.

Leiser, W. L., H. F. W. Rattunde, H.-P. Piepho, E. Weltzien, A. Diallo, A. Toure, and B. I. G. Haussmann. 2015. Phosphorous efficiency and tolerance traits for selection of sorghum for performance in phosphorous-limited environments. *Crop Science* 55 (3):1152–1162.

Lerner, B. R., and J. D. Utzinger. 1986. Wood ash as soil liming material. *Hortscience* 21 (1):76–78.

Liao, H., X. Yan, G. Rubio, S. E. Beebe, M. W. Blair, and J. P. Lynch. 2004. Genetic mapping of basal root gravitropism and phosphorus acquisition efficiency in common bean. *Functional Plant Biology* 31 (10):959–970.

Liu, A., C. Hamel, T. Spedding, T.-Q. Zhang, R. Mongeau, G. R. Lamarre, and G. Tremblay. 2008. Soil microbial carbon and phosphorus as influenced by phosphorus fertilization and tillage in a maize-soybean rotation in south-western Quebec. *Canadian Journal of Soil Science* 88 (1):21–30.

Liverpool-Tasie, L. S. O., B. T. Omonona, A. Sanou, and W. Ogunleye. 2015. Is increasing inorganic fertilizer use in sub-Saharan Africa a profitable proposition? Evidence from Nigeria (February 1, 2015). World Bank Policy Research Working Paper (7201). World Bank Group, Washington, DC.

López-Arredondo, D. L., M. A. Leyva-González, S. I. González-Morales, J. López-Bucio, and L. Herrera-Estrella. 2014. Phosphate nutrition: Improving low-phosphate tolerance in crops. *Annual Review of Plant Biology* 65 (1):95–123.

Lynch, J. P. 2011. Root phenes for enhanced soil exploration and phosphorus acquisition: Tools for future crops. *Plant Physiology* 156 (3):1041–1049.

Lynch, J. P. 2013. Steep, cheap and deep: An ideotype to optimize water and N acquisition by maize root systems. *Annals of Botany* 112 (2):347–357.

Lynch, J. P., and T. Wojciechowski. 2015. Opportunities and challenges in the subsoil: Pathways to deeper rooted crops. *Journal of Experimental Botany* 66:2199–2210.

Lynch, J., A. Läuchli, and E. Epstein. 1991. Vegetative growth of the common bean in response to phosphorus nutrition. *Crop Science* 31 (2):380–387.

Mackay, A. D., J. K. Syers, R. W. Tillman, and P. E. H. Gregg. 1986. A simple model to describe the dissolution of phosphate rock in soils. *Soil Science Society of America Journal* 50 (2):291–296.

Maheshwari, D. K. 2011. *Bacteria in Agrobiology: Plant Nutrient Management.* Springer, Amsterdam.

Mahimairaja, S., N. S. Bolan, and M. J. Hedley. 1994. Dissolution of phosphate rock during the composting of poultry manure: An incubation experiment. *Fertilizer Research* 40 (2):93–104.

Malik, M. A., P. Marschner, and K. Saifullah Khan. 2012. Addition of organic and inorganic P sources to soil—Effects on P pools and microorganisms. *Soil Biology and Biochemistry* 49 (0):106–113.

Margenot, A. J., and R. Sommer. 2014. Amplifying investments in phosphorus by site-specific soil fertility management. In *Visions for a Sustainbale Phosphorus Tomorrow.* European Sustainable Phosphorus Platform, Brussels, Belgium.

Margenot, A. J., R. Sommer, N. Jelinski, J. Mukulama, and S. J. Parikh. 2014. Soil phosphorus (P) fractions and lability in western Kenya Oxisols after a decade of P inputs of different quality and response to simulated liming: Implications for cost-effective P management. Paper read at *Soil Science Society of America Annual Meeting, November 2, 2014, at Long Beach, CA.*

Marschner, H. 1995. *Mineral Nutrition of Higher Plants.* Academic Press, Cambridge, MA.

Materechera, S. A. 2012. Using wood ash to ameliorate acidity and improve phosphorus availability in soils amended with partially decomposed cattle or chicken manure. *Communications in Soil Science and Plant Analysis* 43 (13):1773–1789.

Materechera, S. A., and T. S. Mkhabela. 2002. The effectiveness of lime, chicken manure and leaf litter ash in ameliorating acidity in a soil previously under black wattle (*Acacia mearnsii*) plantation. *Bioresource Technology* 85 (1):9–16.

Mehlich, A. 1984. Mehlich 3 soil test extractant: A modification of Mehlich 2 extractant. *Communications in Soil Science and Plant Analysis* 15 (12):1409–1416.

Mendes, F. F., L. J. M. Guimarães, J. C. Souza, P. E. O. Guimarães, J. V. Magalhaes, A. A. F. Garcia, S. N. Parentoni, and C. T. Guimaraes. 2014. Genetic architecture of phosphorus use efficiency in tropical maize cultivated in a low-P soil. *Crop Science* 54 (4):1530–1538.

Menon, R. G., S. H. Chien, and L. L. Hammond. 1989. Comparison of Bray I and Pi tests for evaluating plant-available phosphorus from soils treated with different partially acidulated phosphate rocks. *Plant and Soil* 114 (2):211–216.

Mike, J. H., and S. B. Nanthi. 2003. Role of carbon, nitrogen, and sulfur cycles in soil acidification. In *Handbook of Soil Acidity*: CRC Press, Boca Raton, FL.

Minasny, B., G. Tranter, Al. B. McBratney, D. M. Brough, and B. W. Murphy. 2009. Regional transferability of mid-infrared diffuse reflectance spectroscopic prediction for soil chemical properties. *Geoderma* 153 (1–2):155–162.

Miyasaka, S. C., and M. Habte. 2001. Plant mechanisms and mycorrhizal symbioses to increase phosphorus uptake efficiency. *Communications in Soil Science and Plant Analysis* 32 (7–8):1101–1147.

Mnkeni, P. N. S., J. M. R. Semoka, and J. B. B. S. Buganga. 1991. Effectiveness of Minjingu phosphate rock as a source of phosphorus for maize in some soils of Morogoro, Tanzania. *Zimbabwe Journal of Agricultural Research* 29 (1):27–37.

Mollier, A., P. de Willigen, S. Pellerin, and M. Heinen. 2001. A two dimensional simulation model of phosphorus uptake including crop growth and P response. In *Plant Nutrition*, edited by Horst, W. J., M. K. Schenk, A. Bürkert, N. Claassen, H. Flessa, W. B. Frommer, H. Goldbach et al. Springer, Amsterdam.

Mollier, A., P. De Willigen, M. Heinen, C. Morel, A. Schneider, and S. Pellerin. 2008. A two-dimensional simulation model of phosphorus uptake including crop growth and P-response. *Ecological Modelling* 210 (4):453–464.

Morris, M. L. 2007. *Fertilizer Use in African Agriculture: Lessons Learned and Good Practice Guidelines: Directions in Development—Agriculture and Rural Development 39037*. The World Bank, Washington, DC.

Mowo, J. G. 2000. *Effectiveness of Phosphate Rock on Ferralsols in Tanzania and the Influence of Within-Field Variability. Landbouwuniversiteit Wageningen* (Wageningen Agricultural University), Wageningen.

Msolla, M. M., J. M. R. Semoka, and O. K. Borggaard. 2005. Hard Minjingu phosphate rock: An alternative P source for maize production on acid soils in Tanzania. *Nutrient Cycling in Agroecosystems* 72 (3):299–308.

Mubiru, D. N., and M. S. Coyne. 2009. Legume cover crops are more beneficial than natural fallows in minimally tilled Ugandan soils. *Agronomy Journal* 101 (3):644–652.

Mukuralinda, A., J. S. Tenywa, L. Verchot, J. Obua, and S. Namirembe. 2009. Decomposition and phosphorus release of agroforestry shrub residues and the effect on maize yield in acidic soils of Rubona, southern Rwanda. *Nutrient Cycling in Agroecosystems* 84 (2):155–166.

Mulla, D. J. 2013. Twenty five years of remote sensing in precision agriculture: Key advances and remaining knowledge gaps. *Biosystems Engineering* 114 (4):358–371.

Murwira, H. K., and H. Kirchmann. 1993. Nitrogen dynamics and maize growth in a Zimbabwean sandy soil under manure fertilisation. *Communications in Soil Science and Plant Analysis* 24 (17–18):2343–2359.

Mustonen, P. J., M. Oelbermann, and D. C. L. Kass. 2011. Using *Tithonia diversifolia* (Hemsl.) gray in a short fallow system to increase soil phosphorus availability on a Costa Rican Andosol. *Journal of Agricultural Science* 4 (2):91.

Mwangi, W. M. 1996. Low use of fertilizers and low productivity in sub-Saharan Africa. *Nutrient Cycling in Agroecosystems* 47 (2):135–147.

Naab, J. B., K. J. Boote, J. W. Jones, and C. H. Porter. 2015. Adapting and evaluating the CROPGRO-peanut model for response to phosphorus on a sandy-loam soil under semi-arid tropical conditions. *Field Crops Research* 176 (0):71–86.

Nakamura, S., M. Fukuda, F. Nagumo, and S. Tobita. 2013. Potential utilization of local phosphate rocks to enhance rice pin sub-Saharan Africa. *Japan Agricultural Research Quarterly: JARQ* 47 (4):353–363.

Nannipieri, P., L. Giagnoni, L. Landi, and G. Renella. 2011. Role of phosphatase enzymes in soil. In *Phosphorus in Action*. Springer, Amsterdam.

Ndung'u, K. W., J. R. Okalebo, C. O. Othieno, M. N. Kifuko, A. K. Kipkoech, and L. N. Kimenye. 2006. Residual effectiveness of minjingu phosphate rock and fallow biomass on crop yields and financial returns in western Kenya. *Experimental Agriculture* 42 (03):323–336.

Nduwumuremyi, A., V. Ruganzu, J. N. Mugwe, and A. C. Rusanganwa. 2013. Effects of unburned lime on soil pH and base cations in acidic soil. *ISRN Soil Science* 2013:7.

Negassa, W., and P. Leinweber. 2009. How does the Hedley sequential phosphorus fractionation reflect impacts of land use and management on soil phosphorus: A review. *Journal of Plant Nutrition and Soil Science* 172 (3):305–325.

Nekesa, P., H. K. Maritim, J. R. Okalebo, and P. L. Woomer. 1999. Economic analysis of maize-bean production using a soil fertility replenishment product (PREP-PAC) in western Kenya. *African Crop Science Journal* 7 (4):585–590.

Nekesa, A. O., J. R. Okalebo, C. O. Othieno, M. N. Thuita, M. Kipsat, A. Bationo, and N. Sanginga. 2005. The potential of Minjingu phosphate rock from Tanzania as a liming material: Effect on maize and bean intercrop on acid soils of western Kenya. *African Crop Science Conference Proceedings* 7:1121–1128.

Nimoh, F., K. Ohene-Yankyera, K. Poku, F. Konradsen, and R. C. Abaidoo. 2014. Farmers' perception on excreta reuse for peri-urban agriculture in southern Ghana. *Journal of Development and Agricultural Economics* 6 (10):421–428.

Njui, N. A., and A. A. O. Musandu. 1999. Response of maize to phosphorus fertilisation at selected sites in western Kenya. *African Crop Science Journal* 7 (4):397–406.

Notholt, A. J. G. 1994. Phosphate rock: Factors in economic and technical evaluation. *Geological Society, London, Special Publications* 79 (1):53–65.

Nwoke, O. C., B. Vanlauwe, J. Diels, N. Sanginga, O. Osonubi, and R. Merckx. 2003. Assessment of labile phosphorus fractions and adsorption characteristics in relation to soil properties of West African savanna soils. *Agriculture, Ecosystems & Environment* 100 (2–3):285–294.

Nye, P. H., and F. H. C. Marriott. 1969. A theoretical study of the distribution of substances around roots resulting from simultaneous diffusion and mass flow. *Plant and Soil* 30 (3):459–472.

Nyenje, P. 2014. *Fate and Transport of Nutrients in Groundwater and Surface Water in an Urban Slum Catchment Kampala, Uganda.* Delft University of Technology, Delft.

Nziguheba, G. 2007. Overcoming phosphorus deficiency in soils of Eastern Africa: Recent advances and challenges. In *Advances in Integrated Soil Fertility Management in Sub-Saharan Africa: Challenges and Opportunities.* Springer, Amsterdam.

Nziguheba, G., R. Merckx, C. A. Palm, and P. Mutuo. 2002. Combining *Tithonia diversifolia* and fertilizers for maize production in a phosphorus deficient soil in Kenya. *Agroforestry Systems* 55 (3):165–174.

Nziguheba, G., R. Merckx, C. A. Palm, and M. R. Rao. 2000. Organic residues affect phosphorus availability and maize yields in a Nitisol of western Kenya. *Biology and Fertility of Soils* 32 (4):328–339.

Nziguheba, G., C. A. Palm, R. J. Buresh, and P. C. Smithson. 1998. Soil phosphorus fractions and adsorption as affected by organic and inorganic sources. *Plant and Soil* 198 (2):159–168.

Oberson, A., and E. J. Joner. 2005. Microbial turnover of phosphorus in soil. In *Organic Phosphorus in the Environment.* CABI, Wallingford, pp. 133–164.

Oberson, A., D. K. Friesen, C. Morel, and H. Tiessen. 1997. Determination of phosphorus released by chloroform fumigation from microbial biomass in high P sorbing tropical soils. *Soil Biology and Biochemistry* 29 (9–10):1579–1583.

Oberson, A., D. K. Friesen, I. M. Rao, P. C. Smithson, B. L. Turner, and E. Frossard. 2006. Chapter 37: Improving phosphorus fertility in tropical soils through biological interventions. In *Biological Approaches to Sustainable Soil Systems*, edited by Uphoff, N. (ed.). Marcel Dekker, New York.

Oberson, A., P. Pypers, E. K. Bünemann, and E. Frossard. 2011. Management impacts on biological phosphorus cycling in cropped soils. In *Phosphorus in Action*, edited by Bünemann, E., A. Oberson, and E. Frossard. Springer, Berlin Heidelberg.

Oehl, F., A. Oberson, M. Probst, A. Fliessbach, H.-R. Roth, and E. Frossard. 2001. Kinetics of microbial phosphorus uptake in cultivated soils. *Biology and Fertility of Soils* 34 (1):31–41.

Ohno, T., and M. S. Erich. 1990. Effect of wood ash application on soil pH and soil test nutrient levels. *Agriculture, Ecosystems & Environment* 32 (3–4):223–239.

Ojo, O. D., E. A. Akinrinde, and M. O. Akoroda. 2010. Residual effects of phosphorus sources in grain amaranth production. *Journal of Plant Nutrition* 33 (5):770–783.

Okalebo, J. R., K. W. Gathua, and P. L. Woomer. 2002. *Laboratory Methods of Soil and Plant Analysis: A Working Manual.* Second ed. Sustainable Agriculture Centre for Research Extension and Development in Africa (SACRED), Nairobi.

Okalebo, J. R., C. O. Othieno, A. O. Nekesa, K. W. Ndungu-Magiroi, and M. N. Kifuko-Koech. 2009. Potential for agricultural lime on improved soil health and agricultural production in Kenya. *Ninth African Crop Science, Conference Proceedings, Cape Town, South Africa, 28 September–2 October 2009*, pp. 339–341.

Okalebo, J. R., P. L. Woomer, C. O. Othieno, N. K. Karanja, S. Ikerra, A. O. Esilaba, A. O. Nekesa et al. 2007. Potential for agricultural lime on improved soil health and agricultural production in Kenya. Paper read at *African Crop Science Proceedings, Egypt.*

Oliveira, S. M., and A. S. Ferreira. 2014. Change in soil microbial and enzyme activities in response to the addition of rock-phosphate-enriched compost. *Communications in Soil Science and Plant Analysis* 45 (21):2794–2806.

Olson, J. S. 1963. Energy storage and the balance of producers and decomposers in ecological systems. *Ecology* 44 (2):322–331.

Osorio, N. W., and M. Habte. 2013a. Synergistic effect of a phosphate-solubilizing fungus and an arbuscular mycorrhizal fungus on leucaena seedlings in an Oxisol fertilized with rock phosphate. *Botany* 91 (4):274–281.

Osorio, N. W., and M. Habte. 2013b. Phosphate desorption from the surface of soil mineral particles by a phosphate-solubilizing fungus. *Biology and Fertility of Soils* 49 (4):481–486.

Osorio, N. W., and M. Habte. 2015. Effect of a phosphate-solubilizing fungus and an arbuscular mycorrhizal fungus on *Leucaena* seedlings in tropical soils with contrasting phosphate sorption capacity. *Plant and Soil* 389 (1–2):375–385.

Owusu-Bennoah, E., and C. Visker. 1994. Organic wastes hijacked. *ILEIA Newsletter* 10 (3):12–13.

Parton, W. J. 1996. The CENTURY model. In *Evaluation of Soil Organic Matter Models*, edited by Powlson, D., P. Smith, and J. Smith. Springer, Berlin Heidelberg.

Pearson, R. W. 1975. *Soil Acidity and Liming in the Humid Tropics.* New York State College of Agriculture and Life Sciences, Cornell University, Ithaca, NY.

Peters, J. B. 1996. Choosing between liming materials. In *University of Wisconsin Extension.* University of Wisconsin-Madison, Madison, WI.

Pitman, R. M. 2006. Wood ash use in forestry—A review of the environmental impacts. *Forestry* 79 (5):563–588.

Probert, M. E. 2004. A capability in APSIM to model phosphorus responses in crops. Paper read at *Nutrient Management in Tropical Cropping Systems, Canberra, Australia.*

Probert, M. E., and B. A. Keating. 2000. What soil constraints should be included in crop and forest models? *Agriculture, Ecosystems & Environment* 82 (1–3):273–281.

Probert, M. E., J. R. Okalebo, J. R. Simpson, and R. K. Jones. 1992. The role of boma manure for improving soil fertility. Paper read at *A Search for Strategies for Sustainable Dryland Cropping in Semi-arid Eastern Kenya. ACIAR Proceedings.* Nairobi, Kenya.

Pur'Homme, M. 2010. World phosphate rock flows, losses, and uses. In *Phosphates International Conference.* International Fertilizer Industry Association, Brussels.

Pypers, P., S. Verstraete, C. P. Thi, and R. Merckx. 2005. Changes in mineral nitrogen, phosphorus availability and salt-extractable aluminium following the application of green manure residues in two weathered soils of South Vietnam. *Soil Biology and Biochemistry* 37 (1):163–172.

Qian, P., and J. J. Schoenau. 2002. Practical applications of ion exchange resins in agricultural and environmental soil research. *Canadian Journal of Soil Science* 82 (1):9–21.

Raison, R. J., P. K. Khanna, and P. V. Woods. 1985. Transfer of elements to the atmosphere during low-intensity prescribed fires in three Australian subalpine eucalypt forests. *Canadian Journal of Forest Research* 15 (4):657–664.

Rajan, S. S. S., J. H. Watkinson, and A. G. Sinclair. 1996. Phosphate rocks for direct application to soils. *Advances in Agronomy* 57:77–159.

Ramaekers, L., R. Remans, I. M. Rao, M. W. Blair, and J.Vanderleyden. 2010. Strategies for improving phosphorus acquisition efficiency of crop plants. *Field Crops Research* 117 (2–3):169–176.

Rao, I. M., and N. Terry. 1989. Leaf phosphate status, photosynthesis, and carbon partitioning in sugar beet: I. Changes in growth, gas exchange, and calvin cycle enzymes. *Plant
Physiology* 90 (3):814–819.

Rao, I. M., D. K. Friesen, and M. Osaki. 1999. Plant adaptation to phosphorus-limited tropical
soils. *Handbook of Plant and Crop Stress*. Marcel Dekker, Inc., New York, pp. 61–96.

Richardson, A. E., and R. J. Simpson. 2011. Soil microorganisms mediating phosphorus availability update on microbial phosphorus. *Plant Physiology* 156 (3):989–996.

Robinson, J. S., and J. K. Syers. 1990. A critical evaluation of the factors influencing the dissolution of Gafsa phosphate rock. *Journal of Soil Science* 41 (4):597–605.

Robinson, J. S., and J. K. Syers. 1991. Effects of solution calcium concentration and calcium
sink size on the dissolution of Gafsa phosphate rock in soils. *Journal of Soil Science* 42
(3):389–397.

Romanya, J., P. K. Khanna, and R. J. Raison. 1994. Effects of slash burning on soil phosphorus
fractions and sorption and desorption of phosphorus. *Forest Ecology and Management*
65 (2):89–103.

Romney, D. L., P. J. Thorne, and D. Thomas. 1994. Some animal-related factors influencing
the cycling of nitrogen in mixed farming systems in sub-Saharan Africa. *Agriculture,
Ecosystems & Environment* 49 (2):163–172.

Rose, T. J., M. T. Rose, J. Pariasca-Tanaka, S. Heuer, and M. Wissuwa. 2011. The frustration
with utilization: Why have improvements in internal phosphorus utilization efficiency in
crops remained so elusive? *Frontiers in Plant Science* 2:73.

Roy, A. H., and G. H. McClellan. 1986. Processing phosphate ores into fertilizers. In
*Management of Nitrogen and Phosphorus Fertilizers in Sub-Saharan Africa*, edited by
Mokwunye, A. U., and P. G. Vlek. Springer, Amsterdam.

Rusinamhodzi, L., M. Corbeels, S. Zingore, J. Nyamangara, and K. E. Giller. 2013. Pushing
the envelope? Maize production intensification and the role of cattle manure in recovery
of degraded soils in smallholder farming areas of Zimbabwe. *Field Crops Research* 147
(0):40–53.

Rutto, E. C., R. Okalebo, C. O. Othieno, M. J. Kipsat, A. Bationo, and K. Girma. 2011. Effect
of PREP-PAC application on soil properties, maize, and legume yields in a Ferralsol of
Western Kenya. *Communications in Soil Science and Plant Analysis* 42 (20):2526–2536.

Ryan, J., H. Ibrikci, A. Delgado, J. Torrent, R. Sommer, and A. Rashid. 2012. Significance of
phosphorus for agriculture and the environment in the West Asia and North Africa region.
In *Advances in Agronomy*, edited by Donald, L. S. Academic Press, Cambridge, MA.

Saa, A. M., C. Trasar-Cepeda, F. Gil-Sotres, and T. Carballas. 1993. Changes in soil phosphorus and acid phosphatase activity immediately following forest fires. *Soil Biology and
Biochemistry* 25 (9):1223–1230.

Sabiha, J., T. Mehmood, M. M. Chaudhry, M. Tufail, and N. Irfan. 2009. Heavy metal pollution from phosphate rock used for the production of fertilizer in Pakistan. *Microchemical
Journal* 91 (1):94–99.

Sale, P. W. G., and A. U. Mokwunye. 1993. Use of phosphate rocks in the tropics. *Fertilizer
Research* 35 (1–2):33–45.

Sanchez, P. A. 1999. Improved fallows come of age in the tropics. *Agroforestry Systems* 47
(1–3):3–12.

Sanchez, P. A. 2002. Soil fertility and hunger in Africa. *Science (Washington)* 295
(5562):2019–2020.

Sanchez, P. A., and T. J. Logan. 1992. Myths and science about the chemistry and fertility of
soils in the tropics. In *Myths and Science of Soils of the Tropics*, edited by Lal, R., and
P. A. Sanchez. Soil Science Society of America and American Society of Agronomy,
Madison, WI.

Sánchez, P. A., and J. G. Salinas. 1981. Low-input technology for managing Oxisols and Ultisols in tropical America. *Advances in Agronomy* 34:279–406.

Sanchez, P. A., S. Ahamed, F. Carré, A. E. Hartemink, J. Hempel, J. Huising, P. Lagacherie et al. 2009. Digital soil map of the world. *Science* 325 (5941):680–681.

Sanchez, P. A., K. D. Shepherd, M. J. Soule, F. M. Place, R. J. Buresh, A.-M. N. Izac, A. U. Mokwunye, F. R. Kwesiga, C. G. Ndiritu, and P. L. Woomer. 1997. Soil fertility replenishment in Africa: An investment in natural resource capital. In *Replenishing Soil Fertility in Africa*, edited by Buresh, R. J., P. A. Sanchez, and F. Calhoun. Soil Science Society of America, Madison, WI. 1–46.

Sanginga, N., and P. L. Woomer. 2009. *Integrated Soil Fertility Management in Africa: Principles, Practices, and Developmental Process*: International Center for Tropical Agriculture, Palmira.

Sarabèr, A., M. Cuperus, and J. Pels. 2011. Ash from combustion of cacao residues for nutrient recycling: A case study. In *Recycling of Biomass Ashes*, edited by Insam, H., and B. A. Knapp. Springer, Berlin Heidelberg.

Satter, M. A., M. M. Hanafi, T. M. M. Mahmud, and H. Azizah. 2006. Role of arbuscular mycorrhiza and phosphorus in acacia mangium-peanut agroforestry system for rejuvenation of tin tailings. *Journal of Sustainable Agriculture* 28 (4):55–68.

Scalenghe, R., A. C. Edwards, E. Barberis, and F. Ajmone-Marsan. 2014. Release of phosphorus under reducing and simulated open drainage conditions from overfertilised soils. *Chemosphere* 95 (0):289–294.

Schiemenz, K., J. Kern, H.-M. Paulsen, S. Bachmann, and B. Eichler-Löbermann. 2011. Phosphorus fertilizing effects of biomass ashes. In *Recycling of Biomass Ashes*, edited by Insam, H., and B. A. Knapp. Springer, Berlin Heidelberg.

Seelan, S. K., S. Laguette, G. M. Casady, and G. A. Seielstad. 2003. Remote sensing applications for precision agriculture: A learning community approach. *Remote Sensing of Environment* 88 (1–2):157–169.

Sharpley, A. N. 1995. Soil phosphorus dynamics: Agronomic and environmental impacts. *Ecological Engineering* 5 (2–3):261–279.

Sheldon, R. P., and D. F. Davidson. 1987. Discovery and development of phosphate rock resources in phosphate-deficient third world countries. *Natural Resources Forum* 11 (1):93–98.

Sibbesen, E. 1978. An investigation of the anion-exchange resin method for soil phosphate extraction. *Plant and Soil* 50 (1–3):305–321.

Sikora, F. J. 2002. Evaluating and quantifying the liming potential of phosphate rocks. *Nutrient Cycling in Agroecosystems* 63 (1):59–67.

Simons, A., D. Solomon, W. Chibssa, G. Blalock, and J. Lehmann. 2014. Filling the phosphorus fertilizer gap in developing countries. *Nature Geoscience* 7 (1):3–3.

Simpson, J. R., J. R. Okalebo, and G. A. S. Lubulwa. 1996. The problem of maintaining soil fertility in Eastern Kenya: A review of relevant research. In *ACIAR Monograph*. Australian Centre for International Agricultural Research, Canberra.

Sinsabaugh, R. L., and J. J. Follstad Shah. 2012. Ecoenzymatic stoichiometry and ecological theory. *Annual Review of Ecology, Evolution, and Systematics* 43 (1):313–343.

Sinsabaugh, R. L., B. H. Hill, and J. J. Follstad Shah. 2009. Ecoenzymatic stoichiometry of microbial organic nutrient acquisition in soil and sediment. *Nature* 462 (7274):795–798.

Six, J., C. Feller, K. Denef, S. Ogle, J. C. De Moraes Sa, and A. Albrecht. 2002. Soil organic matter, biota and aggregation in temperate and tropical soils—Effects of no-tillage. *Agronomie* 22 (7–8):755–775.

Smaling, E. M. A., and B. H. Janssen. 1993. Calibration of quefts, a model predicting nutrient uptake and yields from chemical soil fertility indices. *Geoderma* 59 (1–4):21–44.

Smaling, E. M. A., S. M. Nandwa, and B. H. Janssen. 1997. Soil fertility in Africa is at stake. In *Replenishing Soil Fertility in Africa*, edited by Buresh, R. J., P. A. Sanchez, and F. Calhoun. Soil Science Society of America and American Society of Agronomy, Madison, WI.

Smithson, P., B. Jama, R. Delve, P. van Straaten, R. Buresh, S. S. S. Rajan, and S. H. Chien. 2003. East African phosphate resources and their agronomic performance. Paper read at *Direct Application of Phosphate Rock and Related Appropriate Technology—Latest Developments and Practical Experiences. Proceedings of an International Meeting, Kuala Lumpur, Malaysia, 16–20 July 2001.*

Soil Survey Staff. 2015. *Web Soil Survey*, edited by Natural Resources Conservation Service, U. S. Department of Agriculture, Washington, DC.

Sombroek, W. G., H. M. H. Braun, and B. J. A. van der Pouw. 1982. Exploratory soil map and agro-climatic zone map of Kenya, 1980. National Agricultural Laboratories, Ministry of Agriculture, Nairobi.

Sommer, R., D. Bossio, L. Desta, J. Dimes, J. Kihara, S. Koala, N. Mango, D. Rodriguez, C. Thierfelder, and L. Winowiecki. 2013. Profitable and sustainable nutrient management systems for east and southern African smallholder farming systems–Challenges and opportunities. International Centre for Tropical Agriculture, Valle del Cauca.

Sommer, R., P. L. G. Vlek, T. D. de Abreu Sá, K. Vielhauer, R. de Fátima Rodrigues Coelho, and H. Fölster. 2004. Nutrient balance of shifting cultivation by burning or mulching in the Eastern Amazon—Evidence for subsoil nutrient accumulation. *Nutrient Cycling in Agroecosystems* 68 (3):257–271.

Soriano-Disla J. M., L. J. Janik, R. A. V. Rossel, L. M. Macdonald, and M. J. McLaughlin. 2013. The performance of visible, near-, and mid-infrared reflectance spectroscopy for prediction of soil physical, chemical, and biological properties. *Applied Spectroscopy Reviews* 49 (2):139–186.

Stafford, J. V. 2000. Implementing precision agriculture in the 21st century. *Journal of Agricultural Engineering Research* 76 (3):267–275.

Stoorvogel, J. J., E. M. A. Smaling, and B. H. Janssen. 1993. Calculating soil nutrient balances in Africa at different scales. *Fertilizer Research* 35 (3):227–235.

Stursova, M., and R. L. Sinsabaugh. 2008. Stabilization of oxidative enzymes in desert soil may limit organic matter accumulation. *Soil Biology and Biochemistry* 40 (2):550–553.

Su, J.-Y., Q. Zheng, H.-W. Li, B. Li, R.-L. Jing, Y.-P. Tong, and Z.-S. Li. 2009. Detection of QTLs for phosphorus use efficiency in relation to agronomic performance of wheat grown under phosphorus sufficient and limited conditions. *Plant Science* 176 (6):824–836.

Sugihara, S., S. Funakawa, T. Nishigaki, M. Kilasara, and T. Kosaki. 2012. Dynamics of fractionated P and P budget in soil under different land management in two Tanzanian croplands with contrasting soil textures. *Agriculture, Ecosystems & Environment* 162 (0):101–107.

Suriyagoda, L. D. B., M. H. Ryan, M. Renton, and H. Lambers. 2014. Plant responses to limited moisture and phosphorus availability: A meta-analysis. In *Advances in Agronomy*, edited by Donald, L. S. Academic Press, Cambridge. MA.

Szilas, C., J. M. R. Semoka, and O. K. Borggaard. 2007. Establishment of an agronomic database for Minjingu phosphate rock and examples of its potential use. *Nutrient Cycling in Agroecosystems* 78 (3):225–237.

Tabo, R., A. Bationo, B. Gerard, J. Ndjeunga, D. Marchal, B. Amadou, M. Garba Annou et al. 2007. Improving cereal productivity and farmers' income using a strategic application of fertilizers in West Africa. In *Advances in Integrated Soil Fertility Management in sub-Saharan Africa: Challenges and Opportunities*, edited by Bationo, A., B. Waswa, J. Kihara, and J. Kimetu. Springer, Amsterdam.

Tabu, I. M., A. Bationo, R. K. Obura, and J. K. Masinde. 2007. Effect of rock phosphate, lime and green manure on growth and yield of maize in a non productive niche of a rhodic Ferralsol in farmer's fields. In *Advances in Integrated Soil Fertility Management in sub-Saharan Africa: Challenges and Opportunities*, edited by Bationo, A., B. Waswa, J. Kihara, and J. Kimetu: Springer, Amsterdam.

Terhoeven-Urselmans, T., O. Spaargaren, T.-G. Vagen, and K. D. Shepherd. 2010. Prediction of soil fertility properties from a globally distributed soil mid-infrared spectral library. *Soil Science Society of America Journal* 74 (5):1792–1799.

Tian, J., and H. Liao. 2015. The role of intracellular and secreted purple acid phosphatases in plant phosphorus scavenging and recycling. In *Annual Plant Reviews*, vol. 28. John Wiley & Sons, Inc, New York.

Tiessen, H., I. H. Salcedo, and E. V. S. B. Sampaio. 1992. Nutrient and soil organic matter dynamics under shifting cultivation in semi-arid northeastern Brazil. *Agriculture, Ecosystems & Environment* 38 (3):139–151.

Tiessen, H., J. W. B. Stewart, and J. O. Moir. 1983. Changes in organic and inorganic phosphorus composition of two grassland soils and their particle size fractions during 60–90 years of cultivation. *Journal of Soil Science* 34 (4):815–823.

Timmer, L., and C. Visker. 1998. Possibilities and impossibilities of the use of human excreta as fertiliser in agriculture in sub-Saharan Africa. Royal Tropical Institute and University of Amsterdam, Amsterdam.

Tisdale, S. L., W. L. Nelson, and J. D. Beaton. 1985. *Soil Fertility and Fertilizers*. Collier Macmillan Publishers, New York.

Tittonell, P., and K. E. Giller. 2013. When yield gaps are poverty traps: The paradigm of ecological intensification in African smallholder agriculture. *Field Crops Research* 143 (0):76–90.

Tittonell, P., A. Muriuki, C. J. Klapwijk, K. D. Shepherd, R. Coe, and B. Vanlauwe. 2013. Soil heterogeneity and soil fertility gradients in smallholder farms of the East African highlands. *Soil Science Society of America Journal* 77 (2):525–538.

Tungani, J. O., E. J. Mukhwana, and P. L. Woomer. 2002. MBILI is number 1: A handbook for innovative maize-legume intercropping. Sustainable Agriculture Centre for Research Extension and Development in Africa (SACRED), Nairobi.

Turner, B. L. 2010. Variation in pH optima of hydrolytic enzyme activities in tropical rain forest soils. *Applied and Environmental Microbiology* 76 (19):6485–6493.

Turner, B. L., and M. S. A. Blackwell. 2013. Isolating the influence of pH on the amounts and forms of soil organic phosphorus. *European Journal of Soil Science* 64 (2):249–259.

Turner, B. L., and B. M. J. Engelbrecht. 2011. Soil organic phosphorus in lowland tropical rain forests. *Biogeochemistry* 103 (1–3):297–315.

Turner, B. L., E. Frossard, and D. S. Baldwin. 2005. *Organic Phosphorus in the Environment*. CAB International, Wallingford.

Turner, B. L., M. J. Papházy, P. M. Haygarth, and I. D. Mckelvie. 2002. Inositol phosphates in the environment. *Philosphical Transactions of the Royal Society of London B* 357 (1420):449–469.

Turner, B. L., A. E. Richardson, and E. J. Mullaney. 2007. *Inositol Phosphates: Linking Agriculture and the Environment*. CAB International, Wallingford.

Twomlow, S., D. Rohrbach, J. Dimes, J. Rusike, W. Mupangwa, B. Ncube, L. Hove, M. Moyo, N. Mashingaidze, and P. Mahposa. 2011. Micro-dosing as a pathway to Africa's Green Revolution: Evidence from broad-scale on-farm trials. In *Innovations as Key to the Green Revolution in Africa*, edited by Bationo, A., B. Waswa, J. M. Okeyo, F. Maina, and J. M. Kihara. Springer, Amsterdam.

Ulery, A. L., R. C. Graham, and C. Amrhein. 1993. Wood-ash composition and soil ph following intense burning. *Soil Science* 156 (5):358–364.

Ussiri, D. A., P. N. S. Mnkeni, A. F. MacKenzie, and J. M. R. Seraoka. 1998. Soil test calibration studies for formulation of phosphorus fertilizer recommendations for maize in Morogoro district, Tanzania: II. Estimation of optimum fertilizer rates. *Communications in Soil Science and Plant Analysis* 29 (17–18):2815–2828.

Vågen, T. G., K. D. Shepherd, M. G. Walsh, L. Winowiecki, L. T. Desta, and J. E. Tondoh. 2010. AfSIS technical specifications: Soil health surveillance. *World Agroforestry Centre*, Nairobi, Kenya.

van der Eijk, D., Bert. H. Janssen, and O. Oenema. 2006. Initial and residual effects of fertilizer phosphorus on soil phosphorus and maize yields on phosphorus fixing soils: A case study in south-west Kenya. *Agriculture, Ecosystems & Environment* 116 (1–2):104–120.

van der Velde, M., C. Folberth, J. Balkovič, P. Ciais, S. Fritz, I. A. Janssens, M. Obersteiner et al. 2014. African crop yield reductions due to increasingly unbalanced nitrogen and phosphorus consumption. *Global Change Biology* 20 (4):1278–1288.

van der Waals, J. H., and M. C. Laker. 2008. Micronutrient deficiencies in crops in Africa with emphasis on Southern Africa. In *Micronutrient Deficiencies in Global Crop Production*, edited by Alloway, B. Springer, Amsterdam.

Van Kauwenbergh, S. J. 1991. Overview of phosphate deposits in East and Southeast Africa. *Fertilizer Research* 30 (2–3):127–150.

van Straaten, P.. 2002. *Rocks for Crops: Agrominerals of sub-Saharan Africa*. International Center for Research in Agroforestry, Nairobi.

Vance, C. P., C. Uhde-Stone, and D. L. Allan. 2003. Phosphorus acquisition and use: Critical adaptations by plants for securing a nonrenewable resource. *New Phytologist* 157 (3):423–447.

Vanlauwe, B., and K. E. Giller. 2006. Popular myths around soil fertility management in sub-Saharan Africa. *Agriculture, Ecosystems & Environment* 116 (1–2):34–46.

Vanlauwe, B., D. Coyne, J. Gockowski, S. Hauser, J. Huising, C. Masso, G. Nziguheba, M. Schut, and P. Van Asten. 2014. Sustainable intensification and the African smallholder farmer. *Current Opinion in Environmental Sustainability* 8 (0):15–22.

Vanlauwe, B., N. Sanginga, J. Diels, and R. Merckx. 2006. Case studies related to the management of soil acidity and infertility in the West-African moist savannah. In *Management Practices for Improving Sustainable Crop Production in Tropical Acid Soils*. Results of a coordinated research project organized by the Joint FAO/IAEA Programme of Nuclear Techniques in Food and Agriculture. Vienna, Austria.

Veneklaas, E. J., H. Lambers, J. Bragg, P. M. Finnegan, C. E. Lovelock, W. C. Plaxton, C. A. Price et al. 2012. Opportunities for improving phosphorus-use efficiency in crop plants. *New Phytologist* 195 (2):306–320.

Vitousek, P. M., S. Porder, B. Z. Houlton, and O. A. Chadwick. 2010. Terrestrial phosphorus limitation: Mechanisms, implications, and nitrogen–phosphorus interactions. *Ecological Applications* 20 (1):5–15.

Waigwa, M. W., C. O. Othieno, and J. R. Okalebo. 2003. Phosphorus availability as affected by the application of phosphate rock combined with organic materials to acid soils in western Kenya. *Experimental Agriculture* 39 (4):395–407.

Walbridge, M. R. 1991. Phosphorus availability in acid organic soils of the lower North Carolina coastal plain. *Ecology* 72 (6):2083–2100.

Walker, T. W., and J. K. Syers. 1976. The fate of phosphorus during pedogenesis. *Geoderma* 15 (1):1–19.

Wang, R., Y. Balkanski, O. Boucher, P. Ciais, J. Penuelas, and S. Tao. 2015. Significant contribution of combustion-related emissions to the atmospheric phosphorus budget. *Nature Geoscience* 8 (1):48–54.

Wang, Y.-P., B. Z. Houlton, and C. B. Field. 2007. A model of biogeochemical cycles of carbon, nitrogen, and phosphorus including symbiotic nitrogen fixation and phosphatase production. *Global Biogeochemical Cycles* GB1018. Available at http://onlinelibrary .wiley.com/doi/10.1029/2006GB002797/full

Wang, X., X. Yan, and H. Liao. 2010. Genetic improvement for phosphorus efficiency in soybean: A radical approach. *Annals of Botany* 106 (1):215–222.

Waring, B. G., S. Rose. Weintraub, and R. L. Sinsabaugh. 2014. Ecoenzymatic stoichiometry of microbial nutrient acquisition in tropical soils. *Biogeochemistry* 117 (1):101–113.

Watanabe, F. S., and S. R. Olsen. 1965. Test of an ascorbic acid method for determining phosphorus in water and NaHCO3 extracts from soil. *Soil Science Society of America Journal* 29 (6):677–678.

White, P. J., T. S. George, L. X. Dupuy, A. J. Karley, T. A. Valentine, L. Wiesel, and J. Wishart. 2013. Root traits for infertile soils. *Frontiers in Plant Science* 4:1–7.

Withers, P. J. A., R. Sylvester-Bradley, D. L. Jones, J. R. Healey, and P. J. Talboys. 2014. Feed the crop not the soil: Rethinking phosphorus management in the food chain. *Environmental Science & Technology* 48 (12):6523–6530.

Wolf, A. M., and D. E. Baker. 1985. Comparisons of soil test phosphorus by Olsen, Bray P1, Mehlich I and Mehlich III methods. *Communications in Soil Science and Plant Analysis* 16 (5):467–484.

Woomer, P. L. 2007. Costs and returns of soil fertility management options in Western Kenya. In *Advances in Integrated Soil Fertility Management in sub-Saharan Africa: Challenges and Opportunities*, edited by Bationo, A., B. Waswa, J. Kihara, and J. Kimetu. Springer, Amsterdam.

Yamoah, C., M. Ngueguim, C. Ngong, and D. K. W. Dias. 1996. Reduction of P fertilizer requirement using lime and Mucuna on high P-sorption soils of NW Cameroon. *African Crop Science Journal* 4 (4):441–451.

Yampracha, S., T. Attanandana, A. Sidibe, A. Srivihok, and R. S. Yost. 2006. Predicting the dissolution of four rock phosphates in flooded acid sulfate soils of Thailand. *Soil Science* 171 (3):200–209.

Yang, Z.-B., I. M. Rao, and W. J. Horst. 2013. Interaction of aluminium and drought stress on root growth and crop yield on acid soils. *Plant and Soil* 372 (1–2):3–25.

Yusiharni, E., and R. J. Gilkes. 2012. Short term effects of heating a lateritic podzolic soil on the availability to plants of native and added phosphate. *Geoderma* 191 (0):132–139.

Zhang, Z., H. Liao, and W. J. Lucas. 2014. Molecular mechanisms underlying phosphate sensing, signaling, and adaptation in plants. *Journal of Integrative Plant Biology* 56 (3):192–220.

Zingore, S., R. J. Delve, J. Nyamangara, and K. E. Giller. 2008. Multiple benefits of manure: The key to maintenance of soil fertility and restoration of depleted sandy soils on African smallholder farms. *Nutrient Cycling in Agroecosystems* 80 (3):267–282.

Zingore, S., P. Tittonell, M. Corbeels, M. T. van Wijk, and K. E. Giller. 2011. Managing soil fertility diversity to enhance resource use efficiencies in smallholder farming systems: A case from Murewa District, Zimbabwe. *Nutrient Cycling in Agroecosystems* 90 (1):87–103.

**FIGURE 3.1** A mature lowland rainforest limited by nutrient availability located in northern French Guiana. The ecosystem is characterized by a high diversity of species, low P concentration in tree leaves, a highly weathered soil depleted in nutrients and a tight coupling between cycles of C, N, and P (Barantal et al. 2012). (Courtesy of Oriol Grau.)

(a)

(b)

**FIGURE 5.1** Phosphate deficiency symptoms in some crops: (a) symptoms of P deficiency in lettuce (*Lactuca sativa*) and (b) symptoms of P deficiency in barley (*Hordeum vulgare*).

(*Continued*)

(c)

(d)

**FIGURE 5.1 (CONTINUED)** Phosphate deficiency symptoms in some crops: (c) symptoms of P deficiency in corn (*Zea mays*) and (d) P deficiency in guava (*Psidium gujava*). (Courtesy of International Plant Nutrition Institute, Peachtree Corners, Georgia.)

(a)

(b)

**FIGURE 8.1** (a) Phosphorus-deficient 20-day maize exhibiting characteristic purple vein-ing and stunting consistent with low available soil P (1.9 mg kg$^{-1}$ soil, Olsen test). (Courtesy of International Plant Nutrition Institute, Peachtree Corners, Georgia, http://media.ipni.net/.) (b) Phosphorus omission trial demonstrating severe stunting in the check (no P added; left) as compared to a 33 kg P ha$^{-1}$ treatment (right). (Courtesy of Dr. Thomas Morris.)

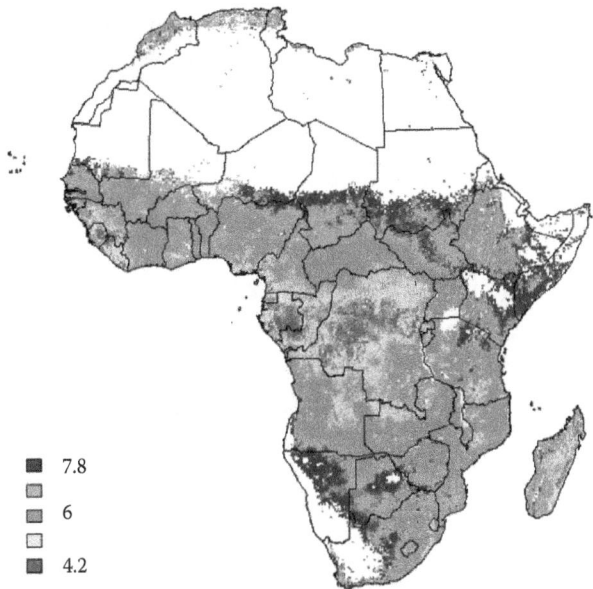

|   | 7.8 |
|---|-----|
|   | 6   |
|   | 4.2 |

**FIGURE 8.3** Working map of predicted pH of surface soil (0–5 cm) developed by AfSIS. Note the prevalence of acidic (pH < 7) soils in SSA. (Courtesy of AfSIS, Wageningen, Netherlands, http://soilgrids.org.)

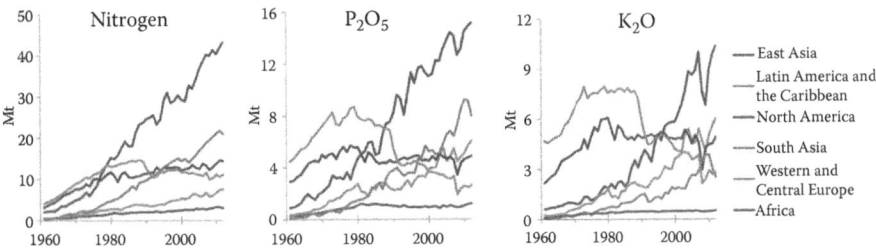

**FIGURE 8.4** Total annual nitrogen, phosphate ($P_2O_5$ equivalent), and potassium ($K_2O$ equivalent) fertilizer consumption (Mt) from 1961–2015 for various regions of the world. (Courtesy of International Fertilizer Industry Association, Paris, France, http://ifadata.fertilizer .org/ucSearch.aspx, 2015.)

**FIGURE 9.5** Algal bloom in Lake Erie extended from Toledo to Cleveland during the summer of 2015 and also covered the length of 800 miles along the Ohio River. (Courtesy of NASA Earth Observatory, Washington, DC, 2015.)

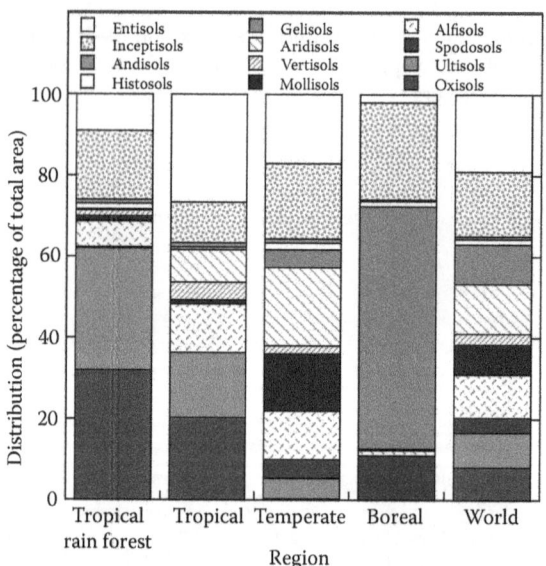

**FIGURE 11.1** Distribution of soil orders among tropical/subtropical moist broadleaf forests (rain forests), all tropical, temperate, and boreal ecosystems, as well as globally. Estimates exclude areas not covered by soils (e.g., rocks, water bodies, shifting sands, ice), and soil order classifications are based on the USDA soil taxonomy system (Soil Survey Staff 1998). The sequence of the soil orders is arranged from less well-developed soils (upper part of the graph; e.g., Entisols) to more well-developed soils (lower part of the graph; e.g., Oxisols). Tropical areas are defined as ≤23.5°, temperate areas as 23.6°–60°, and the boreal zone as >60°. All data are from Palm et al. (2007), who determined the extent of the 12 soil orders of soil taxonomy and their geographical distribution using the USDA's Global Soil Regions dataset.

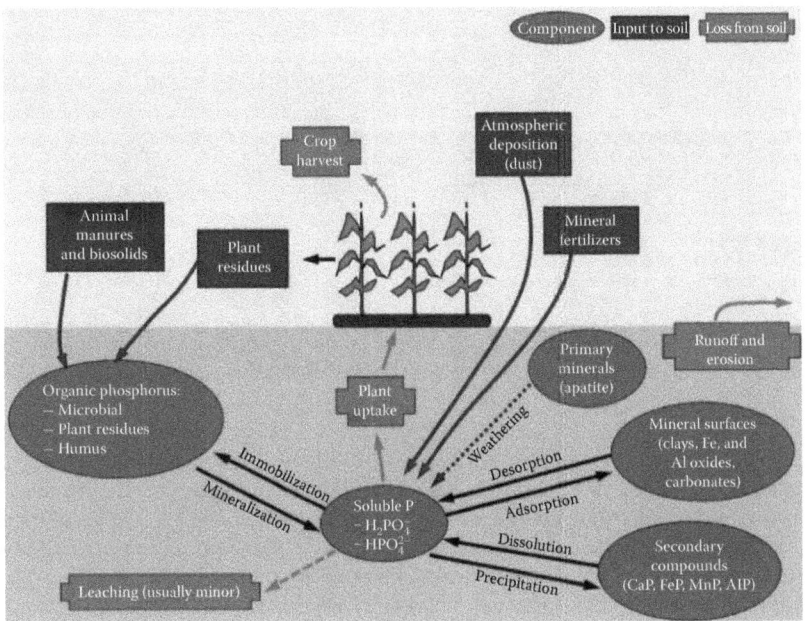

**FIGURE 11.4** The P cycle, focusing on P pools (green ovals), inputs (blue rectangles), and loss (orange polygons). (Reproduced with permission from Wikipedia Creative Commons, en.wikipedia.org/wiki/Phosphorus_cycle.)

(a)　　　　　　　　　　　　　　　　　(b)

**FIGURE 11.6** Plant nutrient–acquisition strategies. (a) Mycorrhizal fungi extend beyond the root zone and greatly increase the volume of soil that can be explored for P uptake. All mycorrhizas (arbuscular mycorrhizas, ectomycorrhizas, and ericoid mycorrhizas) are involved in the uptake of soluble inorganic P. Photo shows the hyphae of *Glomus caledonium*, an arbuscular mycorrhizal fungus that in the picture is growing into the soil from a host root of the plant *Trifolium repens*. (Photo by Iver Jakobsen, reprinted with kind permission from Springer Science+Business Media. *Mycorrhizal Ecology*, Foraging and resource allocation strategies of mycorrhizal fungi in a patchy environment, 2003, 93–115, Olsson, P. A., Jakobsen, I., and Wallander, H.; reprinted from *Trends in Ecology and Evolution*, 23, Lambers, H., Raven, J. A., Shaver, G. R., and Smith, S. E., Plant nutrient-acquisition strategies change with soil age, 95–103, Copyright (2008), with permission from Elsevier.) (b) Root morphology helps plants access P from soil; for example, Proteaceae species that grow under extremely low P supply in highly weathered Australian soils have a number of root morphological adaptations. Shown here is a proteoid root cluster of *Banksia grandis*. (Photo by Michael W. Shane, reprinted from *Trends in Ecology and Evolution*, 23, Lambers, H., Raven, J. A., Shaver, G. R., and Smith, S. E., Plant nutrient-acquisition strategies change with soil age, 95–103, Copyright (2008), with permission from Elsevier.)

(a)                                    (b)                          (c)

**FIGURE 11.7**   Photos showing the effect of animal activity on soil structure. (a) *Atta cephalotes* leaf cutter ant nest in Cockscomb Basin Wildlife Sanctuary, Belize. The photo shows the massive soil translocation ants create in building their nests in highly weathered tropical soils, as they move massive amounts of soil from deeper layers to the surface. (Courtesy of Matthew Meier.) (b) An earthworm cast produced by the tropical earthworm species *Amynthas khami* (which can grow up to 50 cm in length) in a rain forest in northeastern Vietnam. Casts are clumps of digested organic matter excreted by earthworms that aggregate into large and distinctive structures. (Reprinted from *Soil Biology and Biochemistry*, 43, Hong, H. N., Rumpel, C., Henry des Tureaux, T., Bardoux, G., Billou, D., Tran Duc, T., and Jouquet, P., How do earthworms influence organic matter quantity and quality in tropical soils? 223–230, Copyright (2011), with permission from Elsevier.) (c) Tall termite hills such as this one are common in highly weathered soils in West Africa. These hills represent large changes to soil structure and, likely, biogeochemical cycling. (Reproduced with permission from Wikipedia Creative Commons, https://commons.wikimedia.org/wiki/File:Termite_hill_in_forest.jpg.)

(a)                                    (b)

**FIGURE 11.9**   Phosphorus availability on highly weathered soils can be a major constraint to agriculture in the tropics. (a) Adding P to nutrient-poor tropical soils can dramatically enhance crop production, and the picture shows the effects of P fertilization on plant growth in Brazil's cerrado. Maize plants in the background were fertilized with P, and they are much larger than the unfertilized maize plants in the foreground. (Courtesy of D.M.G. de Sousa.) (b) Some crops are much more commonly grown in the tropics than anywhere else on earth. Shown here is cassava farming in Nigeria where the International Institute for Tropical Agriculture. (Courtesy of Tunga Media, Niger State, Nigeria, http://tungamediang.com.)

# 9 Phosphorus and the Environment

*Rattan Lal*

## CONTENTS

## 9.1 INTRODUCTION

Phosphorus (P) is essential to all life and has a limited supply. Isaac Asimov (1974) stated, "Life can multiply until all the phosphorus has gone and then there is an inexorable halt which nothing can prevent" (Ashley et al. 2011). Thus, its availability limits the net primary productivity (NPP) such as in old and highly weathered soils (Buendía et al. 2014). Yet, its misuse has created an environmental disaster around the world including anoxia or dead zones, and toxic algal blooms.

The problem of low reserves is exacerbated by low efficiency of agroecosystems, losses of P through water runoff, drainage, sediment transport (Cordell et al. 2011), and seemingly squandering of a scarce commodity (Ferro et al. 2015). The enhanced interest in soil P and its dynamics is also due to the debate on global P reserves and the effects of the transport from agroecosystems into natural waters leading to eutrophication and severe contamination of water resources. The term *eutrophication* refers to the overenrichment of natural waters by nutrient loading (P, N, etc.) especially by transport from agroecosystems (Carpenter 2005). The need for limiting the nonpoint source P loading of natural waters is highlighted by severe degradation of the quality of natural waters (e.g., algal bloom) around the world in both developed and developing countries. The problem is likely to be aggravated, such as that observed in Lake Erie during 2014 and 2015, by the increase in agricultural activities to feed the world's growing population, the heavy use of fertilizers, and the cultivation of marginal lands. Among the examples of severe degradation in the quality of natural waters observed since 1990s include Great Lakes in North America, Lake Taihu in China, Baltic Sea, Chesapeake Bay, and Gulf of Mexico (Sharpley and Wang 2014). The Baltic Sea is a large brackish sea ($0.4 \times 10^6$ km$^2$) with a fourfold large drainage area with a population of about 85 million (Elmgren and Larsson 2001). Yet, nutrient-rich coastal areas are prone to toxic phytoplankton blooms and filamentous macroalgae, which jeopardize the quality of water.

Ecologically, P cycling is closely linked to that of C and N through allometric relations across species (Niklas 2008). Because of its finite reserves and ecological significance, understanding the fate of P during pedogenesis is critical. While the availabilities of C and N are increasing because of anthropogenic activities, the NPP or the rate of biomass production is limited by the availability of P. Anthropogenic changes in C and N relative to P since the second half of the twentieth century is a unique phenomenon in earth's history (Peñuelas et al. 2014). In addition to N-limited plant growth in most terrestrial ecosystems, P limitation is also becoming an important factor (Aerts and Chapin 2000). The limitation of P in plants is indicated by the foliar N:P mass ratio in wild plants. Plant growth is N limited at N:P ratio of <14, P limited at N:P ratio of >16, and colimited at N:P ratio of intermediate values (Peñuelas et al. 2014).

The land–sea link via rivers is an important component of the global P cycle and the biogeochemistry (See Chapters 1, 11, and 13; Alvarez-Cobelas et al. 2009). However, the link is strongly impacted by human activities. Indeed, P loading, eutrophication, and attendant algal bloom depend on land use and management (Turner et al. 2003). P-laden sediments, transported by water runoff and wind erosion, make lakes eutrophic leading to algal blooms. The overenrichment of agricultural soils can perpetuate the problem for a long time even if significant changes are made in soil, crop, and animal management systems (Carpenter 2005).

Therefore, the objectives of this chapter are to describe the global P cycle, discuss the impacts of land use and management on the transport of P to the aquatic ecosystems, and explain the importance of improved management on use efficiency of P.

## 9.2 IMPACT OF INDISCRIMINANT AGRICULTURAL INTENSIFICATION

Anthropogenic perturbations have increased the global cycle of P by 400% (Falkowski et al. 2000), along with the doubling of the net flux of bioreactive (dissolved) P from land to waters (Filipelli 2008; Whitehead and Crossman 2012) leading to algal blooms and hypoxia. An impressive increase in agronomic production since the middle of the twentieth century (1960s) has been achieved by the increase in fertilizer inputs (especially of N and P; see Table 13.1), the increase in the area under irrigation (Figure 9.1), and the use of input-responsive varieties. The global use of reactive N (as fertilizer) increased from <10 million Mg in 1950 to 123 million Mg in 2012 and is projected to increase to 135 million Mg by 2020. Thus, N use over the 70-year period has increased by a factor of 13.5. Similarly, P use as fertilizer has increased from 11 million Mg in 1960 to 46 million Mg in 2012, and it is projected to be 47.6 million Mg by 2020. Thus, P use over the 70-year period has increased by a factor of 4.37 (Table 13.1). The impact of agricultural intensification on nutrient loading is aggravated by simultaneous increase in the expansion of irrigation. Irrigated land area has increased from 50 million ha in 1900 to >300 million ha by 2015 and is projected to be about 350 million ha by 2050 (Figure 9.1). In addition to increasing

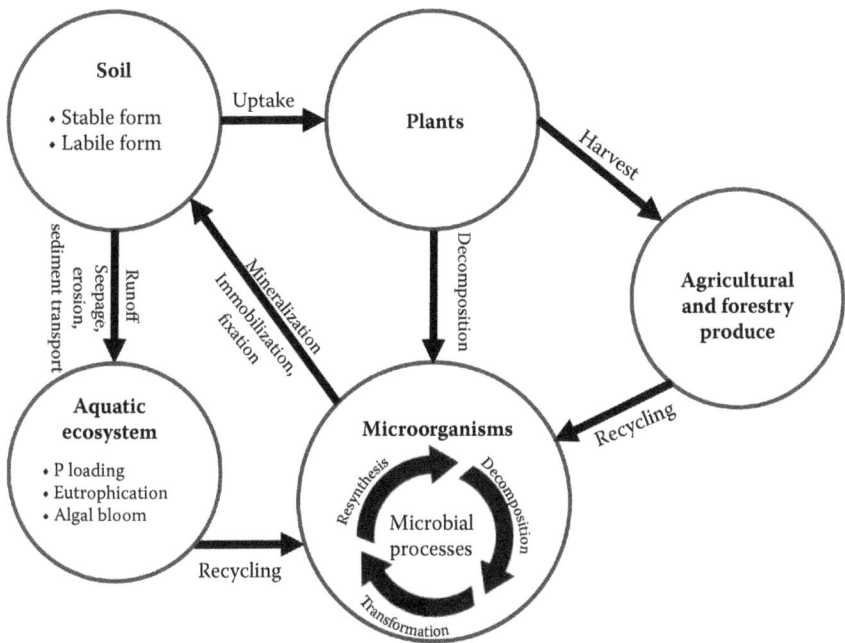

**FIGURE 9.1**   Phosphorus dynamics affecting the transport from soil to aquatic ecosystems.

the risks of secondary salinization, the addition of water in the soils of agroecosystems also accelerates the biogeochemical cycling of P and N because of accelerated nutrient mobilization (Schipanski and Bennett 2012). However, systematic and long-term studies on the impacts of the changes in agroecosystems on P cycling are scanty. In contrast to cycles of N and C, which emit potent greenhouse gases such as $N_2O$ ($NO_X$) along with $CO_2$ and $CH_4$, the transport of P into the environment occurs almost entirely in solid and liquid phrases. Thus, the eutrophication of rivers and lakes is the principal environmental impact of P fertilization. In addition to the use of chemical fertilizers, the spreading of livestock manure on cropland is another source of eutrophication (Bouwman et al. 2013).

Changes in temperature and precipitation because of the projected climate change can affect the P cycling because of the changes in the chemical reaction kinetics and the rate of processes affected by the microbiota (Whitehead and Crossman 2012). If P is a limiting nutrient, the increase of P input may strongly impact the N cycle by increasing the NPP. In addition, emissions of phosphine and polyphosphate gas, although in small amounts, may also impact the P cycle at a local scale (Mackenzie et al. 2002; Whitehead and Crossman 2012). The decrease in P availability can also be caused by the application of N fertilizers and amendments (Vitousek et al. 2010), indicating the need for understanding the coupled cycling of C, N, P, and $H_2O$ in an integrated manner and at a range of spatial scales. Indeed, environmental degradation is an important control of macronutrients cycles. Thus, there is a need to study and understand the interconnectivity between atmosphere, biosphere, pedosphere, lithosphere, and hydrosphere using a nexus approach (Figure 9.2).

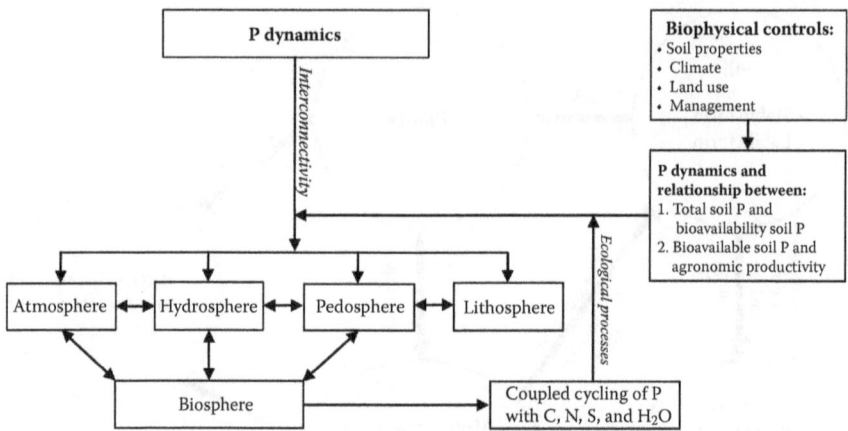

**FIGURE 9.2** Interconnectivity among the ecological processes that control P dynamics and the need for a nexus approach to address the environmental issues.

Ecological processes as controls of P dynamics (Figure 9.2) also influence food and nutritional security at regional and global scales. Agronomic productivity, and thus food and nutritional security, are influenced by (1) the total soil P and bioavailable soil P and (2) the bioavailable soil P and the agronomic productivity (Dumas et al. 2011), both of which are strongly affected by soil properties, climate, and land use and management. The strategies of achieving food security in soils with a limited P reserve include (1) evaluating the vulnerability of diverse regions to increasing scarcity of P and (2) evaluating the deficiency of the total soil P in different regions vis-à-vis the level at which the maximum agronomic productivity is sustained only by recycling and without any external input of P (Dumas et al. 2011).

## 9.3 TRANSPORT OF SOIL PHOSPHORUS IN SURFACE RUNOFF AND SEDIMENTS

The growth of aquatic weeds and algae, which influence the quality of water in lakes, is mostly limited by the availability of P (Rockwell et al. 2005; Schindler et al. 2008), because most filamentous blue-green algae fix atmospheric N. Thus, there is a critical concentration of P dissolved (0.01 mg/L) and total P (0.02 mg/L) in relation to the eutrophication of the surface water (Sawyer 1947; Vollenweider 1968; Daniel et al. 1993). The total P (TP) concentration of >0.02 mg/L in a runoff aggravates eutrophication (Correll 1998). Cyanobacteria have a TP threshold of 0.01 mg/L and become dominant over phytoplankton when the concentration of TP $\geq$ 0.1 mg/L (Bridgeman et al. 2012). The increase in P concentration also accentuates the growth of toxic compared with nontoxic strains of *Microcystis* (Davis et al. 2009; Bridgeman et al. 2012). These algae feed on P transported through the runoff from ecosystems (e.g., fertilizers and manure from farmlands, human waste, and leakage from septic tanks).

The loading of P and other nutrients in lakes and aquatic bodies occurs from external and internal sources (Figure 9.3). Important among external loading are the

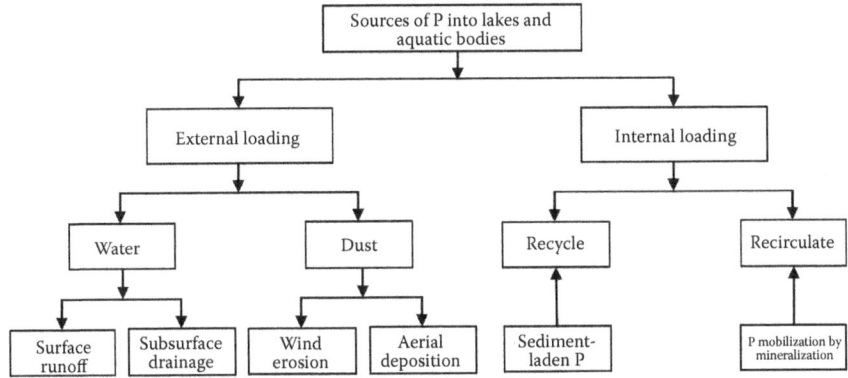

**FIGURE 9.3**  Understanding the external and internal sources of phosphorus in lakes is essential to P management in terrestrial and aquatic ecosystem.

sediments transported from the adjacent watershed. Surface runoff from agricultural watersheds/lands is a major determinant of eutrophication (Daniel et al. 1993). In the context of partial/critical area concept, some regions of the watershed may be more vulnerable to erosional processes than others. Types of P being transported from the watershed may include (1) P dissolved in water, which can be measured as a water-extractable P, and (2) sediment-laden P, which is absorbed on the clay surfaces. For assessing the loading potential of a specific soil, Vadas et al. (2005) proposed a single extraction coefficient for water-extractable P (Mehlich-3 P) and P-sorption saturation data.

Furthermore, the concentration of dissolved P in a surface runoff is also affected by the volume of the runoff and the duration since the application of P fertilizer and the clay content of the soil (Cox and Hendricks 2000). In addition to surface runoff, transport of P can also occur in subsurface drainage. The factors that determine the transport of P in subsurface drainage water include soil properties, drainage design, climatic conditions, and management (King et al. 2015). The transport of P in subsoil drainage is also exacerbated by the downward translocation of P from the surface layer into the subsoil. A field experiment in Florida showed that the high pH of the surface water/sediment systems may cause high P concentrations in streams and lakes (Elrashidi et al. 2001). Macropore flow, directly from the surface to the drain line, also plays a critical role. In Norway, Bechmann (2012) observed that the (TP) and the dissolved reactive P (DRP) losses (g/ha × yr) were 460 and 160, respectively, in the surface runoff compared with 530 and 270 in the subsurface flow.

The loading of P also occurs through P-laden sediments transported by water and wind erosion. Measurements of P transport from a catchment study in eastern Norway indicated that for suspended sediment load of 2140 kg/ha × yr, the transport of P was 2320 g/ha × yr for TP and 240 g/ha × yr for DRP (Bechmann 2012). Based on a village-level study in the west African Sahel, Visser and Sterk (2007) reported that nutrient losses from one field during a wind erosion event can be as high as 73% of the N and 100% of the P needs for crop production. Thus, the deposition of these P-laden sediments in lakes can lead to eutrophication. Notable progress has been made in developing guidelines for reducing the risks of eutrophication from point sources of P (e.g., industrial plant affluent, sewage treatment). However, additional research is needed to reduce P loading from nonpoint sources including agricultural and urban ecosystems. The latter includes home lawns, golf courses, and recreational facilities. Since the 1960s, agroecosystems have been transformed from being the sink of P to the source of P because of the application of fertilizers and animal manure and the surface application of P in no-till systems.

Sediment-laden P transported into streams and rivers is eventually carried into lakes, where it enriches the nutrient concentration in the lake. The transport of P in surface runoff and subsurface drainage flow originates from a few areas (the partial/critical area concept) that are vulnerable to runoff or subsurface drainage (interflow or tile drain) and where the high rate of P application also coincides with those of high runoff or drainage flow (Sharpley et al. 2001). Thus, balancing P inputs and outputs at farm level is essential to minimize the P loading of aquatic ecosystems. Daloğlu et al. (2012) used the soil and water assessment tool to model the DRP in the Sandusky watershed in northern Ohio. Daloğlu et al. observed that the reasons for the increased DRP load since mid-1990s include increase in storm events, change in fertilizer application and timing rate, and increase of soil stratification with concentration of P in the soil surface.

The atmospheric deposition of P is another source of P loading in lakes. Increased P deposition may be related to meteorological conditions. An example of the long-distance transfer includes the transport of dust from northern Africa and the Iberian Peninsula to the Pyrenees and to the Pyrenean Lake (Camarero and Catalan 2012). The Harmattan dust, originating from northwestern Africa every year between November and March, makes its way across the Sahara to the Gulf of Guinea, and across the Atlantic to the Americas (Kaufman et al. 2005).

In addition to P loading by wind-blown sediments, wind wave disturbance can also contribute P from sediments to the overlying water column, such as in Lake Taihu in China (Zhu et al. 2013). Furthermore, Zhu et al. observed that P bound to Fe, Al, and Mn oxides and hydroxides and organic P can also be bioactive and make these sources potentially algal-available P.

## 9.4   ALGAL BLOOMS IN THE GREAT LAKES

Phytoplankton and cyanobacterium comprise lake water. The relative predominance of these depends on nutrient concentration. Phytoplankton predominate in nutrient-poor water and cyanobacteria in nutrient-rich water. The trophic status of the lake water, or its biomass productivity, depends on the nutrient concentration (Figure 9.4). Based on the nutrient-driven biomass productivity, the trophic status of lakes may be (1) eutrophic with high plant growth related to well-nourished and enriched water, (2) mesotrophic with medium-nourished or moderate growth, and (3) oligotrophic or poorly nourished and low plant growth. In addition to the endogenous factors related to nutrient status, the plant growth in lakes is also determined by several exogenous factors including climate, rate of nutrient supply, and lake morphometry (Figure 9.4). Depending upon the amount and the species composition, the growth of some plants in lake water can have severe effects on water quality. Algal bloom in Lake Erie is an example of such an

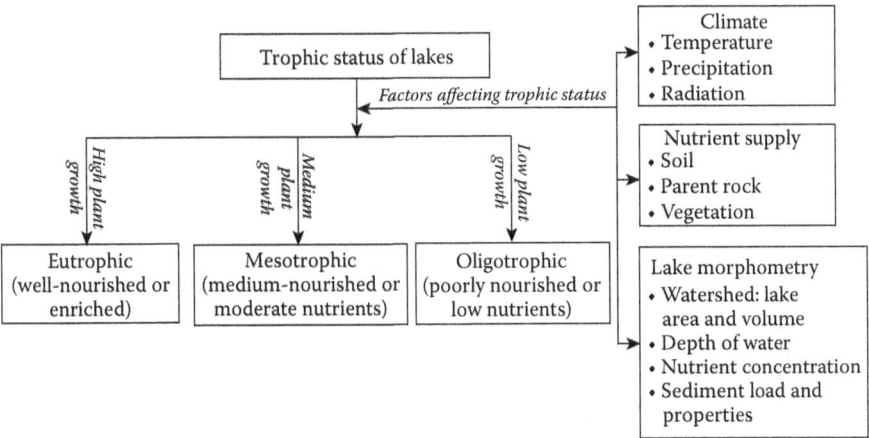

**FIGURE 9.4** Factors affecting the trophic status of lake water.

adverse effect related to nutrient loading. During the summer of 2015, the algal bloom also spread to the Ohio rivers over 650 miles (1040 km) from near Pittsburgh to Tell City, covering more than two-thirds of the river's length (Arenschield 2015).

Algal blooms are caused by the explosive growth of microorganisms, including dinoflagellates and bacteria that produce and exude highly toxic biomolecules. Prominent among bacteria are cyanobacteria, which are the most ancient phytoplankton on earth and form harmful algal blooms in freshwater. Cyanobacteria from freshwater are *Microcystis, Anabaena, Synechococcus*, and *Trichodesmium*; those from estuarine are *Nodularia, Aphanizomenon*; and from marine ecosystems are *Lyngbya, Synechococcus*, and *Trichodesmium* (O'Neil et al. 2012).

Among the dinoflagellates, *Pfiesteria piscicida* is an important causative organism in the estuaries of the middle and southern Atlantic coast (Silbergeld et al. 2000). *Pfiesteria* is a single-cell dinoflagellate, microscopic, and free-swimming microorganism. Rather than being plant-like, *Pfiesteria* is more animal-like and acquires most of its energy by eating other organisms rather than through photosynthesis. These organisms are more abundant near sewage treatment outlets (Silbergeld et al. 2000).

Algal blooms in the western Lake Erie have increased since the mid-1990s (Conroy et al. 2005; Bridgeman et al. 2012). Ecosystem services (e.g., water quality) are strongly impaired by algal bloom formation of surface scums of *Microcystis* stretching over large areas (Figure 9.5). Eutrophication adversely affects water use for fisheries, recreation, and industry because of the rapid growth of undesirable algae and aquatic weeds and the attendant anoxia. Harmful algal blooms of surface water render them toxic to human consumption. The development of cyanobacterial blooms adversely impacts many ecosystem services provisioned by freshwater (e.g., recreation, drinking water supply, irrigation, aquaculture). It makes water undrinkable by the production of microcystin toxins whose concentrations may exceed the safety levels established by the World Health Organization (WHO) at 16 µg/L (Bridgeman et al. 2012). Algal blooms occur in warm temperatures (>15°C), and microcystin colonies are deposited at the lake floor during fall.

**FIGURE 9.5** **(See color insert.)** Algal bloom in Lake Erie extended from Toledo to Cleveland during the summer of 2015 and also covered the length of 800 miles along the Ohio River. (Courtesy of NASA Earth Observatory, Washington, DC, 2015.)

There is also a strong link between loadings of N and P. Many phytoplankton causing harmful algal blooms have physiological adaptive capacity that accentuates their growth under elevated N (e.g., urea, ammonium). Glibert et al. (2004) hypothesize that algal bloom is caused by (1) inefficient incorporation of N in the food supply chain, leakage of the N cycle from the crop to the table and the fate of lost N relative to P, and (2) adaptive capacity of many microorganisms to excessive N. Thus, aggressive control is needed for both N and P transport to natural waters.

Among the major types of harmful cyanobacteria in Lake Erie (*Microcystis aeruginosa* and *Lyngbya wollei*) the *Lyngbya* species forms the filamentous benthic mat-forming cyanobacterium. In Lake Erie, *Lyngbya wollei* grows mainly in the Maumee Bay area with the greatest biomass between 1.5 and 3.0 m depths (Bridgeman et al. 2012). Together, *Microcystis* and *Lyngbya* are major environmental and human health hazards in the Maumee Bay region.

Algal blooms are also observed elsewhere due to excessive P loading of surface waters. Using the dataset from 800 European lakes, Carvalho et al. (2013) observed that cyanobacteria exhibit a nonlinear response to P with the sharpest increase in cyanobacterial growth occurring in the TP range of 20–100 µg/L. They further observed that 50% of the lakes in Europe exceed the threshold P levels established by the WHO (1981, 2003, 2004). In China, the surface waters in Lake Taihu are increasingly threatened by P-related eutrophication and promote algal growth (Zhao et al. 2012; Paerl et al. 2014).

Algal bloom is also affected by meteorological parameters and climate change. In a nutrient-rich blackwater river in Florida called St. Johns River, Philips et al. (2007) observed that the eutrophic status of the river was reflected in high concentrations of

N (3100 µg/L) and P (180 µg/L) and to the algal bloom. The nutrient concentrations were strongly correlated to the water replacement rates Thus, algal blooms were sensitive to rainfall patterns (Philips et al. 2007). The cyanobacterial bloom in Florida Bay, United States, may also be associated with the amount of dissolved organic N and P uptake (Glibert et al. 2004). Florida Bay, a shallow, sea grass-dominated bay in southern Florida, receives significant nutrient loading and is vulnerable to algal blooms.

Because climate change may affect rainfall patterns and temperature regimes, an increase in cyanobacteria bloom formation within lakes in Europe may occur as a result of global warming (Wagner and Adrian 2009). Despite the complexity of the processes involved, the magnitude and the frequency of algal blooms are likely to increase with the projected global warming (O'Neil et al. 2012).

## 9.5  MANAGING P LOADING IN SURFACE WATERS

There are two strategies of managing the algal bloom: (1) harvesting the harmful algal bloom and using it for producing biofuels (biogas), soil amendments, and other innovative industrial uses (Sharpley and Wang 2014) and (2) developing long-term strategies of reducing the nutrient (P, N) loading and minimizing the risks of eutrophication (Figure 9.6). The majority of industrialized farms, especially those in developed countries, operate with an annual P surplus. The problem is especially severe in intensively managed livestock production systems in the UK that have a surplus of 10–20 kg P/ha × yr (Edwards and Withers 1998). It is this surplus of P in croplands and grazing lands that is the root cause of the problem leading to the eutrophication of waters. So, the excessive P loading of natural waters caused by

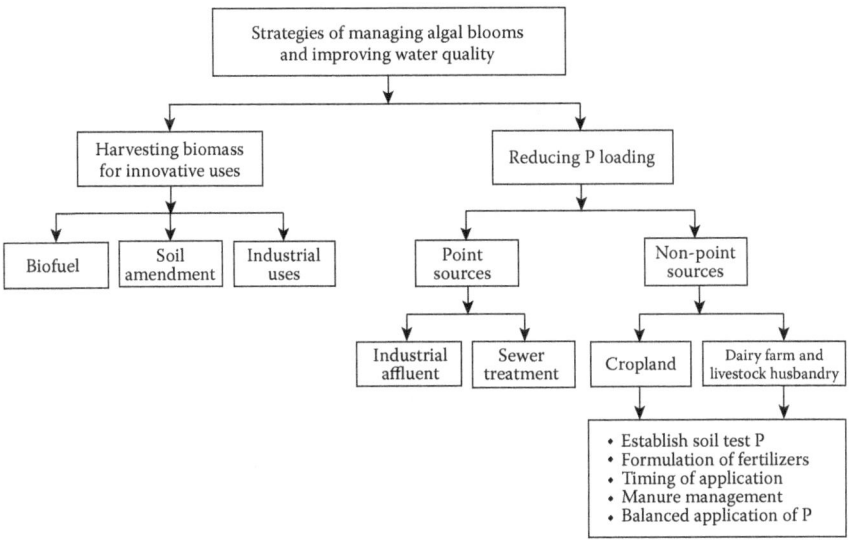

**FIGURE 9.6**  Short-term and long-term strategies of P management for improving water quality and reducing risks of algal blooms.

the heavy use of fertilizer and animal manure on agricultural lands must be systematically addressed. Therefore, it is important to establish an environmental soil test critical level above which either no or only enough P to replace the crop removal should be applied (Cox and Hendricks 2000; Gartley and Sims 1994; Sharpley et al. 1994, 1996). In addition to applying P at the rate determined by the soil test P, the time of the application of fertilizers and manure may also be an important factor.

The accumulated P in the surface layer, because of the heavy rate of the application of fertilizer and animal manure, is often in labile and semilabile P fraction (Sun et al. 2015). While these fractions can be absorbed by the subsequent crops, these are also a serious eutrophication threat to adjacent water bodies. Thus, appropriate formulation (slow release) may also be important. The balanced application of nutrients (Yan et al. 2008), at the appropriate rate based on soil test and applied at the right time while using the correct formulation, is important to minimize the P loading of natural waters. Yuan et al. (2013) recommended that applying P fertilizer in the spring may reduce P loss during the fallow winter season. Similar to cropland, there is also a strong need to identify cost-effective mitigation strategies to reduce P loss from dairy farms. In New Zealand and Australia, McDowell and Nash (2012) recommended the following strategies to reduce P loss from dairy farms: (1) management to decrease soil test P and application of low water-soluble P, (2) amendments such as alum or red mud, and (3) natural or constructed wetlands as edge-of-field mitigation devices. Improving the PUE in agroecosystems is a critical requirement to reducing P loading (Schröder et al. 2011). Therefore, a better understanding of the soil test P and the loss of dissolved soluble P from soil to water is critical to identifying management strategies (Sims et al. 2000; Styles et al. 2006).

Furthermore, the pattern of P cycling and the dominant loss pathways differ between livestock and arable farming systems (Edwards and Withers 1998) and must be understood. Pathways of lateral and vertical transports of P must be understood in relation to soil properties (macropore or bypass flow) and water table dynamics. Above all, farmers, land managers, and policy makers can make an important contribution through effective stewardship of natural resources.

## 9.6    PHOSPHORUS DYNAMICS DURING PEDOGENESIS

Being a finite resource, P availability and dynamics must be understood on short-term (ecological timescale) and long-term (geological timescale) bases (Buendía et al. 2014). The availability of P depends on the mineralization and the recycling of SOM on ecological timescales (Cuevas and Medina 1988) and on the weathering of the parent rock on geological timescales (Buendía et al. 2010; Schlesinger 1997). Because of low P reserves, old or highly weathered soils have a lower biomass production rate when P becomes a limiting factor under unfertilized condition (Buendía et al. 2014). Furthermore, the accumulation of SOM in soil is an important indicator of the degree of soil formation. The accumulation of SOM also depends on litter chemistry, including the concentration of P vis-à-vis that of N and other elements. The weathering of rocks by factors of soil formation (Jenny 1941), influenced by biotic agents, also depends on the bioavailability of P. Further, P is a major element in the biochemical conversion of biomass-C into humus (Lal 2014; Walker and Syers

1976). Forms and availability of soil P also change with the age of chronosequences. Similar trends are observed in P fractions over time in soils of volcanic origin (Crews et al. 1995; Schlesinger et al. 1998). The soil C pool differs among world's life zones (Post et al. 1982) and has a possible correlation with P reserves because it must be entirely supplied by the parent material of the unfertilized soil. The rapid accumulation of P in organic forms (and fixation with other minerals) can lead to P deficiency in crops (Schlesinger et al. 1998). Knowledge about the distribution of P among different organic and inorganic fractions at a given stage of soil development is important information in understanding the cycling and the availability of P and other elements (N, S, etc.) in diverse ecosystem processes (Crews et al. 1995).

## 9.7 CONCLUSION

Phosphorus, an essential but finite resource, is an important cause of the eutrophication of natural waters leading to algal blooms. The loading of P in natural waters may be exacerbated by the increase in the application of P fertilizers and animal manure. Accelerated soil erosion from croplands and transport of P by surface runoff and drainage flow are other sources. Therefore, the strategies of reducing eutrophication include (1) carefully designed rates of P application that do not exceed that of the crop uptake of P, (2) application of slow-release formulation of P fertilizers, (3) adoption of conservation-effective measures (conservation agriculture) that reduce surface runoff and sediment transport, (4) diversion of surface runoff and drainage effluent into wetlands prior to overflow into lakes, and (5) establishment of hydromorphic vegetation in lakes, which can effectively absorb P. In addition, the biomass of algal blooms can be harvested for industrial uses (e.g., biofuel) and composted for use as a soil amendment. Judicious use of fertilizers and amendments, adoption of conservation-effective measures, use of slow-release formulations are among the strategies to minimize risks of P loading of natural waters.

## REFERENCES

Aerts, R., and F. S. Chapin. 2000. The mineral nutrition of wild plants revisited. *Advances in Ecological Research* 30:1–67.

Alvarez-Cobelas, M., S. Sanchez-Carrillo, D. G. Angeler, and R. Sanchez-Andres. 2009. Phosphorus export from catchments: A global view. *Journal of the North American Benthological Society* 28 (4):805–820.

Arenschield, L. 2015, October 3. Algae along the Ohio. *Columbus Dispatch*, Columbus, OH.

Ashley, K., D. Cordell, and D. Mavinic. 2011. A brief history of phosphorus: From the philosopher's stone to nutrient recovery and reuse. *Chemosphere* 84 (6):737–746.

Asimov, I. 1974. *Asimov on Chemistry*. Doubleday Garden City, New York, p. 267.

Bechmann, M. 2012. Effect of tillage on sediment and phosphorus losses from a field and a catchment in south eastern Norway. *Acta Agriculturae Scandinavica Section B—Soil and Plant Science* 62:206–216.

Bouwman, L., K. K. Goldewijk, K. W. Van Der Hoek, A. H. W. Beusen, D. P. Van Vuuren, J. Willems, M. C. Rufino, and E. Stehfest. 2013. Exploring global changes in nitrogen and phosphorus cycles in agriculture induced by livestock production over the 1900–2050 period. *Proceedings of the National Academy of Sciences of the United States of America* 110 (52):20882–20887.

Bridgeman, T. B., J. D. Chaffin, D. D. Kane, J. D. Conroy, S. E. Panek, and P. M. Armenio. 2012. From river to lake: Phosphorus partitioning and algal community compositional changes in Western Lake Erie. *Journal of Great Lakes Research* 38 (1):90–97.

Buendía, C., S. Arens, T. Hickler, S. I. Higgins, P. Porada, and A. Kleidon. 2014. On the potential vegetation feedbacks that enhance phosphorus availability—Insights from a process-based model linking geological and ecological timescales. *Biogeosciences* 11 (13):3661–3683.

Buendía, C., A. Kleidon, and A. Porporato. 2010. The role of tectonic uplift, climate, and vegetation in the long-term terrestrial phosphorous cycle. *Biogeosciences* 7 (6):2025–2038.

Camarero, L., and J. Catalan. 2012. Atmospheric phosphorus deposition may cause lakes to revert from phosphorus limitation back to nitrogen limitation. *Nature Communications* 3:1–5.

Carpenter, S. R. 2005. Eutrophication of aquatic ecosystems: Bistability and soil phosphorus. *Proceedings of the National Academy of Sciences* 102:10002–10005.

Carvalho, L., C. McDonald, C. de Hoyos, U. Mischke, G. Phillips, G. Borics, S. Poikane et al. 2013. Sustaining recreational quality of European lakes: Minimizing the health risks from algal blooms through phosphorus control. *Journal of Applied Ecology* 50 (2):315–323.

Conroy, J. D., D. D. Kane, D. M. Dolan, W. J. Edwards, M. N. Charlton, and D. A. Culver. 2005. Temporal trends in Lake Erie plankton biomass: Roles of external phosphorus loading and dreissenid mussels. *Journal of Great Lakes Research* 31:89–110.

Cordell, D., A. Rosemarin, J. J. Schroder, and A. L. Smit. 2011. Towards global phosphorus security: A systems framework for phosphorus recovery and reuse options. *Chemosphere* 84 (6):747–758.

Correll, D. L. 1988. The role of phosphorus in the eutrophication of receiving waters: A review. *Journal of Environmental Quality* 27:261–266.

Cox, F. R., and S. E. Hendricks. 2000. Soil test phosphorus and clay content effects on runoff water quality. *Journal of Environmental Quality* 29 (5):1582–1586.

Crews, T. E., K. Kitayama, J. H. Fownes, R. H. Riley, D. A. Herbert, D. Mueller-Dombris, and P. M. Vitousek. 1995. Changes in soil phosphorus fractions and ecosystem dynamics across a long chronosequence in Hawaii. *Ecology* 76:1407–1424.

Cuevas, E., and E. Medina. 1988. Nutrient dynamics within Amazonian forests: 2. Fine root-growth, nutrient availability and leaf litter decomposition. *Oecologia* 76 (2):222–235.

Daloğlu, I., K. H. Cho, and D. Scavia. 2012. Evaluating causes of trends in long-term dissolved reactive phosphorus loads to Lake Erie. *Environmental Science & Technology* 46 (19):10660–10666.

Daniel, T. C., D. R. Edwards, and A. N. Sharpley. 1993. Effect of extractable soil surface phosphorus on runoff water-quality. *Transactions of the ASAE* 36 (4):1079–1085.

Davis, T. W., D. L. Berry, G. L. Boyer, and C. J. Gobler. 2009. The effects of temperature and nutrients on the growth and dynamics of toxic and non-toxic strains of *Microcystis* during cyanobacteria blooms. *Harmful Algae* 8 (5):715–725.

Dumas, M., E. Frossard, and R. W. Scholz. 2011. Modeling biogeochemical processes of phosphorus for global food supply. *Chemosphere* 84 (6):798–805.

Edwards, A. C., and P. J. A. Withers. 1998. Soil phosphorus management and water quality: A UK perspective. *Soil Use and Management* 14:124–130.

Elmgren, R., and U. Larsson. 2001, December 18. Nitrogen and the Baltic Sea: Managing nitrogen in relation to phosphorus. *The Scientific World Journal* 1(52):371–377.

Elrashidi, M. A., A. K. Alva, Y. F. Huang, D. V. Calvert, T. A. Obreza, and Z. L. He. 2001. Accumulation and downward transport of phosphorus in Florida soils and relationship to water quality. *Communications in Soil Science and Plant Analysis* 32 (19–20):3099–3119.

Falkowski, P., R. J. Scholes, E. Boyle, J. Canadell, D. Canfield, J. Elser, N. Gruber et al. 2000. The global carbon cycle: A test of our knowledge of earth as a system. *Science* 290 (5490):291–296.

Ferro, C. J., E. Ritz, and J. N. Townend. 2015. Phosphate: Are we squandering a scarce commodity? *Nephrology Dialysis Transplantation* 30 (2):163–168.

Filippelli, D. M. 2008. The global phosphorus cycle: Past, present, future. *Elements* 4:89–95.

Gartley, K. L., and J. T. Sims. 1994. Phosphorus soil testing—Environmental uses and implications. *Communications in Soil Science and Plant Analysis* 25 (9–10):1565–1582.

Glibert, P. M., C. A. Heil, D. Hollander, M. Revilla, A. Hoare, J. Alexander, and S. Murasko. 2004. Evidence for dissolved organic nitrogen and phosphorus uptake during a cyanobacterial bloom in Florida Bay. *Marine Ecology Progress Series* 280:73–83.

Jenny, H. 1941. *Factors of soil formation.* McGraw Hill, New York.

Kaufman, Y. J., I. Koren, L. A. Remer, D. Tanré, P. Ginoux, and S. Fan. 2005. Dust transport and decomposition observed from the Terra-MODIS spacecraft over the Atlantic Ocean. *Journal of Geophysical Research* 110:1–16.

King, K. W., M. R. Williams, M. L. Macrae, N. R. Fausey, J. Frankenberger, D. R. Smith, P. J. A. Kleinman, and L. C. Brown. 2015. Phosphorus transport in agricultural subsurface drainage: A review. *Journal of Environmental Quality* 44 (2):467–485.

Lal, R. 2014. Societal value of carbon. *Journal of Soil and Water Conservation* 69 (6):186A–192A.

Mackenzie, F. T., L. M. Vera, and A. Lerman. 2002. Century-scale nitrogen and phosphorus controls of the carbon cycle. *Chemical Geology* 190 (1–4):13–32.

McDowell, R. W., and D. Nash. 2012. A review of the cost-effectiveness and suitability of mitigation strategies to prevent phosphorus loss from dairy farms in New Zealand and Australia. *Journal of Environmental Quality* 41 (3):680–693.

NASA (National Aeronautics and Space Administration) Earth Observatory. 2015, July 28. Algae Bloom on Lake Erie. Available at http://earthobservatory.nasa.gov/IOTD/view.php?id=86327.

Niklas, K. L. 2008. Carbon/nitrogen, phosphorus allometric relations across species. In White, P. J., and J. P. Hamond (Eds.) *The Ecophysiology of Plant Phosphorus Interactions.* Springer, Dordrecht, pp. 9–30.

O'Neil, J. M., T. W. Davis, M. A. Burford, and C. J. Gobler. 2012. The rise of harmful cyanobacteria blooms: The potential roles of eutrophication and climate change. *Harmful Algae* 14:313–334.

Paerl, H. W., W. S. Gardner, M. J. McCarthy, B. L. Peierls, and S. W. Wilhelm. 2014. Algal blooms: Noteworthy nitrogen. *Science* 346 (6206):175.

Peñuelas, J., B. Poulter, J. Sardans, P. Ciais, M. van der Velde, L. Bopp, O. Boucher et al. 2013. Human-induced nitrogen-phosphorus imbalances alter natural and managed ecosystems across the globe. *Nature Communications* 4:1–10.

Philips, E. J., J. Hendrickson, E. L. Quinlan, and M. Cichra. 2007. Meteorological influences on algal bloom potential in a nutrient-rich blackwater river. *Freshwater Biology* 52 (11):2141–2155.

Post, W. M., W. R. Emanuel, P. J. Zinke, and A. G. Stangenberger. 1982. Soil carbon pools and world life zones. *Nature* 298 (5870):156.

Rockwell, D. C., G. J. Warren, P. E. Bertram, D. K. Salisbury, and N. M. Burns. 2005. The US EPA Lake Erie indicators monitoring program 1983–2002: Trends in phosphorus, silica, and chlorophyll a in the central basin. *Journal of Great Lakes Research* 31:23–34.

Sawyer, C. N. 1947. Fertilization of lakes by agricultural and urban drainage. *National English Water Works Association* 61:109–127.

Schindler, D. W., R. E. Hecky, D. L. Findlay, M. P. Stainton, B. R. Parker, M. J. Paterson, K. G. Beaty, M. Lyng, and S. E. M. Kasian. 2008. Eutrophication of lakes cannot be controlled by reducing nitrogen input: Results of a 37-year whole-ecosystem experiment. *Proceedings of the National Academy of Sciences of the United States of America* 105 (32):11254–11258.

Schipanski, M. E., and E. M. Bennett. 2012. The influence of agricultural trade and livestock production on the global phosphorus cycle. *Ecosystems* 15 (2):256–268.

Schlesinger, W. 1997. In Waltham, M. A. (Ed.) *Biogeochemistry: An Analysis of Global Change*, 2nd ed. Academic Press, Cambridge, MA.

Schlesinger, W. H., L. A. Bruijnzeel, M. B. Bush, E. M. Klein, K. A. Mace, J. A. Raikes, and R. J. Whittaker. 1998. The biogeochemistry of phosphorus after the first century of soil development on Rakata Island, Krakatau, Indonesia. *Biogeochemistry* 40 (1):37–55.

Schröder, J. J., A. L. Smit, D. Cordell, and A. Rosemarin. 2011. Improved phosphorus use efficiency in agriculture: A key requirement for its sustainable use. *Chemosphere* 84 (6):822–831.

Sharpley, A., and X. Wang. 2014. Managing agricultural phosphorus for water quality: Lessons from the USA and China. *Journal of Environmental Sciences—China* 26 (9):1770–1782.

Sharpley, A., T. C. Daniel, J. T. Sims, and D. H. Pote. 1996. Determining environmentally sound soil phosphorus levels. *Journal of Soil and Water Conservation* 51 (2):160–166.

Sharpley, A. N., R. W. McDowell, and P. J. A. Kleinman. 2001. Phosphorus loss from land to water: Integrating agricultural and environmental management. *Plant and Soil* 237 (2):287–307.

Sharpley, A. N., J. T. Sims, G. M. Pierzynski, J. L. Havlin, and J. S. Jacobsen. 1994. Innovative soil phosphorus availability indices—Assessing inorganic phosphorus. *Soil Testing: Prospects for Improving Nutrient Recommendations* (40):115–142.

Silbergeld, E. K., L. Grattan, D. Oldach, and J. G. Morris. 2000. *Pfisesteria*: Harmful algal blooms as indicators of human: Ecosystem interactions. *Environment Research Section* A82:97–105.

Sims, J. T., A. C. Edwards, O. F. Schoumans, and R. R. Simard. 2000. Integrating soil phosphorus testing into environmentally based agricultural management practices. *Journal of Environmental Quality* 29 (1):60–71.

Styles, D., I. Donohue, C. Coxon, and K. Irvine. 2006. Linking soil phosphorus to water quality in the Mask catchment of western Ireland through the analysis of moist soil samples. *Agriculture Ecosystems & Environment* 112 (4):300–312.

Sun, W. X., B. Huang, M. K. Qu, K. Tian, L. P. Yao, M. M. Fu, and L. P. Yin. 2015. Effect of farming practices on the variability of phosphorus status in intensively managed soils. *Pedosphere* 25 (3):438–449.

Turner, R. E., N. N. Rabalais, D. Justic, and Q. Dortch. 2003. Global patterns of dissolved N, P and Si in large rivers. *Biogeochemistry* 64 (3):297–317.

Vadas, P. A., P. J. A. Kleinman, A. N. Sharpley, and B. L. Turner. 2005. Relating soil phosphorus to dissolved phosphorus in runoff: A single extraction coefficient for water quality modeling. *Journal of Environmental Quality* 34 (2):572–580.

Visser, S. M., and G. Sterk. 2007. Nutrient dynamics—Wind and water erosion at the village scale in the sahel. *Land Degradation & Development* 18 (5):578–588.

Vitousek, P. M., S. Porder, B. Z. Houlton, and O. A. Chadwick. 2010. Terrestrial phosphorus limitation: Mechanisms, implications, and nitrogen-phosphorus interactions. *Ecological Applications* 20 (1):5–15.

Vollenweider, R. A. 1968. Scientific fundamentals of the eutrophication of lakes and flowing waters with particular reference to nitrogen and phosphorus as factors in eutrophication. Paris Report DAS/CSI/68.27. Organization for Economic Cooperation and Development, Paris.

Wagner, C., and R. Adrian. 2009. Cyanobacteria dominance: Quantifying the effects of climate change. *Limnology and Oceanography* 54 (6):2460–2468.

Walker, T. W., and J. K. Syers. 1976. The fate of phosphorus during pedogenesis. *Geoderma* 15:1–19.

Whitehead, P. G., and J. Crossman. 2012. Macronutrient cycles and climate change: Key science areas and an international perspective. *Science of the Total Environment* 434:13–17.

WHO (World Health Organization). 1981. Eutrophication: A global problem. *World Water Quality Bulletin* Part 1, 6(3), Part 2, 6(4). WHO, Geneva.

WHO. 2003. Guidelines for safe recreational water environments. WHO, Geneva.

WHO. 2004. Guidelines for drinking-water quality. WHO, Geneva.

Yan, X., J. Y. Jin, H. E. Ping, and M. Z. Liang. 2015. Recent advances on the technologies to increase fertilizer use efficiency. *Agricultural Science in China* 7:469–479.

Yuan, Y., M. A. Locke, R. L. Bingner, and R. A. Rebich. 2013. Phosphorus losses from agricultural watersheds in the Mississippi Delta. *Journal of Environmental Management* 115:14–20.

Zhao, G., J. Du, Y. Jia, Y. Lv, G. Han, and X. Tian. 2012. The importance of bacteria in promoting algal growth in eutrophic lakes with limited available phosphorus. *Ecological Engineering* 42:107–111.

Zhu, M., G. Zhu, W. Li, Y. Zhang, L. Zhao, and Z. Gu. 2013. Estimation of the algal-available phosphorus pool in sediments of a large, shallow eutrophic lake (Taihu, China) using profiled SMT fractional analysis. *Environmental Pollution* 173:216–223.

WHO, 2006. Guidelines for safe recreational water environments, WHO, Geneva.

WHO, 2011. Guidelines for drinking-water quality, WHO, Geneva.

# 10 Enhancing Efficiency of Phosphorus Fertilizers through Formula Modifications

*Ruiqiang Liu and Rattan Lal*

## CONTENTS

## 10.1 INTRODUCTION

The rapidly growing world population is projected to reach 9.6 billion by the year 2050 (United Nations Department of Economic and Social Affairs, 2015), signifying a great pressure on the world agriculture for food production. Meanwhile, the people's preference for meat-based diet and the increasing requirement for bioenergy crops are driving an ever-increasing demand for global agricultural output. The world grain demand is thus projected to be increased by 70% by 2050 (Food and Agriculture Organization of the United Nations [FAO], 2009). Given the limited arable lands and scarce water resources, a significant increase in fertilizer application is a practical approach to achieving the required massive increase in global crop production. Consequently, the demand for phosphorus (P) fertilizers is predicted to increase by 144% and that for nitrogen (N) by 170% by 2050, due to the globally growing food demands (Tilman et al., 2001). However, as compared with those of N fertilizers, the

P reserves are relatively limited with an amount of about 67 billion t in the form of PRs (Scholz et al., 2014). A complete depletion of this critical resource would occur in 300–400 years even if the current mining rate of 200 million t per year was maintained (Lal, 2014). Therefore, a high dependence on the traditional P fertilizers of low use efficiencies would be a major bottleneck for global agriculture to feed the future world population and to achieve global sustainable development.

The use efficiency of the conventional P fertilizers is only 10–20% due to the high chemical affinity of phosphates with soil particles. Thus, a high percentage (80–90%) is lost to the environment, which not only leads to a loss of the valuable resource but also pollutes the environment. As a matter of fact, the agricultural use of P fertilizers is a major anthropogenic factor exacerbating the eutrophication of surface freshwater bodies and degrading water quality. For example, a serious algal bloom occurred in Lake Erie in 2014, which forced one of Ohio's largest cities (Toledo with 282,000 population) to ban the use of drinking water for 2 days (Bacon, 2014). The problem worsened in 2015 (Sutherland, 2015).

Therefore, as far as sustainable agriculture and environmental protection are concerned, the traditional P fertilizers and the associated manufacturing technology are inadequate, and further improvements of fertilizer properties are required. Thus, there exists an imperative research need on innovative P fertilizers for enhancing fertilizer use efficiency, increasing crop yields, and minimizing environmental pollution. In fact, several research endeavors have been made in the development of new P fertilizers featuring much higher quality in recent years. Thus, the primary objective of this chapter is to synthesize the available research and development in formula-modified P fertilizers, to identify the related knowledge gaps, and to prioritize the research needs. The ultimate goal is to improve P use efficacy for agronomic production and to minimize the associated environmental risks through P fertilizer formula modifications. Four major types of the formula-modified P fertilizers are reviewed in this work: soluble P fertilizers with coatings (controlled P release), solid P fertilizers (PRs) with enhanced release rates, recycled struvite as an alternative P fertilizer, and P nanofertilizers.

## 10.2  SOLUBLE P FERTILIZERS WITH COATINGS (CONTROLLED-RELEASE P FERTILIZERS)

In order to reduce the high mobility of P associated with the conventional P fertilizers, an additional barrier layer (or coating) can be added around the fertilizer granules to prevent the direct contact of the soluble salts with the soil solution so that the P release can be somewhat controlled (controlled-release P fertilizers, CRPs). Compared with the conventional formulations, CRPs release the nutrient(s) in a gradual fashion, so that their fertilizing effects last longer, and the loss of nutrient(s) to the environment is substantially reduced. The slow release of the nutrients can also reduce the osmotic stress to plants, prevent root and leaf scorching (Qian and Schoenau, 2010), and avoid the risks of soil salinization (Davidson and Gu, 2011), which are often associated with heavy applications of traditional fertilizers. Additionally, coated fertilizers are more convenient to transport, store, and handle than the conventional types.

Organic polymers are the most commonly used barrier or coating materials for CRPs. Based on the sources of these polymers, CRPs are further categorized into two classes based on coating with (1) synthetic polymers and (2) natural polymers.

### 10.2.1 CONTROLLED RELEASE OF SOLUBLE P FERTILIZERS WITH SYNTHETIC POLYMER COATINGS

Synthetic organic polymers are traditional coating materials in CRP research and manufacture and have been widely used in various commercial CRPs. These polymer coatings form semipermeable or impermeable layers with small porous structures through which nutrients can diffuse. When in contact with a moist soil, the pores in the coating allow water to diffuse into the core, dissolving the water-soluble compounds inside. The infusion of the water increases the osmotic pressure and causes the coating to stretch and the pore size to increase, which allows the nutrients to gradually diffuse back out through the pores (Carson and Ozores-Hampton, 2014) for crops to absorb.

Many types of synthetic polymers have been applied as coating materials for commercial CRPs. These compounds include polyesters, polyolefines, polyurethane, polyacrylic acid, polyacrylamide (PAM), polysulfonate, polyvinyl chloride, polystyrene, polylactide, polyacetate and polydopamine. For example, Osmocote (Everris Inc., Dublin, Ohio) is a CRP with a polyester (alkyd resin) coating. Polyon is the trade name given to a polyurethane-coated slow-release fertilizer. The coating material in Nutricote is a thermoplastic resin such as polyolefins, poly(vinylidene chloride), and copolymers.

Research shows that these CRPs indeed release P and other nutrients in a gradual fashion when compared with the conventional soluble fertilizers without coatings, thus potentially reducing the risks of eutrophication. For example, 80% of the total applied P quickly leached out of a soil column with the leaching water in a short period of 2.5 days when a conventional P fertilizer (DAP, $(NH_4)_2HPO_4$) was used. This fast-leaching rate was relatively constant at approximately 18% $d^{-1}$ in the first 5 days (Jin et al., 2011). In comparison, commercial CRPs showed much lower P release rates of ~0.15% $d^{-1}$ with only ~40% of ATP leached out in 300 days (Adams et al., 2013). The releasing patterns of CRPs depend on the coating types (polymer type, processing method, etc.), the polymer coating thickness, and the P type. In general, thinner coatings or coated MAP ($NH_4H_2PO_4$) respond better than the thicker ones or coated DAP (Pauly et al., 2002). But, unlike coated N and K fertilizers, the P release pattern is relatively independent of the ambient temperature (Adams et al., 2013).

However, there has been no unanimous evidence to support that CRPs are always superior to the conventional P fertilizers on improving agronomic production. Chien et al. (2009) cited a set of unpublished experimental data showing that applications of a polymer-coated P fertilizer (MAP, trade name AVAIL) produced better corn (*Zea mays*) yields (9.2 and 10.1 Mg ha⁻¹) than the conventional MAP fertilizer (8.5 and 7.0 Mg ha⁻¹) at a rate of 11.6 kg P ha⁻¹. The yields reported earlier resulted from the different fertilizer placement methods: band and broadcast, respectively. In a greenhouse study, Pauly et al. (2002) reported that the PUE of CRPs by barley (*Hordeum vulgare*) was 6–16% higher than that of MAP. Their field trials also demonstrated that CRPs increased the barley yield by 4% over MAP for the same application rates. Under laboratory conditions, Qian and Schoenau (2010) compared the effects of a conventional P fertilizer and a CRP (Agrium Inc., Denver, Colorado) on the growth of a range of common crops: wheat (*Triticum aestivum*), canola (*Brassica napus*), mustard (*Brassica juncea*), flax (*Linum usitatissimum*), yellow pea (*Pisum sativum*), and alfalfa (*Medicago sativum*) in soil for 1 month (starting from germination). They found that the CRP had no significant negative impacts (fertilizer burns) on the seed germinations at application rates of up to 80–100 kg $P_2O_5$, while plant injury occurred in the case of the conventional P fertilizer at a rate of 20–40 kg $P_2O_5$. However, the plant biomass and the P contents were not significantly different for these two types of fertilizers used at the same rate. Conventional MAP sometimes even resulted in slightly higher yields and P uptake rates than those from CRPs at this early growth stage of crops, but CRPs performed better at later stages (Pauly et al., 2002). However, many field studies have shown that commercial CRPs did not result in a significant difference in crop growth, crop yields, biomass production, or P bioavailability (Cahill et al., 2013; Kleinschmidt and Prill, 2008; Murdock et al., 2007; McGrath and Binford, 2012; Randall and Vetsch, 2004) in comparison to other conventional P fertilizers.

Additionally, there is a lack of theoretical basis for CRPs in enhancing the P bioavailability (Chien et al., 2009). Since the P bioavailability in the soil is controlled by phosphate adsorption processes, a reduced rate of phosphates entering the soil solution would not significantly affect or reverse these P fixation processes. Thus, innovative technology for solving the P fixation problem is still needed for enhancing the PUE in agricultural practices. However, considering the environmental benefits, the application of CRPs is advantageous over the conventional P fertilizers in reducing the risks of P transport into the environment while maintaining similar or higher crop yields.

## 10.2.2 CONTROLLED RELEASE OF SOLUBLE P FERTILIZERS WITH NATURAL POLYMER COATINGS

Although synthetic polymers possess better mechanical and controlled-release properties, the accumulation of residual or nondegradable plastics in the soil creates a new type of pollution in ecosystems. Natural polymers are preferred over synthetic polymers as CRP coating materials because of their nontoxicity, water solubility, and biodegradability. In addition to their porous structure for nutrient diffusion, the coatings of natural polymers are also able to release the nutrients through dissolution and/ or biodegradation of the polymer barriers. Natural polymers (e.g., starch, cellulose, lignin, chitosan, alginate, wheat gluten, rubber, and latex) in their original or modified form have been studied for their suitability in enhancing the efficiency of different fertilizer formulations (Table 10.1). For example, Wu and Liu (2008a) reported a chitosan-coated NPK compound fertilizer with properties of controlled release and high water retention. The fertilizer core was water-soluble NPK fertilizer granules, the inner coating was chitosan, and the outer coating was P(AA-co-AM) superabsorbent polymer. As much as 60% of the TP was released in 30 days with an average rate of 2% P $d^{-1}$. And the water-holding capacity (WHC) of a sandy soil increased by 10% from 30.2 to 40.4% when the CRP was applied to the soil at a rate of 1% (by weight).

**TABLE 10.1**
**Recent Studies on Slow-Release P Fertilizers with Natural Polymer Coatings**

| Coating Formula | Effective P Ingredient | Other Macronutrients | Corp and Soil Type | P Release Effectiveness | Crop Yield | Other Benefits | References |
|---|---|---|---|---|---|---|---|
| Chitosan, polyacrylic acid, polyacrylamide | An NPK compound fertilizer | N, K, | nr | Slow release, 69% P released in 30 days | nr | 10% higher water retention | Wu and Liu, 2008a |
| Cellulose acetate, polyacrylic acid, polyacrylamide, vermiculite | An NPK compound fertilizer | N, K, Ca, Mg | nr | Slow release, 69% P released in 30 days while 100% in 15 days for a regular compound fertilizer | nr | 10% higher water retention | Wu and Liu, 2008b |
| Carboxymethyl-chitosan, polyacrylic acid, polyacrylamide, activated carbon | A polyphosphate-based fertilizer | N, Fe | nr | 40–60% P released in 30 days while 90% released within 5 days for a regular P fertilizer | nr | 10% higher water retention | Jin et al., 2011 |
| Starch, polyacrylic acid, polyacrylamide | $NH_4H_2PO_4$ | N | nr | 60% P released in 30 days while 90% released within 5 days for a regular P fertilizer | nr | 10% higher water retention | Jin et al., 2013 |
| Chitosan, polyacrylic acid, diatomite | An NPK compound fertilizer | N, K | nr | ~60% P released in 30 days while 90% released within 5 days for a regular P fertilizer | nr | 10% higher water retention | Wu et al., 2008 |
| $Al(SO_4)_3,H_2O$, $Fe_2(SO_4)_3$, starch, chitosan, lignin | $NH_4H_2PO_4$, $Ca(H_2PO_4)_2$, $Fe(P_2O_7)$ | N | Loam, sandy loam, sand, spring wheat | Reduced P leaching by 200% in 112 days compared to a commercial CRP | Biomass 15–55% lower than that under the commercial CRPF | nr | Entry and Sojka, 2008 |

*Note:* nr = not reported.

Starch and cellulose were also found to be useful as coating materials for slow-release fertilizers with similar patterns (Wu and Liu, 2008b; Jin et al., 2013). Another bio-polymer, polydopamine, inspired by the composition of adhesive proteins in mussels (Lee et al., 2007), has also been tested as a coating material for CRPs (Jia et al., 2013). This polymer exhibited a unique property as a CRP coating: the thickness could be simply manipulated through repeating of a single coating process, thereby control-ling the slow releasing properties of the synthesized CRPs. For example, one time of the deposition process resulted in a coating thickness of 8 nm; three times, 11 nm; and five times, 15 nm (Jia et al., 2013). Another property was that the coating was biodegradable, and thus, the nutrient release was also affected by the microorganisms present in the soils (Jia et al., 2013). Sometimes, clay minerals such as diatomite are added as the CRP compositions with an intention to enhance nutrient and water reten-tion (Wu et al., 2008). Entry and Sojka (2010) applied a rather complicated recipe for a CRP, including fertilizers (nutrient source), Al and Fe salts (for binding P), lignin, cellulose, and starch (as biodegradable coating materials). Although the CRP leached 93% less P in comparison with a commercial CRP, their 210-day greenhouse study of Kentucky bluegrass (*Poa pratensis*) showed that the biomass under this newly syn-thesized CRP was only 25% of that under the commercial CRP, suggesting that their new CRPs were much less effective than the latter in providing P. As a matter of fact, most of the newly reported CRPs (Table 10.1) were not tested for their effects on crop growth, making it difficult to evaluate these new fertilizers. Simply showing a reduction of the nutrient release rate is not enough for an effective CRP since the ultimate goal of fertilizer development is to enhance agronomic production and to achieve food security. Without any evidence of supporting crop growth or not being able to enhance the crop growth as effective as the conventional fertilizers, new CRPs would be less competitive or attractive in agronomical practices. Thus, evaluating the capacity of a newly developed CRP to improve crop growth and yield is a research-able priority. In addition, some reports (Table 10.1) indicated that a few hydrophilic coatings have high water storage capacity, potentially increasing the plant-available WHC of soil and supplying the crops with more water during drought. However, high water content in the polymers does not necessarily mean that all the water is available for crop use. Indeed, plant-available water content (AWC) in a soil is the difference in the water content between the field capacity and the permanent wilting point. There is a specific method (Cassel and Nielsen, 1986; Liu and Lal, 2013) to measure the AWC in a soil, which was not used in these studies. Moreover, higher cost and poor con-trol over nutrient release are also notable limiting factors for these newly developed CRPs with natural polymer coatings. Therefore, enhancing the agronomic produc-tion, reducing the cost, and improving the controls of nutrient and water releases are among high researchable priorities.

## 10.3 ENHANCING BIOAVAILABILITY OF SOLID P FERTILIZERS

### 10.3.1 Phosphate Rocks with Inorganic Acids

PRs (apatites, $Ca_5(PO_4)_3OH$, $Ca_5(PO_4)_3F$, or $Ca_5(PO_4)_3Cl$) are the major natural forms of P and the raw materials for traditional mineral P fertilizers such as MAP, DAP,

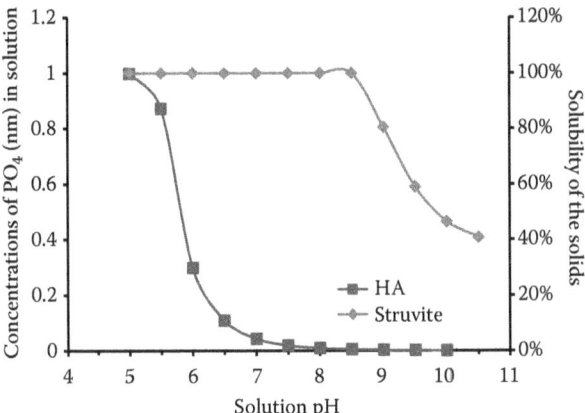

**FIGURE 10.1**   A comparison of HA ($Ca_5(PO_4)_3OH$) and struvite ($NH_4MgPO_4$) solubility under different pH (estimated with Visual Minteq 3.0, initial solid concentrations equivalent to 1 mM $PO_4$).

phosphoric acid, and others. PRs have for long been applied in the natural form as a cheap P fertilizer for agronomic production (FAO, 2003), but the fertilizing efficacy is low in comparison with the soluble P fertilizers because P is tightly locked in the solid apatite matrices (solubility products of apatites is $<10^{-50}$) (Bengtsson, 2007) at neutral or alkaline pH ranges (Figure 10.1). However, due to their relatively low costs (Kpomblekou-A and Tabatabai, 2003) and environmental friendliness (PRs could be regarded as a slow P-releasing fertilizer), PRs have been considered as a valuable candidate for P fertilizer developments for many years (FAO, 2003). Further efforts have been made to enhance the P release rate of the rocks or increase their fertilizer efficiency. Since acids can increase the P release from PRs (Figure 10.1), several inorganic acids have been experimented on with related research. For example, by reacting PRs, sulfuric acid ($H_2SO_4$) or phosphoric acid ($H_3PO_4$) has been used to prepare the so-called partially acidulated phosphate rocks (PAPRs). The use of PAPRs has been widespread in Europe and South America (FAO, 2003). PAPRs offer an economic means of enhancing the agronomic effectiveness of indigenous PR sources that may otherwise be unsuited for direct application. Research on the agronomic effectiveness of PAPRs has been reviewed by FAO (2003), and the general conclusion is that PAPRs of 40–50% acidulation with $H_2SO_4$ or 20–30% acidulation with $H_3PO_4$ are as effective as fully acidulated TSP ($Ca(H_2PO_4)_2$). However, the presence of cadmium (Cd) in PRs may limit the direct use of PAPRs (Lal, 2014).

### 10.3.2   Phosphate Rocks with Naturally Occurring Organic Acids

Some plants are able to increase P acquisition under P-deficit conditions by excreting some organic acids of carboxylates (mainly of citrate and oxalate) (Kpomblekou-A and Tabatabai, 2003; Erro et al., 2010; Gerke, 2015). Therefore, these organic acids of small molecular weight could potentially help in mobilizing the fixed P in PRs or in soils. The mechanisms involved include acidic dissolution of the insoluble P species and/or enhanced desorption of phosphate ions by citrate and oxalate ligands through anion exchange processes.

Kpomblekou-A and Tabatabai (2003) reported that these organic acids increased the release rate of P from PRs of different origins, with oxalic and citric acids being the most efficient in mobilizing solid P. Their data indicated that the oxalic acid performed similarly to or even better than sulfuric acid in improving P release. For instance, by simply mixing a PR and 10 mM acid solutions at a ratio of 1 g to 25 mL, the P release rate from the PR in a 24-hour period was 90 mM P kg$^{-1}$ PR by citric acid, about 160 mM P kg$^{-1}$ PR by oxalic acid, and 150 mM P kg$^{-1}$ PR by sulfuric acid. The release rates differed with the PR origins and the acid types, and a presence of carbonate impurities depressed the P-mobilizing ability of these acids. Moreover, their data showed that the addition of oxalic and citric acids to soil–PR mixtures for a month resulted in a 50% increase in available P from PRs in comparison with those mixtures without acid amendments, clearly indicating an enhancement of P release from the PRs by these organic acids. In addition, through a 60-day green-house study, amendments of organic acids with PRs were also found to enhance the aboveground biomass production of corn when compared to those amended by PRs alone. Specifically, at a PR application rate of 200 mg P kg$^{-1}$ soil, oxalic acid and citric acid produced a yield (dry aboveground biomass) of about 6.3 g pot$^{-1}$ and 4.9 g pot$^{-1}$, respectively. In comparison, PR-only amendment just produced a yield of 3.5 g pot$^{-1}$, evidencing the enhanced fertilizing effects of PRs through organic acids with high crop yields (Kpomblekou-A and Tabatabai, 2003). Erro et al. (2007) added humic acid to a synthesized P-metallic (Mg and Zn) fertilizer (a type of CRP) at a rate of 11% by weight and observed that the acid treatment enhanced the water solubility of P by 2% as compared with the same fertilizer but without humic acid addition. These trends suggest that mixing humic acids with low-soluble P fertilizer can improve P bioavailability. Follow-up studies by Erro et al. (2010) showed that biomass production and P contents of corn, chickpea (*Cicer arietinum*), and white lupin (*Lupinus albus*) with humic acid were better than those with no-P or CRP-only treatment but were similar to that of a soluble P fertilizer treatment (KH$_2$PO$_4$) after 24 days of incubation in a growth chamber.

Some bacteria and fungi have long been found to be able to solubilize PRs or P in soil by excreting similar organic acids (Narsian and Patel, 2000; Ramirez et al., 2009; Arcand and Schneider, 2006; Mohammadi, 2012). Microbiological means of improving the agronomical effectiveness of PR have been reviewed by FAO (2003). Those methods included composting PRs with organic wastes and inoculating the seeds or the PRs with PSMs. Interested readers are encouraged to refer to the study of FAO (2003) for additional details.

## 10.3.3 Siderophores

Siderophores (Figure 10.2) are a group of organic compounds with low molecular weight produced by microorganisms and plants growing under low iron (Fe) conditions (Ahmed and Holmström, 2014). The unique property of this group of biochemicals is their extreme high affinity with ferric iron (Fe (III)) in a wide pH range

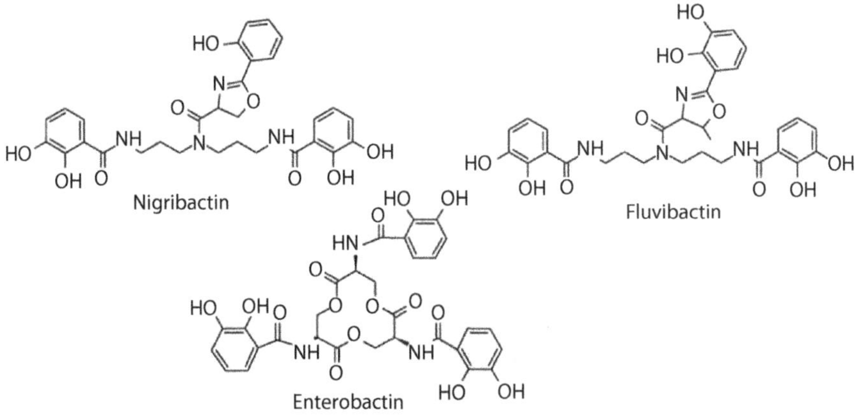

**FIGURE 10.2** Example of some siderophores. (From Nielsen, A., M. Mansson, M. Wietz, A. N. Varming, R. K. Phipps, T. O. Larsen, L. Gram, and H. Ingmer, *Mar. Drugs*, 10, 2584–2595, 2012, http://creativecommons.org/licenses/by/3.0/. With permission.)

commonly observed in soils. Specifically, siderophores form 1:1 complexes with Fe (III) with stability constants ($K$) ranging between $10^{30}$ and $10^{52}$, while the constants of oxalic and citric acids with Fe (III) are $10^8$ and $10^{12}$, respectively (Ahmed and Holmström, 2014). The primary biochemical function of these compounds to microorganisms and plants is to help the requisition of Fe nutrients from the environment, especially under unfavorable conditions (e.g., in calcareous soils of high pH). Some important crops such as barley, rice (*Oryza sativa*), and wheat are able to produce siderophores, while oats (*Avena sativa*) are able to use microbial siderophores to assimilate iron. More than 500 different types of siderophores have been found in various ecosystems (e.g., soils, fresh water, sediments, and seawater) and 270 of these have been structurally characterized (Ahmed and Holmström, 2014).

Studies have also shown that siderophores play a significant role in mineral-weathering processes by accelerating the dissolution of Fe-containing minerals such as ferrihydrate and goethite (e.g., Hiradate and Inoue, 1998; Sokolova et al., 2010; Gómez-Galera et al., 2012). Since a large proportion of P fertilizers applied to the soil (especially the acidic soils) is sequestrated by iron or aluminum (hydro) oxides, preventing it from plant use (Fageria, 2009); the dissolution of these iron- or aluminum-containing minerals by siderophores can also help the release of the fixed P and increase its bioavailability. In addition to Fe, plants might have used siderophores since time immemorial to acquire nutrient P from the environment by dissolution of P-fixing agents in the soil.

Kearns et al. (2004) observed an increased P dissolution rate with a siderophore added to some sparingly soluble iron phosphate minerals (e.g., $FePO_2$ $2H_2O$ with a solubility product constant of $9.91 \times 10^{-16}$), thereby making both Fe and P bioavailable. Ghosh et al. (2015) reported that a type of arsenic-resistant bacteria could produce a significant amount of siderophores (6.2 times more than that from a soil bacterium isolated from non-As-contaminated soils). These bacteria grow better in alkaline soils, which are characterized by low P and Fe availabilities and/or high concentrations of As (Lessl and Ma, 2013). Phosphorus and As anions compete for uptake by bacterial cells, so bacteria have developed an effective system to increase P solubility from soils by secreting siderophores. These bacteria showed high capacity for solubilizing P from the $FePO_4$ solid. Further, a 7-day greenhouse test on 1-month-old tomato (*Solanum lycopersicum*) seedlings showed positive responses to the combined amendments of $FePO_4$ and the liquid medium containing siderophores produced by the strain. The root and shoot biomass under siderophore treatment were 2.3–4.0-fold higher than those in the control (a low P medium without siderophores) and a medium treated by a nonsiderophore-producing bacterial strain. The shoots and roots with siderophore treatment had 1.5–2.7-fold higher P concentration (7.46–11.6 mg $g^{-1}$) than that in the control and 1.3–1.7-fold higher than that of the nonsiderophore-producing bacterial strain. The Fe concentrations in the tomato tissues also increased tenfold (10.8 mg $kg^{-1}$) in the shoots and twofold (270 mg $kg^{-1}$) in the roots with siderophore treatment compared to nonsiderophore-producing bacterial strain. This study also indicated that siderophores were more effective in P solublization from iron phosphate–containing compounds than from calcium phosphate or PRs since siderophores are very strong in chelating Fe (III) but not Ca (II).

## 10.4 USES OF RECYCLED STRUVITES AS AN ALTERNATIVE P FERTILIZER

Given the limited natural reserves for P and the large quantities of industrial, municipal, and agricultural waste streams containing considerable amounts of P, recovery from these wastes and reuse of these as P fertilizers have long been regarded as a sustainable approach for P fertilizer industry and for the world agriculture and food production. Moreover, P removal from waste streams has also been required by federal regulations to reduce the risks of eutrophication to the surface water bodies. Thus, technology for reducing P levels in industrial or municipal wastewaters have been intensively researched. For example, anion exchange resins and other sorbents (e.g., iron minerals) are reportedly effective means for this purpose (Rittmann et al., 2011; Karunanithi et al., 2015). However, studies on the recovery of the removed P and its reuses as P fertilizers are scanty. Data on the agronomic evaluation of these recovered P are extremely scarce, indicating a missed link on the P recycling chain. The recovery of P in the form of struvite (Equation 10.1) from waste streams is the only well-studied and promising technology for alternative P fertilizer productions. The related mechanisms, production techniques, and economical feasibility have been widely reviewed (e.g., Karunanithi et al., 2015). Thus, this section focuses on the latest studies regarding the fertilizing effect of the recovered struvite as an alternative P fertilizer on the growth and the yields of different crops in diverse soils (Table 10.2). The primary objective of this section is to create a bridge between environmental scientists, water-treatment engineers, soil scientists, agronomists, and farmers, regarding the agronomically, economically, and environmentally beneficial roles of using the recovered struvite as alternative P fertilizers. Indeed, without the practical reuse of the waste P in agriculture, sustainable agriculture or sustainable economy would not be possible.

**TABLE 10.2**

**Effects of Recycled Struvite as an Alternative P Fertilizer on Plant Growth**

| Crop Type | Source of P | Source of Mg | P Fertilizer Control | Soil Type, Test Method and Duration | Observations | Note | Reference |
|---|---|---|---|---|---|---|---|
| Maize (Zea mays L.), soybean (Glycine max [L.] Merr.) | Wastewater effluent of corn fiber processing for bioenergy | nr[a] | TSP fertilizer $(Ca(H_2PO_4)_2 \cdot H_2O)$ | Silt loam of pH 6, loam of pH 6, silty clay of pH 7; field study, corn–soybean– corn rotations; three-years | Biomass, grain yield, P availability similar to or higher than TSP | | Thompson, 2013 |
| Maize, barley (Hordeum vulgare L.), sorghum (Sorghum bicolor Moench), amaranth (Amaranthus hypochondracus L.), sunflower (Helianthus annuus L.) | Anaerobic digestion effluent | $MgCl_2 \cdot 6H_2O$ | TSP and seven other types of organic fertilizers derived from sewage sludge | Loamy sand of pH 5.2, greenhouse test; 56 days | Dry biomass yields constantly higher or similar to TSP or other sludge-derived P fertilizers | | Vogel et al., 2015 |
| Spring wheat (Triticum aestivum L.) | Dairy manure supernatant | nr[a] | TSP | Loam sand of pH 6.5 and 7.6; greenhouse test; 120 days | Dry biomass and P uptakes similar or higher than TSP | Struvite suitable for both acidic and slightly alkaline soils | Massey et al., 2009 |

(Continued)

## TABLE 10.2 (CONTINUED)
### Effects of Recycled Struvite as an Alternative P Fertilizer on Plant Growth

| Crop Type | Source of P | Source of Mg | P Fertilizer Control | Soil Type, Test Method and Duration | Observations | Note | Reference |
|---|---|---|---|---|---|---|---|
| Canola (*Brassica napus* L.) | Swine manure supernatant | Swine manure supernatant | MAP | Sandy loam of pH 7.7; greenhouse study: 56 days | Yields and P uptakes lower 18% and 15% than those of controls, respectively | Alkaline soil reduced struvite solubility and bio-availability | Ackerman et al., 2013 |
| Lettuce (*Lactuca sativa* L.) | Anaerobic digestion effluent | Low-grade MgO by-product | Commercial P fertilizer | Loamy sand of pH 5.9; greenhouse study; ~90 days | Fresh weight and nutrient uptake similar or better than those of the control at the same application rates | | González-Ponce et al., 2009 Plaza et al., 2007 |
| Ryegrass (*Lolium multiflorum* L.), maize | Human urine | MgO | Commercial P fertilizer | nr*; greenhouse test; 82 days | Dry biomass and P uptakes similar or higher than those of the control | Low heavy metals | Antonini et al., 2012 |
| Ryegrass | Synthetic urine solutions | $MgCl_2 \cdot 6H_2O$ | $NH_4NO_3$ and $KH_2PO_4$ | Sandy loam of pH 5.4; greenhouse test, ~72 days | Dry biomass and P uptakes similar or higher than those of the control | Crop recovered 26% P from struvite | Bonvin et al., 2015 |

*(Continued)*

**TABLE 10.2 (CONTINUED)**
**Effects of Recycled Struvite as an Alternative P Fertilizer on Plant Growth**

| Crop Type | Source of P | Source of Mg | P Fertilizer Control | Soil Type, Test Method and Duration | Observations | Note | Reference |
|---|---|---|---|---|---|---|---|
| Ryegrass (*Lolium perenne* L.) | Different municipal or industrial wastewaters | nr[a] | $CaHPO_4$ | Sandy loam of Ph 6.6, a candy clay loam of pH 7.1; pot test; ~100 days | Dry biomass and P uptakes similar or higher than those of the control | | Johnston and Richards, 2003 |
| Chinese flowering cabbage (*Brassica para chinensis*), Chinese chard (*Brassica rapa* var. *chinensis*), Water spinach (*Ipomoea aquatica*), | P salt, landfill leachate as nitrogen source | Mg salt | $NH_4NO_3$ and $Ca(H_2PO_4)_2$ | Sandy clay of pH 6.2; pot test, outdoor; ~35 days | Growth rates similar to the control; no heavy metal accumulation, 106–146% more Mg and 77–172% more P in plants than the controls | | Li and Zhao, 2003 |
| Chinese cabbage | Wastewater from a semiconductor factory | $MgCl_2 \cdot 6H_2O$ | Composite commercial fertilizer: one compost and one manure | Loam of pH 5.3; laboratory tests; 32 days | Thirty-two-days fresh weight 3.5 times higher than organic fertilizers but 2 times lower than that under the chemical fertilizer | | Ryu et al., 2012 |
| Lilac (*Syringa vulgaris*; Mademoiselle Lemoine), rhododendron (*Rhododendron catawbiense*; Roseum Elegans) | Municipal or industrial wastewaters | nr[a] | TSP | Wood chips and peat moss; greenhouse test; ~224 days | Plant growth (shoot and root growth, visual quality, leaf color) similar to that of the control | | Fish, 2015 |

[a] nr: not reported.

Generally, phosphate and ammonium contained in the waste streams (e.g., municipal wastewater, supernatant of sludge digesters, landfill leachate, cattle manure, human urine, wastewater discharge from fertilizer industry, etc.) can be recovered in the form of struvite precipitate as described in Equation 10.1. In practice, an external application of Mg salts and a pH adjustment to ~9 or higher are necessary to completely precipitate phosphates (sometimes, additional phosphate is needed when N recovery is the target). Several full-scale struvite recovery systems have been in operation in wastewater treatment plants around the world (Karunanithi et al., 2015).

$$Mg^{2+} + NH_4^+ + PO_4^{3-} + 6H_2O = MgNH_4PO_4\,6H_2O\ (struvite) \downarrow \qquad (10.1)$$

Figure 10.1 shows the dissolution of struvite under various soil solution pH values as compared with that of one type of PRs (HA). These data show that struvite is a better P fertilizer than the latter since a complete dissolution of struvite occurs across a wider soil pH range of 5.0–8.0. In comparison, soluble P is less than 20% from HA at pH 6 and close to zero when the pH is higher than 8.0. Indeed, greenhouse and field studies have shown that the recovered struvite could be as effective as the conventional P fertilizers in enhancing crop yields (Table 10.2) in acidic or slightly alkaline soils. Moreover, research data have shown that the recovered struvite does not contain high levels of heavy metals and other organic toxicants (Antonini et al., 2012). Thus, the recovered struvite is a more effective and safer fertilizer than the biosolids. The latter may contain a significant amount of heavy metals (Liu and Lal, 2013) and other contaminants and may not be suitable as a soil amendment.

Shu et al. (2006) indicated that approximately 1 kg of struvite can be produced from 100 m³ of wastewater. If struvite were to be recovered from wastewater treatment plants worldwide, 0.63 Tg phosphorus (as $P_2O_5$) could be harvested annually, possibly reducing PR mining by 1.6%.

## 10.5  NANOSIZED P FERTILIZERS (NANOFERTILIZERS)

Research efforts have been aimed at innovative P fertilizer formulations for enhancing the nutrient use efficiencies. Nanotechnology has been a promising field to explore new types of nutrient delivery schemes and initiate new fertilizer development (nanofertilizers) (Liu and Lal, 2014, 2015; Bindraban et al., 2015; Subramanian et al., 2015; Sharonova et al, 2015; Montalvo et al., 2015a,b). Nanotechnology is associated with particles measuring a dimension of 1–100 nm. This enables atom-by-atom manipulation, and thus, the processes or the products evolved from nanotechnology are different from those achieved through conventional methods.

Nanoparticles possess unique characteristics in delivering nutrient elements due to their small particle sizes and large specific surface areas. The nutrient release pattern from the particle surfaces to the solution occurs in a slow-release fashion (at moderate release rates for longer periods) in comparison to soluble nutrient salts (at high release rates but in extremely short time) or conventional particles (at extremely low release rate) (Casals et al., 2014). Thus, it renders the nanoparticles as a promising carrier for delivering crop nutrients. For example, Casals et al. (2014) observed that Fe nanoparticles of 7 nm in size were able to enhance biogas production from experimental anaerobic digesters by significantly threefold as compared with the controls of regular Fe salt or Fe solid treatments. The underlying mechanisms were that, superior to other Fe sources, the nanoparticulate Fe provides enough Fe nutrient for a long time to the microorganisms for metabolism. Santner et al. (2012) tested the uptake of P by canola (*Brassica napus*) through a nutrient solution containing an $Al_2O_3$ nanoparticle–loaded P nanofertilizer (particle size <40 nm) in a greenhouse study. They reported that the nutrient (P) uptake in this case was eight times as much as that in the control (with nutrient solution of $KH_2PO_4$ but without nano-$Al_2O_3$). The authors believed that nanoparticle-associated P were able to penetrate the diffusion layer (a stagnant water film) on the root surfaces and continuously supply P for the root cells to uptake, while free phosphate ions were blocked out of the film, supplying limited P to the root cells. This research indicates another advantage of nanofertilizers that nanoparticles are superior to other chemical forms in transporting nutrients to crop roots in the rhizosphere. Montalvo et al. (2015a) tested the uptake of P by wheat from a solution containing natural colloid-bound P (in sizes of 1–1000 nm) and a solution without particles. They reported that the P uptake was up to fivefold from the former than the latter, again emphasizing the important role of nanoparticulate P to crops.

Liu and Lal (2014) synthesized a new type of HA ($Ca_5(PO_4)_3OH$) nanoparticles of ~16 nm in size (Figure 10.3) and assessed the fertilizing effect of the nanoparticles

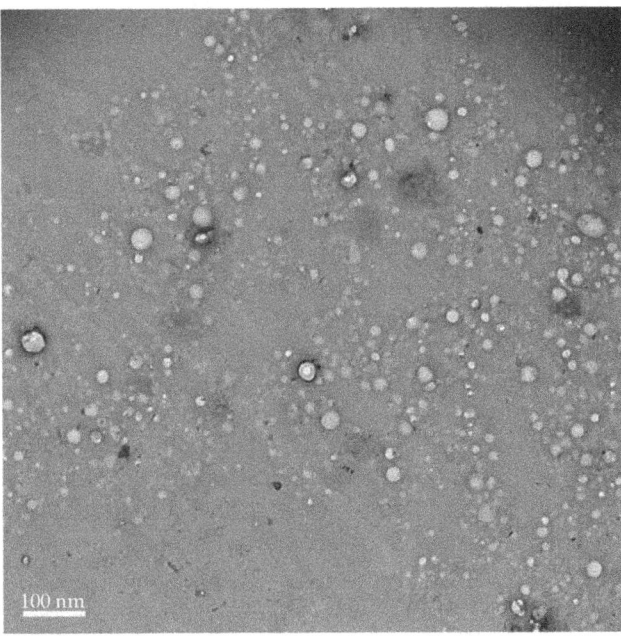

**FIGURE 10.3** A transmission electron microscopic image of HA nanoparticles prepared as a new type of P nanofertilizer. (Reprinted by permission from Macmillan Publishers Ltd. *Sci. Rep.* [Liu and Lal, 2014], copyright [2014].)

on soybean in an inert growing medium (50% perlite and 50% peat moss) through a greenhouse experiment. The data showed that the application of the nanoparticles increased the growth rate and the seed yield by 33% and 20%, respectively, compared to those with a regular P fertilizer (TSP). The production of biomass was enhanced by 18% for the aboveground and 41% for the belowground components. The data indicated that soybean roots can absorb HA nanoparticles as an effective P nutrient source and maintain healthy growth and high yield. In addition to P, the nanoparticles may have also supplied the plants with nutrient Ca. Liu and Lal (2014) indicated that the apatite nanoparticles had much weaker interactions with soil components than the charged ions of $PO_4^{3-}$, $HPO_4^{2-}$, $H_2PO_4^-$, or $Ca^{2+}$. Therefore, a significant portion of nanoparticles remained in the soil solution for roots to absorb, while most of the charged phosphate ions from conventional P fertilizers were often absorbed by the soil particles and were not available for plant uptake.

The porous structures of the plant cell wall have openings ranging from 5 to 20 nm. Hence, nanoparticles with sizes less than the pore diameters could easily pass through the cell wall and reach the plasma membrane. There is also a chance for the enlargement of the pores or the induction of new cell wall pores upon interaction with engineered nanoparticles, which in turn enhances nanoparticle uptake. Further internalization occurs during endocytosis with the help of a cavity-like structure that forms around the nanoparticles by the plasma membrane. They may also cross the membrane using embedded transport carrier proteins or through ion channels. These might be another advantage of P nanofertilizer to be easily absorbed by plant cells. However, the particle size is critical (<20 nm) in this regard. Montalvo et al. (2015b) reported that the wheat uptake of P from HA nanoparticles was higher (40–61%) than that from the bulk HA (12–18%), but it is lower than that from TSP (64–88%). But they measured the size of the HA nanoparticles as large as 280 nm, out of the size range for nanoparticles or for cell wall openings, suggesting that the particle size might be a critical parameter in determining the effectiveness of a nanofertilizer. Further, there was also no indication of any phytotoxicity of the P nanofertilizers observed through a seed germination test of crops and vegetables (Liu and Lal, 2014; Sharonova et al., 2015).

In general, nanotechnology opens a new field to explore and address long-standing problems in global agriculture—increasing P availability to the crops, improving P fertilizer efficiencies, and reducing eutrophication. Research on P nanofertilizers is an opportunity to utilize this new nutrient-delivery approach, enhance P uptakes by plants, and reduce P losses to the environment. The potential is promising, although the related studies are just emerging. The hypotheses behind nanofertilizer developments include direct entrance into the cell walls for cells to absorb P nutrient, slow release of P nutrient from the nanoparticle surface, and enhanced diffusion of the nutrient from the soil solution to the root surfaces. More research is needed both in the theoretical and practical aspects (e.g., field studies) through a collaboration of researchers in a variety of disciplines (e.g., nanochemistry, plant biology, soil science, and agronomy). In chemistry, the syntheses of nanoparticles with different sizes and compositions for optimum P uptakes are important to P nanofertilizer developments. In botany, research is needed to study how plant cells interact with nanoparticulate nutrients and preferentially absorb nanonutrients. In soil science, research is needed

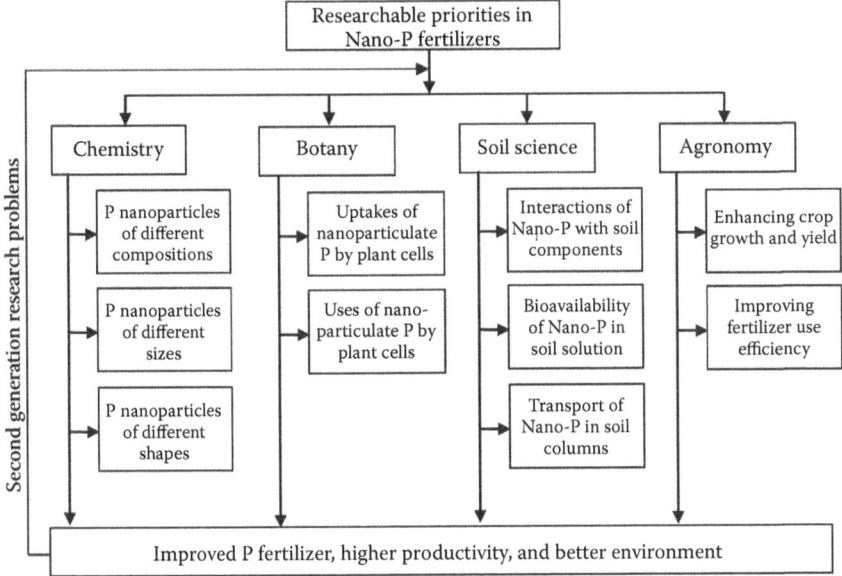

**FIGURE 10.4**    Researchable priorities in nano-P fertilizers.

to understand the transport and the transformation of the P nanofertilizers within the soil and to elucidate the interactions of the particles with the complicated soil components. In agronomy, greenhouse and field studies are critical to nanofertilizer research on evaluating the effectiveness of P nanofertilizers on different crops (Figure 10.4).

## 10.6   SUMMARY

The low use efficiency of P fertilizers (10–15%) and the related eutrophication risks to ecosystems have been a challenge to the scientific community for a long time but with little success. Further, finite and unevenly distributed P resources have been increasingly recognized as a major bottleneck to sustainable agriculture and economy. Therefore, enhancing the efficiency of P fertilizers through the modification of their formulation is important in increasing global agronomic productions, minimizing the risks of widely-spreading eutrophication issues, and ensuring future food security. The synthesis of the available information presented herein shows the related potential of slow-release P fertilizers with natural polymer coatings, improved PR fertilizers with organic acids and siderophores, recycled struvite as an alternative P fertilizer, and P nanofertilizers with innovative fertilizing mechanisms. Theoretical bases and studies at laboratory or pilot scales have shown that all these fertilizers are effective in increasing P uptakes by plants and/or reducing P loss to the environment as compared with the conventional P fertilizers. But more research data at the field scales (especially on improving crop yields) are needed to encourage the adoption of these new P fertilizers in agricultural practices.

## REFERENCES

Ackerman, J. N., F. Zvomuya, N. Cicek, and D. Flaten. 2013. Evaluation of manure-derived struvite as a phosphorus source for canola. *Can. J. Plant Sci.* 93: 419–424.

Adams, C., J. Frantz, and B. Bugbee. 2013. Macro- and micronutrient-release characteristics of three polymer-coated fertilizers: Theory and measurements. *J. Plant Nutr. Soil Sci.* 176: 76–88.

Ahmed, E., and S. J. M. Holmström. 2014. Siderophores in environmental research: Roles and applications. *Microb. Biotechnol.* 7: 196–208.

Antonini, S., M. A. Arias, T. Eichert, and J. Clemens. 2012. Greenhouse evaluation and environmental impact assessment of different urine-derived struvite fertilizers as phosphorus sources for plants. *Chemosphere* 89: 1202–1210.

Arcand, M. M., and K. D. Schneider. 2006. Plant- and microbial-based mechanisms to improve the agronomic effectiveness of phosphate rock: A review. *An. Acad. Bras. Ciênc.* 78: 791–807.

Bacon, J. 2014. Third day is the charm: Toledo can drink its water. *USA Today.* Available at http://www.usatoday.com/story/news/usanow/2014/08/04/ohio-toledo-water-ban/13562707 (accessed September 20, 2014).

Bengtsson, A. 2007. Solubility and surface complexation studies of apatite. PhD dissertation, Umeå University, Umeå.

Bindraban, P. S., C. Dimkpa, L. Nagarajan, A. Roy, and R. Rabbinge. 2015. Revisiting fertilisers and fertilisation strategies for improved nutrient uptake by plants. *Biol. Fertil. Soils.* 51: 897–911.

Bonvin, C., B. Etter, K. M. Udert, E. Frossard, S. Nanzer, F. Tamburini, and A. Oberson. 2015. Plant uptake of phosphorus and nitrogen recycled from synthetic source-separated urine. *AMBIO* 44: S217–S227.

Cahill, S., R. J. Gehl, D. Osmond, and D. Hardy. 2013. Evaluation of an organic copolymer fertilizer additive on phosphorus starter fertilizer response by corn. *Crop Manage.* 12: 1–11 (accessed June 10, 2015).

Carson, L. C., and M. Ozores-Hampton. 2014. *Description of Enhanced-Efficiency Fertilizers for Use in Vegetable Production.* Publication No. HS1247. Institute of Food and Agricultural Sciences (IFAS) Extension Service, University of Florida, Gainesville, FL. Available at http://edis.ifas.ufl.edu/pdffiles/HS/HS124700.pdf (accessed June 10, 2015).

Casals, E., R. Barrena, A. García, E. González, L. Delgado, M. Busquets-Fité, X. Font et al. 2014. Programmed iron oxide nanoparticles disintegration in anaerobic digesters boosts biogas production. *Small* 10: 2801–2808.

Cassel, D. K., and D. R. Nielsen. 1986. Field capacity and available water capacity. In *Methods of Soil Analysis (Part 1): Physical and Mineralogical Methods* (Second ed.), Klute, A. (ed.), pp. 635–662. American Society of Agronomy and Soil Science Society of America, Madison, WI.

Chien, S. H., L. I. Prochnow, and H. Cantarella. 2009. Recent developments of fertilizer production and use to improve nutrient efficiency and minimize environmental impacts. *Adv. Agron.* 102: 267–322.

Davidson, D., and F. X. Gu. 2011. Materials for sustained and controlled release of nutrients and molecules to support plant growth. *J. Agric. Food Chem.* 60: 870–876.

Entry, J. A., and R. E. Sojka. 2010. Matrix-based fertilizers reduce nutrient leaching while maintaining Kentucky bluegrass growth. *Water Air Soil Poll.* 207: 181–193.

Erro, J., O. Urrutia, S. San Francisco, and J. M. García-Mina. 2007. Development and agronomical validation of new fertilizer compositions of high bioavailability and reduced potential nutrient losses. *J. Agric. Food Chem.* 66: 7831–7839.

Erro, J., A. M. Zamarreño, and J. M. García-Mina. 2010. Ability of various water-insoluble fertilizers to supply available phosphorus in hydroponics to plant species with diverse phosphorus-acquisition efficiency: Involvement of organic acid accumulation in plant tissues and root exudates. *J. Plant Nutr. Soil Sci.* 173: 772–777.

Fageria, N. K. 2009. *The Use of Nutrients in Crop Plants.* CRC Press Boca Raton, FL.

FAO (Food and Agriculture Organization of the United Nations). 2003. *Use of Phosphate Rocks for Sustainable Agriculture.* Publishing Management Service, Information Division, FAO, Rome.

FAO. 2009. How to feed the world in 2050. *Proceedings of the Expert Meeting on How to Feed the World in 2050, June 24–26 2009. FAO Headquarters, Rome.*

Fish, M. C. 2015. Developing struvite as a fertilizer for container grown nursery crops. MS thesis, Washington State University, Pullman, WA.

Gerke, J. 2015. The acquisition of phosphate by higher plants: Effect of carboxylate release by the roots: A critical review. *J. Plant Nutr. Soil Sci.* 178: 351–364.

Ghosh, P., B. Rathinasabapathi, and L. Q., Ma, 2015. Phosphorus solubilization and plant growth enhancement by arsenic-resistant bacteria. *Chemosphere* 134: 1–6.

Gómez-Galera, S., D. Sudhakar, A. M. Pelacho, T. Capell, and P. Christou. 2012. Constitutive expression of a barley Fe phytosiderophore transporter increases alkaline soil tolerance and results in iron partitioning between vegetative and storage tissues under stress. *Plant Physiol. Biochem.* 53: 46–53.

González-Ponce, R., E. G. López-de-Sá, and C. Plaza. 2009. Lettuce response to phosphorus fertilization with struvite recovered from municipal wastewater. *Hortscience* 44: 426–430.

Hiradate, S., and K. Inoue. 1998. Dissolution of iron from iron (hydr)oxides by mugineic acid. *Soil Sci. Plant Nutr.* 44: 305–313.

Jia, X., Z. Ma, G. Zhang, J. Hu, Z. Liu, H. Wang, and F. Zhou. 2013. Polydopamine film coated controlled-release multielement compound fertilizer based on mussel-inspired chemistry. *J. Agric. Food Chem.* 61: 2919–2924.

Jin, S., Y. Wang, J. He, Y. Yang, X. Yu, and G. Yue. 2013. Preparation and properties of a degradable interpenetrating polymer network based on starch with water retention, amelioration of soil, and slow release of nitrogen and phosphorus fertilizer. *J. Appl. Polym. Sci.* 128: 407–415.

Jin, S., G. Yue, L. Feng, Y. Han, X. Yu, and Z. Zhang. 2011. Preparation and properties of a coated slow-release and water-retention biuret phosphoramide fertilizer with superabsorbent. *J Agric Food Chem.* 59: 322–327.

Johnston, A. E., and R. Richards. 2003. Effectiveness of different precipitated phosphates as phosphorus sources for plants. *Soil Use Manage.* 19: 45–49.

Karunanithi, R., A. A. Szogi, N. Bolan, R. Naidu, P. Loganathan, P. G. Hunt, M. B. Vanotti, C. P. Saint, Y. S. Ok, and S. Krishnamoorthy. 2015. Phosphorus recovery and reuse from waste streams. *Adv. Agron.* 131: 173–250.

Kearns, J., J. Cervini-Silva, and J. Banfield. 2004. Siderophores may simultaneously influence iron and phosphorus bioavailability in soils. *Geochim. Cosmochim. Ac.* 68: A395–A395.

Kleinschmidt, A., and G. Prill. 2008. *Evaluations of Starter Fertilizers for Field Corn.* Ohio State University Extension, Van Wert, OH. Available at http://agcrops.osu.edu/on-farm -research/archive%20pages/2008/corn%20pop-up%20starter%20fertilizer%20FINAL .pdf (accessed June 10, 2015).

Kpomblekou-A, K., and M. A. Tabatabai. 2003. Effect of low-molecular weight organic acids on phosphorus release and phytoavailabilty of phosphorus in phosphate rocks added to soils. *Agr. Ecosyst. Environ.* 100: 275–284.

Lal, R. 2014. Book Review: R. W. Scholz, A. H. Roy, F. S. Brand, D. T. Hellums, and A. E. Ulrich (eds.): Sustainable phosphorus management: A global transdisciplinary roadmap. *J. Plant Nutr. Soil Sci.* 177: 934–935.

Lee, H., S. M. Dellatore, W. M. Miller, and P. B. Messersmith. 2007. Mussel-inspired surface chemistry for multifunctional coatings. *Science* 318: 426–430.

Lessl, J. T., and L. Q. Ma. 2013. Sparingly soluble phosphate rock induced significant plant growth and arsenic uptake by *Pteris vittata* from three contaminated soils. *Environ. Sci. Technol.* 47: 5311–5318.

Li, X. Z., and Q. L. Zhao. 2003. Recovery of ammonium-nitrogen from landfill leachate as a multi-nutrient fertilizer. *Ecol. Eng.* 20: 171–181.

Liu, R., and R. Lal. 2013. A laboratory study on amending mine soil quality. *Water Air Soil Poll.* 224: 1679–1796.

Liu, R., and R. Lal. 2014. Synthetic apatite nanoparticles as a phosphorus fertilizer for soybean (*Glycine max*). *Sci. Rep.* 4: 5686–5691.

Liu, R., and R. Lal. 2015. Potentials of engineered nanoparticles as fertilizers for increasing agronomic productions. *Sci. Total Environ.* 514: 131–139.

Massey, M. S., J. G. Davis, J. A. Ippolito, and R. E. Sheffield. 2009. Effectiveness of recovered magnesium phosphates as fertilizers in neutral and slightly alkaline soils. *Agron. J.* 101: 323–329.

McGrath, J. M., and G. D. Binford. 2012. Corn response to starter fertilizer with and without AVAIL. *Crop Manage.* 11: 1–8 (accessed June 10, 2015).

Mohammadi, K. 2012. Phosphorus solubilizing bacteria: Occurrence, mechanisms and their role in crop production. *Resour. Environ.* 2: 80–85.

Montalvo, D., F. Degryse, and M. J. McLaughlin. 2015a. Natural colloidal P and its contribution to plant P uptake. *Environ. Sci. Technol.* 49: 3427–3434.

Montalvo, D., M. J. McLaughlin, and F. Degryse. 2015b. Efficacy of hydroxyapatite nanoparticles as phosphorus fertilizer in Andisols and Oxisols. *Soil Sci. Soc. Am. J.* 79: 551–558.

Murdock, L., J. James, and G. Olson. 2007. Effect of AVAIL® polymer applied to phosphorus fertilizers on dry matter production and P uptake of fescue at Princeton, KY. *Soil Science News & Views* 27: 1–6. Available at http://www.uky.edu/Ag/Agronomy/Extension/ssnv /vol27no3.pdf (accessed June 10, 2015).

Narsian, V., and H. H. Patel. 2000. *Aspergillus aculeatus* as rock phosphate solubilizers. *Soil Biol. Biochem.* 32: 559–565.

Nielsen, A., M. Mansson, M. Wietz, A. N. Varming, R. K. Phipps, T. O. Larsen, L. Gram, and H. Ingmer. 2012. Nigribactin, a novel siderophore from *Vibrio nigripulchritudo*, modulates *Staphylococcus aureus* virulence gene expression. *Mar. Drugs* 10: 2584–2595.

Pauly, D. G., M. Nyborg, and S. S. Malhi. 2002. Controlled-release P fertilizer concept evaluation using growth and P uptake of barley from three soils in a greenhouse. *Can. J. Soil Sci.* 82: 201–210.

Plaza, C., R. Sanz, C. Clemente, J. M. Fernández, R. González, A. Polo, and M. F. Colmenarejo. 2007. Greenhouse evaluation of struvite and sludges from municipal wastewater treatment works as phosphorus sources for plants. *J Agric. Food Chem.* 55: 8206–8212.

Qian, P., and J. J. Schoenau. 2010. Effects of conventional and controlled release phosphorus fertilizer on crop emergence and growth response under controlled environment conditions. *J. Plant Nutr.* 33: 1253–1263.

Ramirez, R., B. Mendoza, and J. I. Lizaso. 2009. Mycorrhiza effect on maize P uptake from phosphate rock and superphosphate. *Commun. Soil Sci. Plan.* 40: 2058–2071.

Randall, G., and J. Vetsch. 2004. *Effect of AVAIL on Corn production in Minnesota*. Research project. Southern Research and Outreach Center, University of Minnesota, Waseca, MN. Available at http://sroc.cfans.umn.edu/prod/groups/cfans/@pub/@cfans/@sroc /@research/documents/asset/cfans_asset_128198.pdf (accessed June 10, 2015).

Rittmann, B. E., B. Mayer, P. Westerhoff, and M. Edwards. 2011. Capturing the lost phosphorus. *Chemosphere* 84: 846–853.

Ryu, H. D., C. C. Lim, M. K. Kang, and S. I. Lee. 2012. Evaluation of struvite obtained from semiconductor wastewater as a fertilizer in cultivating Chinese cabbage. *J. Hazard. Mater.* 221–222: 248–255.

Santner, J., E. Smolders, W. W. Wenzel, and F. Degryse. 2012. First observation of diffusion-limited plant root phosphorus uptake from nutrient solution. *Plant Cell Environ.* 35: 1558–1566.

Scholz, R. W., A. H. Roy, F. S. Brand, D. Hellums, and A. E. Ulrich (eds.). 2014. *Sustainable Phosphorus Management: A Global Transdisciplinary Roadmap.* Springer Publishers, Dordrecht.

Sharonova, N. L., A. Kh. Yapparov, N. Sh. Khisamutdinov, A. M. Ezhkova, I. A. Yapparov, V. O. Ezhkov, I. A. Degtyareva, and E. V. Babynin. 2015. Nanostructured water-phosphorite suspension is a new promising fertilizer. *Nanotechnol. Russia.* 10: 651–661.

Shu, L., P. Schneider, V. Jegatheesan, and J. Johnson. 2006. An economic evaluation of phosphorus recovery as struvite from digester supernatant. *Bioresource Technol.* 97: 2211–2216.

Sokolova, T. A., I. I., Tolpeshta, and I. V. Topunova. 2010. Biotite weathering in podzolic soil under conditions of a model field experiment. *Eurasian Soil Sci.* 43: 1150–1158.

Subramanian, K. S., A. Manikandan, M. Thirunavukkarasu, and C. S. Rahale. 2015. Nano-fertilizers for balanced crop nutrition. In *Nanotechnologies in Food and Agriculture*, Rai, M., C. Ribeiro, L. Mattoso, and N. Duran (eds.), pp. 69–80. Springer International Publishing, Cham.

Sutherland, S. 2015. *Lake Erie's Toxic Algae Bloom Is Back and It's Spreading.* Available at http://www.theweathernetwork.com/us/news/articles/climate-and-environment/lake-eries-toxic-algae-bloom-is-back-and-its-spreading/55505/ (accessed October 5, 2015).

Thompson, L. B. 2013. Field evaluation of the availability for corn and soybean of phosphorus recovered as struvite from corn fiber processing for bioenergy. MS thesis, Iowa State University, Aimes, IA.

Tilman, D., J. Fargione, B. Wolff, C. D'Antonio, A. Dobson, R. Howarth, D. Schindler, W. H. Schlesinger, D. Simberloff, and D. Swackhame. 2001. Forecasting agriculturally driven global environmental change. *Science* 292: 281–284.

UN (United Nations Department of Economic and Social Affairs, Population Division). 2015. *World Population Prospects: The 2015 Revision.* Available at http://esa.un.org/unpd/wpp/Graphs/ (accessed September 29, 2015).

Vogel, T., M. Nelles, and B. Eichler-Löbermann. 2015. Phosphorus application with recycled products from municipal waste water to different crop species. *Ecol. Eng.* 83: 466–475.

Wu, L., and M. Liu. 2008a. Preparation and properties of chitosan-coated NPK compound fertilizer with controlled-release and water-retention. *Carbohyd. Polym.* 72: 240–247.

Wu, L., and M. Liu. 2008b. Preparation and characterization of cellulose acetate-coated compound fertilizer with controlled-release and water-retention. *Polym. Adv. Technol.* 19: 785–792.

Wu, L., M. Liu, and R. Liang. 2008. Preparation and properties of a double-coated slow-release NPK compound fertilizer with superabsorbent and water-retention. *Bioresource Technol.* 99: 547–554.

# 11 Soil Phosphorus Cycling in Tropical Soils: An Ultisol and Oxisol Perspective

*Sasha C. Reed and Tana E. Wood*

## CONTENTS

## 11.1 INTRODUCTION

Phosphorus (P) is essential for life. It is the backbone of our DNA, provides energy for biological reactions, and is an integral component of cell membranes. As such, it is no surprise that P availability plays a strong role in regulating ecosystem structure and function (Wassen et al. 2005; Elser et al. 2007; Condit et al. 2013) and in determining our capacity to grow food for a burgeoning human population (Sharpley et al. 1997; Sims and Sharpley 2005; Lal 2009). Concerns that P supplies are insufficient to meet growing demands of our species are on the rise (Tiessen 2001; Cordell et al. 2009; Richardson and Simpson 2011), and scientific and media outlets increasingly discuss P as an element worthy of our attention and concern (e.g., Cordell et al. 2009; Lougheed 2011; Edixhoven et al. 2013; Ulrich et al. 2013). Indeed, a number of groups are calling for the explicit stewardship of our planet's P stocks (Schipper 2014; Withers et al. 2015). Yet a focus on P as a vital and limited resource is not new in the tropics, where an abundance of soils characterized by low P has resulted in

**TABLE 11.1**

**Soil Taxonomy Order Distribution[a] within the Tropics[b] and Average Soil P Pool Sizes for Those Orders as Determined by the Hedley Fractionation Method**

| Order | Tropical 10^6 (ha) | Percentage (%) | Resin Inorganic P (mg/kg) | Bicarbonate Inorganic P (mg/kg) | Bicarbonate Organic P (mg/kg) | Apatite P (mg/kg) | Total P (mg/kg) |
|---|---|---|---|---|---|---|---|
| Entisols | 1276 | 26.8 | $40.27 \pm 9.63$ ($n = 9$) | $27.62 \pm 9.93$ ($n = 9$) | $284.19 \pm 46.80$ ($n = 9$) | $18.33 \pm 6.51$ ($n = 8$) | $579.68 \pm 30.05$ ($n = 9$) |
| Inceptisols | 470 | 9.9 | $35.90 \pm 9.88$ ($n = 9$) | $13.98 \pm 3.26$ ($n = 9$) | $63.96 \pm 23.94$ ($n = 9$) | $41.04 \pm 11.68$ ($n = 9$) | $491.71 \pm 89.58$ ($n = 9$) |
| Andisols | 48 | 1.0 | $63.33 \pm 40.92$ ($n = 3$) | $36.62 \pm 6.57$ ($n = 13$) | $179.61 \pm 30.98$ ($n = 13$) | $39.69 \pm 6.52$ ($n = 13$) | $903.07 \pm 107.80$ ($n = 13$) |
| Histisols | 31 | 0.7 | $38.68 \pm 12.45$ ($n = 5$) | $23.06 \pm 4.05$ ($n = 5$) | $28.88 \pm 14.06$ ($n = 5$) | $87.00 \pm 26.55$ ($n = 4$) | $628.60 \pm 134.17$ ($n = 5$) |
| Gelisols | 1 | 0.0 | | | | | |
| Aridisols | 376 | 8.0 | $22.12 \pm 4.78$ ($n = 14$) | $13.52 \pm 2.69$ ($n = 14$) | $244.57 \pm 35.78$ ($n = 14$) | $5.27 \pm 1.49$ ($n = 8$) | $411.54 \pm 64.18$ ($n = 14$) |
| Vertisols | 206 | 4.4 | $19.80 \pm 5.44$ ($n = 5$) | $14.80 \pm 7.07$ ($n = 5$) | $212.60 \pm 75.19$ ($n = 5$) | $6.75 \pm 0.85$ ($n = 4$) | $497.00 \pm 95.00$ ($n = 5$) |
| Mollisols | 48 | 1.0 | $19.33 \pm 1.81$ ($n = 43$) | $11.49 \pm 0.99$ ($n = 43$) | $169.05 \pm 11.53$ ($n = 39$) | $19.12 \pm 3.46$ ($n = 40$) | $552.77 \pm 23.33$ ($n = 43$) |
| Alfisols | 561 | 11.9 | $15.11 \pm 2.99$ ($n = 17$) | $13.15 \pm 2.97$ ($n = 17$) | $78.92 \pm 15.61$ ($n = 17$) | $14.71 \pm 2.43$ ($n = 17$) | $364.74 \pm 37.01$ ($n = 17$) |
| Spodosols | 5 | 0.1 | $18.45 \pm 2.48$ ($n = 11$) | $15.54 \pm 2.12$ ($n = 11$) | $30.84 \pm 13.51$ ($n = 6$) | $27.87 \pm 4.56$ ($n = 11$) | $292.90 \pm 38.86$ ($n = 11$) |
| Ultisols | 757 | 16.0 | $6.80 \pm 1.08$ ($n = 38$) | $5.85 \pm 0.72$ ($n = 38$) | $7.36 \pm 1.64$ ($n = 38$) | $11.66 \pm 1.19$ ($n = 38$) | $225.18 \pm 21.77$ ($n = 38$) |
| Oxisols | 956 | 20.2 | $3.62 \pm 0.47$ ($n = 11$) | $10.15 \pm 5.66$ ($n = 14$) | $0.90 \pm 0.22$ ($n = 14$) | $8.12 \pm 1.69$ ($n = 13$) | $193.35 \pm 35.39$ ($n = 14$) |
| Total | 4726 | 100.0 | | | | | |

*Source:* Palm, C., Sanchez, P., Ahamed, S., and Awiti, A., *Annual Review of Environment and Resources*, 32, 99–129, 2007; Yang, X., and Post, W. M., *Biogeosciences*, 8, 2907–2916, 2011. With permission.

*Note:* Values shown are means ± standard error, and the number of values comprising the mean is shown in parentheses.

[a] Estimates exclude areas not covered by soils (e.g., rocks, water bodies, shifting sands, ice).

[b] Definitions: tropical, <23.5°; temperate zone, 23.6°–60°; boreal, >60°.

a substantial, long-standing reliance on P inputs for tropical ecosystem function in both unmanaged and agricultural settings (Table 11.1, Figure 11.1; Sanchez 1976; Swap et al. 1992; Chadwick et al. 1999; Okin et al. 2004; Lal 2009). Indeed, there is a long history of cultivation in the tropics, where for thousands of years land management practices have included methods that effectively modify P availability for plant growth (e.g., Sanchez 1976; Giardina et al. 2000; Lawrence and Schlesinger 2001; Vitousek et al. 2004; Tiessen et al. 2005; Lewis et al. 2015). Nevertheless, low soil fertility in tropical systems where fertilizer is scarce has long been recognized as a major source of hunger and starvation (Sanchez and Buol 1975; Sanchez 2002; Sanchez and Swaminathan 2005; Vitousek et al. 2009).

In addition to supporting the majority of the world's population (Vira et al. 2015) and biodiversity (Myers et al. 2000), tropical ecosystems store and exchange enormous amounts of energy and carbon (C) with the atmosphere (Brown and Lugo 1982; Foley et al. 2003; Beer et al. 2010; Pan et al. 2013). Accordingly, tropical ecosystems maintain a vast potential to create feedbacks to future climate at the global scale, and there is a growing interest in understanding P cycling interactions with tropical forest responses to environmental change (Silver 1998a; Townsend et al. 2011; U.S. Department of Energy 2012; Yang et al. 2014; Cavaleri et al. 2015; Reed et al. 2015).

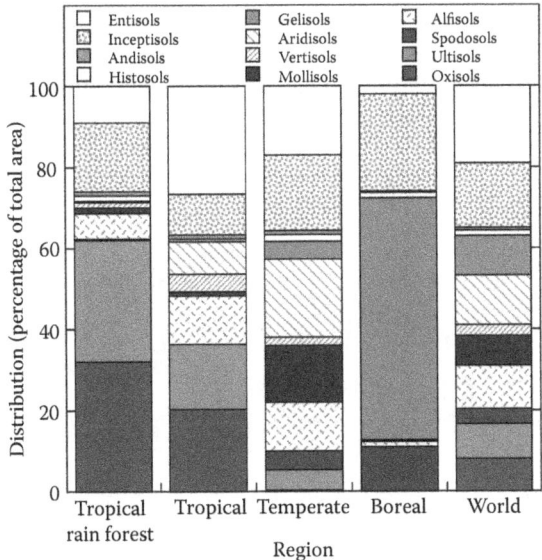

**FIGURE 11.1** **(See color insert.)** Distribution of soil orders among tropical/subtropical moist broadleaf forests (rain forests), all tropical, temperate, and boreal ecosystems, as well as globally. Estimates exclude areas not covered by soils (e.g., rocks, water bodies, shifting sands, ice), and soil order classifications are based on the USDA soil taxonomy system (Soil Survey Staff 1998). The sequence of the soil orders is arranged from less well-developed soils (upper part of the graph; e.g., Entisols) to more well-developed soils (lower part of the graph; e.g., Oxisols). Tropical areas are defined as ≤23.5°, temperate areas as 23.6°–60°, and the boreal zone as >60°. All data are from Palm et al. (2007), who determined the extent of the 12 soil orders of soil taxonomy and their geographical distribution using the USDA's Global Soil Regions dataset.

Now is an exciting time to evaluate our understanding of tropical P cycling in this context of change (e.g., climate change, land use change), and along with highlighting the diversity in tropical P cycling and synthesizing the fundamental aspects of the tropical P cycle in Ultisols and Oxisols, this is our goal for the chapter. In the coming sections, we will discuss the multiplicity of tropical soils, dive into soil chemical P cycling on numerous timescales, consider how P affects and is affected by tropical biota, and highlight considerations of P cycling and the interactions with other biogeochemical cycles in a framework of global change.

## 11.2 SOIL DIVERSITY AND PHOSPHORUS CYCLING IN THE TROPICS

The tropics is a region that encircles the equator, delineated in the north by the Tropic of Cancer (23°26'14.0" N) and in the south by the Tropic of Capricorn (23°26'14.0" S; Figure 11.2). These latitudes correspond with the planet's axial tilt, and the tropics include the areas of earth where the sun reaches a point directly overhead at least once during the solar year. While we often think of a *tropical climate* as being warm and wet, in reality, the tropics maintain a myriad of climates. For example, the tropics comprise not only earth's wettest ecosystems (e.g., rain forests near Lloró, Colombia, which average 12,892 mm of precipitation/year), but also its driest (the Atacama Desert, averaging 15 mm/yr). The temperature is also highly variable, and in conjunction with this large range in climate, biota and edaphic conditions greatly vary across the tropics. In fact,

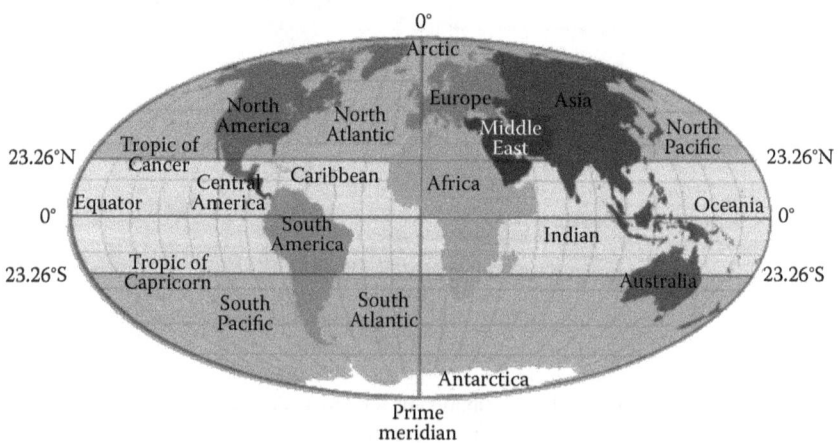

**FIGURE 11.2** Global map outlining the region between 23.26° latitude north (the Tropic of Cancer) and 23.26° latitude south (the Tropic of Capricorn). The lands within these bounds are known as the *Tropics* and include all of Central America, the Caribbean, and large portions of South America, Africa, Asia (particularly India, Southeast Asia, and Indonesia), and Australia.

of the 116 Holdridge life zones (a global bioclimatic classification scheme), the tropics contain more than the sum of all the other geographic regions combined (Holdridge 1967). The tropics contain hyperdiverse wet forests, arid deserts, alpine ecosystems, dry forests, wetlands and peat lands, and many other ecosystem types. In this way, there is no representative tropical ecosystem or tropical soil (Hilgard 1906; Sanchez and Buol 1975; Richter and Babbar 1991) nor any singular tropical P cycle.

Beyond the variations in climate, soil diversity in the tropics is driven by an extremely wide variety of parent materials, biota, landforms, geomorphic characteristics, and soil ages—the variation in these factors, each of which affects soils, is greater in the tropics than in the temperate zone (Sanchez 1976; Vitousek 1984; Richter et al. 1985; McGroddy et al. 2004; Porder et al. 2007; Townsend et al. 2008). The tropics maintain all 12 soil types of the USDA soil taxonomy system (Table 11.1, Figure 11.2; Natural Resources Conservation Service 2005; Palm et al. 2007), although Gelisols are notably rare. Moreover, while the number of soil types does not differ between tropical and temperate ecosystems, the relative abundance of particular soil types is quite distinct (Figure 11.2). For example, the tropics have a much lower abundance of fertile Mollisols (e.g., the soil that supports the "breadbasket" of America), as well as a much higher abundance of highly weathered Oxisols and Ultisols. This chapter focuses on these two soil orders that make up 36% of total tropical land area (16% and 20% of global soils, respectively; Palm et al. 2007). In addition, these two soil types comprise 62% of tropical rain forests, a biome that accounts for 17% of earth's land area and plays a disproportionately large role in global biogeochemical cycles (Palm et al. 2007; Townsend et al. 2011). Interestingly, these two prevalent soil types represent the latest stages of the soil pedogenic spectrum, which makes sense, as the extended time since glaciation and the relatively warm and wet climates found in many tropical forests are factors that result in highly weathered soils (Figure 11.3).

## 11.3  SOIL PEDOGENESIS AND THE DEVELOPMENT OF TROPICAL PHOSPHORUS CYCLES

Because all soil types are represented in the tropics (Table 11.1), models of soil development (pedogenesis) provide a powerful foundation on which to consider tropical soil P cycling and with which to contextualize the unique characteristics of highly weathered soils. In 1941, Jenny proposed five state factors that interact to determine the nature of a soil. This was captured in the iconic equation: $s = f(cl, o, r, p, t)$, where the state of the soil system ($s$) is a function of climate ($cl$), organisms ($o$), relief ($r$), parent material ($p$), and time ($t$) (Jenny 1941). It has been stated that P is perhaps the key element of soil pedogenesis (Walker 1965), because a parent material is P's central source and because of P's ecological significance (Tiessen et al. 1984; Elser et al. 2007; Vitousek et al. 2010). In contrast to the cycles of C and nitrogen (N), which have significant gaseous components, the gaseous phase of P (phosphine gas [$PH_3$]) is notably rare. Phosphine gas has been measured under highly anaerobic conditions,

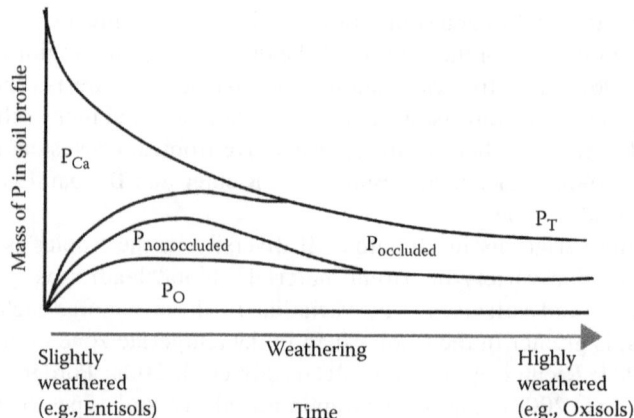

**FIGURE 11.3** Graphical representation of the changes in the relative abundance of different soil P pools over the course of soil development. $P_T$ = total soil P; $P_{Ca}$ = calcium phosphates; $P_O$ = organic P; $P_{occluded}$ = sorbed P (relatively unavailable to organisms); $P_{nonoccluded}$ = P in the soil solution (relatively available to organisms; often called labile P). The extent of weathering is driven by climate, type and abundance of organisms, topography, parent material characteristics, and amount of time weathering has occurred. Disturbances can also affect the rate and the nature of pedogenic change. The graphic is built of data from four soil chronosequences and shows that highly weathered soils (e.g., Oxisols) have less total P, higher relative proportions of $P_{occluded}$ and $P_O$, and very little available P ($P_{nonoccluded}$) relative to soils at earlier stages of soil development (e.g., Entisols). (Recreated from *Geoderma*, 15, Walker, T. W., and Syers, J. K., The fate of phosphorus during pedogenesis, 1–19, Copyright (1976), with permission from Elsevier; with kind permission from Springer Science+Business Media: *Phosphorus in Action*, Phosphorus cycling in tropical forests growing on highly weathered soils, 2011, 339–369, Reed, S., Townsend, A., Taylor, P., and Cleveland, C.)

such as sewage sludge, sediment soils, animal manure, marshes, and landfills; however, this form of P is thermodynamically unfavorable under most biogenic conditions and thus does not significantly contribute to the P cycle (Hanrahan et al. 2005). As such, it is the evolution and cycling of P within the soil system that plays the central role in determining soil P availability in terrestrial ecosystems (Figure 11.4).

Walker and Syers (1976) synthesized data from soil chronosequences to explore patterns in soil P pools in the context of pedogenesis and found that early in soil development, the majority of P is in primary mineral forms (Figure 11.3). As mineral P is weathered, it results in an increase in biologically available inorganic P (i.e., $PO_4^{3-}$; unoccluded P). Some of this P is taken up by plants and microbes, while other P molecules are bound to secondary soil minerals with high P sorption capacities. *Sorption* broadly describes any process that removes a reactant from a solution, and P sorption represents the removal of P from the soil solution into less reactive geochemical sinks. In the highly weathered and acidic soils typical of tropical forests, crystalline and noncrystalline oxides of iron (Fe) and aluminum (Al) are considered the main geochemical sinks of $PO_4^{3-}$ (Parfitt et al. 1975; López-Hernández 1977; Schwertmann and Taylor 1977; Parfitt 1978; Hsu 1989). In contrast,

P mineralization—the conversion of organic P to inorganic P—returns organic P (as created by plants and microbes) to inorganic pools, which can then be readily utilized by the biota, sorbed onto the soil surfaces, or remain within the soil solution (Figures 11.4 and 11.5). Much of the variability in unoccluded P can be accounted for by organic forms of P, which suggests that P mineralization may be a major determinant of P fertility in these soils (Tiessen et al. 1984). Each turn of the P cycle can lead to P being sorbed to soil surfaces, precipitated, and/or lost via leaching in both organic and inorganic forms (e.g., Hedin 1995). Thus, as soil development reaches more advanced stages (i.e., Ultisols and Oxisols), the total amount of P in the system declines, and much of the P that is left is bound, insoluble, or physically protected in occluded forms (Table 11.1; Figure 11.3).

This decline over time in P obtained from parent material increases the relative importance of soil P inputs that occur from atmospheric dust and mineral aerosol deposition (Swap et al. 1992; Chadwick et al. 1999; Okin et al. 2004; Pett-Ridge 2009). Phosphorus can be transferred through the atmosphere over extremely long distances in the form of dust (Swap et al. 1992; Chadwick et al. 1999; Okin et al. 2004; Mahowald et al. 2005; Pett-Ridge 2009) and as ash from local biomass burning (Mahowald et al. 2005; Das et al. 2013). The importance of dust inputs to tropical ecosystems depends on the existing reservoir of P as well as the flux of dust being transported and deposited. Multiple studies suggest that atmospheric P inputs

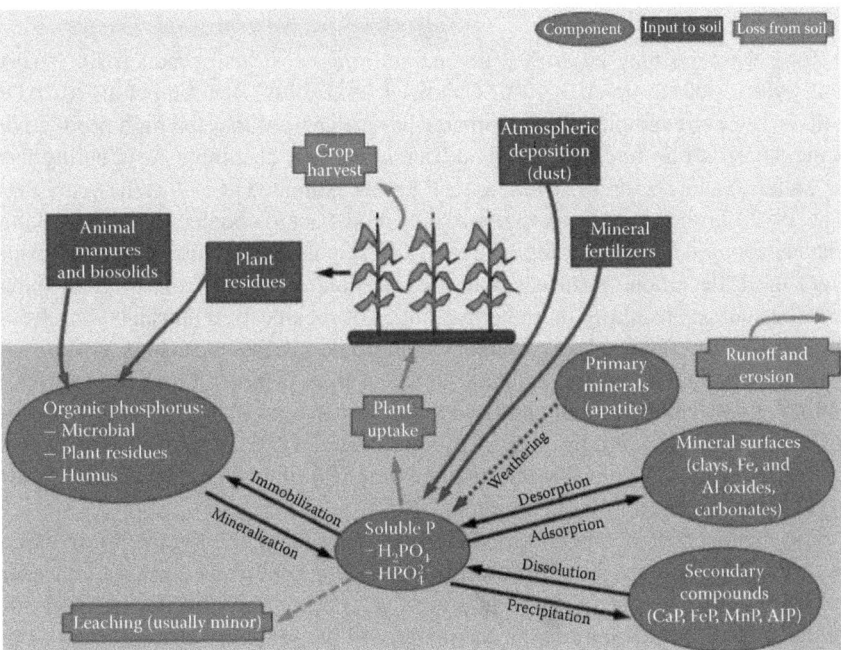

**FIGURE 11.4** (**See color insert.**) The P cycle, focusing on P pools (green ovals), inputs (blue rectangles), and loss (orange polygons). (Reproduced with permission from Wikipedia Creative Commons, en.wikipedia.org/wiki/Phosphorus_cycle.)

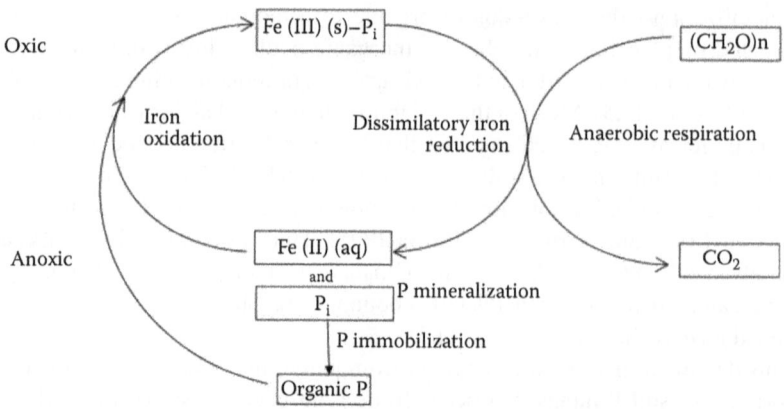

**FIGURE 11.5** Conceptual model from Liptzin and Silver (2009) showing the relationship between Fe, $O_2$, P availability, and carbon oxidation. During more anoxic periods (e.g., following precipitation events), bacteria reduce Fe (III) to Fe (II), oxidizing organic C into $CO_2$. At the same time, Fe-bound inorganic P ($P_i$) is released and may be quickly immobilized by roots and/or microbes. During periods when $O_2$ availability is higher, Fe (II) is oxidized to Fe (III) via biotic and/or abiotic reactions, and the Fe (III) oxides are again available to bind inorganic P. (Reprinted from *Soil Biology and Biochemistry*, 41, Liptzin, D., and Silver, W. L., Effects of carbon additions on iron reduction and phosphorus availability in a humid tropical forest soil, 1696–1702, Copyright (2009), with permission from Elsevier.)

to tropical forests may be critical for maintaining ecosystem productivity (Artaxo et al. 1988, 2002; Swap et al. 1992; Chadwick et al. 1999; Ben-Ami et al. 2010; Das et al. 2013). For example, some estimates have suggested that the high productivity of the Amazon rain forest is fueled, at least in part, by dust inputs originating from the Sahara/Sahel region in Africa (e.g., P inputs estimated at 1–4 kg/ha year; Swap et al. 1992). From a P pool perspective, dust inputs act to counteract the P-depleting effects of long-term weathering (Figure 11.4). Landscape dynamics can also counteract the P depletion of surface soils that are long removed from original parent material sources, resulting in higher quantities of rock-derived nutrients than would otherwise be predicted. In particular, geomorphic processes such as erosion and tectonic uplift have a strong potential to affect soils in general and tropical soil P cycling in particular. Erosion and tectonic uplift can rejuvenate such soils by bringing the parent material to the surface (Vitousek et al. 2003; Bern et al. 2005; Porder et al. 2007), and tropical ecosystems have some of the highest erosion and uplift rates on earth (White et al. 1998; Porder et al. 2007).

The chemistry of weathered tropical soils affects the mobility of soil P in unique ways. Both organic and inorganic forms of P are susceptible to sorption (e.g., Berg and Joern 2006), but the extent of this sorption is primarily influenced by soil mineralogy and the concentration, the chemistry, and the solubility of soil P (Smyth and Sanchez 1982; Singh and Gilkes 1991; Iyamuremye et al. 1996; Guppy et al. 2005; Reed et al. 2011a), as well as the competition with biological sinks (Uehara and Gillman 1981; Sollins et al. 1988; Olander and Vitousek 2004). In particular, the prevalence of 1:1 clays (e.g., kaolinite) and high concentrations of Al and Fe oxides and hydroxides, which have a high affinity for P, are an interesting characteristic of many highly weathered tropical soils (Sanchez 1976; Harrison 1987; Sollins et al. 1988). Variable-charge clays maintain an electric charge that results from the protonation and the deprotonation of surface hydroxyl groups (Gillman 1984; White and Zelazny 1986), which occur at the edge of the 1:1 clays and Al and Fe hydroxides that dominate these soils. The charge of clay soils depends upon the soil pH, whereby small changes in pH can significantly affect the net charge of the soil and thus the mobility of anions and cations (Sollins et al. 1988). Accordingly, the mobility of P in highly weathered soils varies with soil pH, and unoccluded P may be less mobile in Ultisols and Oxisols than in soils dominated by clays with more permanent charge because permanent-charge clays typically maintain net cation exchange capacity (Gillman 1984). Fertilization studies on these soils have found high potential for P to be rapidly and nearly completely sorbed to the soil surfaces over short timescales (e.g., Sanchez 1976; Uehara and Gillman 1981; Oberson et al. 1997). Consequently, the concentrations of sorbed P often exceed that of P in the soil solution by several orders of magnitude (Sanchez 1976). However, while the process of P retention by Fe and Al oxides has been historically thought to exhibit low reversibility in highly weathered tropical soils, fluctuating redox potential common in the humid tropics can reduce Fe oxides when in the presence of labile C and humic substances. Such reduced soil conditions favor the remobilization of the geochemical or occluded pool of P, making this P once again biologically available (Figure 11.5; Chacon et al. 2006; Liptzin and Silver 2009).

The highly weathered, finely textured clay Ultisols and Oxisols that are high in Fe and Al oxides, combined with the high rainfall typical of the humid tropics, can limit the diffusive transport of oxygen ($O_2$). These circumstances lead to rapid fluctuations between reducing (low $O_2$) and oxidizing (high $O_2$) conditions (Silver et al. 1999; Shin 2005), which have the potential to release C and P that has been sorbed to the soil surfaces (Figure 11.5; Chacon et al. 2006; Liptzin and Silver 2009). There are several mechanisms that result in the release of P from geochemical pools under reducing conditions (Ponnamperuma et al. 1967; Gambrell and Patrick 1978; Chacon et al. 2006), which stem from bacteria obtaining energy by coupling the oxidation of organic matter (OM) to the reduction of different electron acceptors. Under anaerobic conditions, Manganese (Mn) and Fe oxides are the primary electron acceptors in soils (Lovley 1991). For example, as Fe oxides are converted to reduced forms, the phosphate anions chemically associated to Fe oxides are released (Figure 11.5; Baldwin and Mitchell 2000). Iron reduction is primarily fueled by the availability of organic acids of low molecular weight such as acetate (Kusel et al. 2002). While more complex organic substrates can also be used by Fe oxide reducers, it has been suggested that a very small percentage of these compounds and their reducing equivalents (less than 5%) are converted to Fe oxide (Lovley 1991). Humic substances also affect the patterns of Fe reduction and reoxidation in soils, whereby humic substances and quinones (oxidized derivatives of aromatic compounds) are used to shuttle electrons between electron donors and acceptors (Lovley et al. 1998). A high availability of labile C (C that is easily biologically available) combined with an abundance of the Fe-reducing bacteria typical of wet tropical forests is likely to stimulate these reactions (Liptzin and Silver 2009; Dubinsky et al. 2010). In summary, the microbial reduction of Fe oxide can be an important mechanism for increased bioavailability of soil P in tropical forest soils (Figure 11.5; Chacon et al. 2006). There are also growing datasets that suggest that plants and soil microbes are able to directly access occluded P (Tiessen et al. 1984; Olander and Vitousek 2004; Richter et al. 2006). Regardless of the cause, the transfer of even modest amounts of P from the geochemical pool (i.e., occluded) to the soil solution can play an important role in maintaining soil P bioavailability (Chacon et al. 2006; Liptzin and Silver 2009).

McGill and Cole (1981) proposed a framework for OM decomposition that suggests P mineralization may also be largely decoupled from the mineralization of C and N. Within OM molecules, C and N are intricately bound (Asner 1997), but P is bound by phosphate ester bonds that phosphatase enzymes can independently mineralize. In this way, whether organisms mineralize organic substrates to obtain energy (in the form of reduced C) or N, mineralization of both elements occurs. In contrast, linkages between P limitation, phosphatase production, and P mineralization may be more direct because P mineralization can be relatively independent from the mineralization of C and N. Importantly, this decoupling may be particularly prevalent on highly weathered P-poor soils (McGill and Cole 1981).

Overall, the spatial heterogeneity of tropical forest soils can be viewed at two levels of organization: the geographic scale of hundreds to thousands of hectares (as described earlier) and the scale of landforms and patches that exist within this larger landscape (Webb et al. 1972; Scatena and Lugo 1995). At this finer spatial scale, local topography and drainage can substantially influence many soil characteristics that influence soil P availability, such as soil pH, soil organic matter (SOM), and exchangeable Fe (Figure 11.5). Research in tropical montane forests and in tropical forest watersheds suggests that higher landscape positions (e.g., ridge tops) are typically characterized as having higher SOM, lower pH, and higher concentrations of extractable Fe than lower topographic positions (e.g., valleys; Scatena and Lugo 1995). This variability in soil characteristics with topographic position can interact to affect soil P availability in complex ways. For example, in a wet tropical forest in Puerto Rico, soil P availability in the surface soils (0–10 cm) is correlated with forest age and not topographic position, and thus at this depth, P is likely primarily controlled by biological processes (Silver 1994; Scatena and Lugo 1995). However, when deeper profiles are considered (0–60 cm), the topographic position is the dominant control, with significantly higher biologically available soil P in valleys relative to the ridge tops (Scatena and Lugo 1995). Correlations between soil P availability and topographic position for the deeper profiles likely occur due to prevention of phosphate adsorption by organic anions (Fox 1982) and more acidic and oxic conditions in the higher topographic positions (ridges). In contrast, the lower topographic positions (valleys) likely experience altered P adsorption in soils with low Fe concentrations, more reducing conditions, and greater transport of P from higher topographic positions (Scatena and Lugo 1995). Taken together, these processes interact to increase the relative size of the P pool in subsoils located at lower topographic positions relative to higher positions (Silver 1994; Scatena and Lugo 1995). In line with the findings in Puerto Rico, biologically available P tends to increase from high to low topographic positions (e.g., Tsui et al. 2004; Porder et al. 2007; Werner and Homeier 2015). Many tropical species are often correlated with specific topographic positions (Scatena and Lugo 1995), and foliar P concentrations also correlate with soil P variation across the landscape (Wood et al. 2006). Teasing apart the controls of geomorphology on the distribution of plants versus the effects of plants on soil P availability remains a significant challenge to the scientific community.

## 11.4 BIOLOGICAL INTERACTIONS WITH THE TROPICAL SOIL PHOSPHORUS CYCLE

### 11.4.1 Biotic Strategies for Acquiring and Utilizing Soil Phosphorus

Soils are not the only component of ecosystems to evolve over long periods: organisms in tropical environments growing on P-poor soils have developed a phenomenal capacity for efficiently using and acquiring P. A central reason tropical organisms are able to maintain high activity in the face of low P availability (Table 11.1) lies in the fact that organisms in general—but especially in the tropics—have evolved a number of key nutrient conservation and acquisition strategies (e.g., Clark et al. 1999; Lambers et al. 2008). These strategies can vary along the pedogenic spectrum (Vance et al. 2003; Lambers et al. 2008; Reed et al. 2011a; Turner et al. 2013a; Ellsworth et al. 2015), as well as among species (Lambers et al. 2008; Reed et al. 2012; Mayor et al. 2014; Steidinger et al. 2015; Zemunik et al. 2015), and conservation strategies include increased growth per unit P (Vitousek 1984), reallocation of internal P (e.g., foliar P resorption prior to leaf senescence; Aerts 1996; Kitayama and Aiba 2002; Yuan and Chen 2009; Wood et al. 2011; Reed et al. 2012), and modifications in metabolism to minimize P-requiring steps (e.g., alternative glycolytic reactions that bypass ATP-requiring steps [Theodorou and Plaxton 1996; Uhde-Stone et al. 2003]). Plants also have mechanisms to maximize soil P acquisition. Phosphorus is much less mobile in the soil solution than most other major plant nutrients (Barber 1984), and P uptake is often assumed to vary in proportion to the surface area of the plant uptake organs involved. Thus, low P availability can be somewhat countered via symbiotic relationships and a number of morphological adaptations that increase root surface area, including mycorrhizal relationships, root hairs, and roots with unusual architecture (Figure 11.6; Lodge 1993; Aldrich-Wolfe 2007; Jansa et al. 2011). For example, some tropical plants make use of lateral roots that scavenge P from litter layers (Herrera et al. 1978; Stark and Jordan 1978; Cuevas and Medina 1986), which contain relatively high P concentrations. Other plants produce specialized root structures (e.g., cluster roots) that allow plants to mine insoluble forms of inorganic P from the soil: cluster roots produce large amounts of carboxylates, which release P from strongly sorbed forms (Figure 11.6; Lambers et al. 2008). Cluster-rooted plants have been observed in many tropical forests, and many of the best-known cluster-rooted species are found in western Australia, in soils with some of the most weathered and P-poor soils on earth (Figure 11.6). The fungal translocation of P through hyphal connections can also play an important role in conserving P in tropical forested ecosystems (Figure 11.6; Lodge et al. 2014).

Microbial organisms, including mycorrhizae, can solubilize P from occluded inorganic pools, transforming it into available P. By releasing strong organic acids, certain bacteria can liberate P from pools that are typically thought of as biologically

**FIGURE 11.6   (See color insert.)** Plant nutrient–acquisition strategies. (a) Mycorrhizal fungi extend beyond the root zone and greatly increase the volume of soil that can explored for P uptake. All mycorrhizas (arbuscular mycorrhizas, ectomycorrhizas, and ericoid mycorrhizas) are involved in the uptake of soluble inorganic P. Photo shows the hyphae of *Glomus caledonium*, an arbuscular mycorrhizal fungus that in the picture is growing into the soil from a host root of the plant *Trifolium repens*. (Photo by Iver Jakobsen, reprinted with kind permission from Springer Science+Business Media. *Mycorrhizal Ecology*, Foraging and resource allocation strategies of mycorrhizal fungi in a patchy environment, 2003, 93–115, Olsson, P. A., Jakobsen, I., and Wallander, H.; reprinted from *Trends in Ecology and Evolution*, 23, Lambers, H., Raven, J. A., Shaver, G. R., and Smith, S. E., Plant nutrient-acquisition strategies change with soil age, 95–103, Copyright (2008), with permission from Elsevier.) (b) Root morphology helps plants access P from soil; for example, Proteaceae species that grow under extremely low P supply in highly weathered Australian soils have a number of root morphological adaptations. Shown here is a proteoid root cluster of *Banksia grandis*. (Photo by Michael W. Shane, reprinted from *Trends in Ecology and Evolution*, 23, Lambers, H., Raven, J. A., Shaver, G. R., and Smith, S. E., Plant nutrient-acquisition strategies change with soil age, 95–103, Copyright (2008), with permission from Elsevier.)

inaccessible (Bonan 1991; Bünemann et al. 2004; Olander and Vitousek 2004; Richter et al. 2006). These results are supported by data suggesting that microbial organisms can effectively compete with geochemical sinks for soil P (Olander and Vitousek 2004). At a global scale, the C:N:P ratios of both the soil (186:13:1) and the soil microbial biomass (60:7:1) are well constrained (Cleveland and Liptzin 2007), and evidence suggests the C:N:P of microbial biomass may lend perspective into questions of nutrient limitation in terrestrial ecosystems. When OM accumulates in an ecosystem, the soil C and N concentrations become increasingly decoupled from the total soil P concentration (McGill and Cole 1981). This observation may reflect a more efficient use of P released from the cycling of organic P pools in the mineral soil and the forest floor, rather than the weathering of inorganic P from primary minerals. Organic P pools are thought to be an important source for labile P in forest soils, especially for tropical forests (Johnson et al. 2003). Measured site-specific microbial N:P ratios that diverge from the calculated average (i.e., 6.9 ± 0.4) may provide insight into the nature of nutrient limitation, at least within lowland tropical ecosystems. For example, relatively high measured microbial N:P ratios (suggesting P limitation) correspond with direct evidence that soil P availability limits microbial biomass, activity, and other ecosystem processes in a wet tropical forest in Costa Rica (Cleveland et al. 2002; Cleveland and Townsend 2006; Reed et al. 2007; Wieder et al. 2009).

### 11.4.2 PLANT AND ANIMAL INTERACTIONS WITH SOIL PHOSPHORUS

We know from temperate ecosystems that different plant species have the capacity to affect the chemistry of soil in species-specific ways: temperate forest research indicates a large interspecific variation in the effects of trees on soil properties (Zinke 1962; Gersper and Holowaychuk 1971; Binkley and Giardina 1998; Finzi et al. 1998; Hobbie et al. 2006). Factors such as volume and chemistry of stemflow, nutritional requirements of trees, organic chemistry of litter, and $N_2$-fixing abilities affect how different tree species alter soils on micro- and mesoscales. An emerging body of work suggests that even in highly diverse tropical forests, tree species can affect the soil in which they live (Reed et al. 2008; Keller et al. 2013; Uriarte et al. 2015; Waring et al. 2015). For example, a recent study in Puerto Rico by Uriarte et al. (2015) found that most tree leaf litter fell less than 5 m from the source tree, which generated fine-scale spatial heterogeneity in leaf litter inputs, with significant implications for soil P cycling. Another study in Costa Rica observed strong correlations between canopy, litter, and soil P concentrations, such that species-specific differences in P were observable along a canopy-to-soil profile, suggesting a species-generated P footprint (Reed et al. 2008). It is important to note the likelihood not only for tree species effects on soil chemistry, but also for soils to dictate the community composition (and thus community-level plant traits) and/or chemistry of trees (John et al. 2007; Condit et al. 2013). The two possibilities are not mutually exclusive nor is it new to recognize how difficult it is to tease apart these controls. For example, in his assessments of the drivers of soil development, Jenny himself recognized the issue, saying "Like everybody else I could see that the vegetation affects the soil and the soil affects the vegetation, the very circulus vititis that I was trying to avoid" (Jenny 1981).

These suggestions of vertical linkages between canopy and soil chemistries are also intriguing in the context of increasing remote-sensing capacity for assessing biogeochemical patterns and functions in diverse tropical forests. If we can use remote-sensing technologies to gain information about forest floor processes, the information would not only provide insight into direct nutrient controls over foliar function (e.g., Walker et al. 2014) but could also help quantify the spatial patterns of biogeochemical cycles occurring beneath the tree crowns. This provides hope that we can continue to distill the tropical forest biogeochemical complexity represented by hundreds of species per hectare (Townsend et al. 2008).

Ecosystem controls and effects are mediated through many trophic levels. For example, in a temperate forest common garden experiment, Hobbie et al. (2006) showed significant tree species effects on litter decomposition, with data suggesting that tree species controls were occurring, in part, through interactions between calcium concentrations and earthworm abundance. We know that animals can be affected by and affect soil P pools in tropical forests (Figures 11.7 and 11.8; McGlynn et al. 2007; Sayer et al. 2010; Hong et al. 2011; Marichal et al. 2011), and animals

(a)                              (b)                              (c)

**FIGURE 11.7   (See color insert.)** Photos showing the effect of animal activity on soil structure. (a) *Atta cephalotes* leaf cutter ant nest in Cockscomb Basin Wildlife Sanctuary, Belize. The photo shows the massive soil translocation ants create in building their nests in highly weathered tropical soils, as they move massive amounts of soil from deeper layers to the surface. (Courtesy of Matthew Meier.) (b) An earthworm cast produced by the tropical earthworm species *Amynthas khami* (which can grow up to 50 cm in length) in a rain forest in northeastern Vietnam. Casts are clumps of digested organic matter excreted by earthworms that aggregate into large and distinctive structures. (Reprinted from *Soil Biology and Biochemistry*, 43, Hong, H. N., Rumpel, C., Henry des Tureaux, T., Bardoux, G., Billou, D., Tran Duc, T., and Jouquet, P., How do earthworms influence organic matter quantity and quality in tropical soils? 223–230, Copyright (2011), with permission from Elsevier.) (c) Tall termite hills such as this one are common in highly weathered soils in West Africa. These hills represent large changes to soil structure and, likely, biogeochemical cycling. (Reproduced with permission from Wikipedia Creative Commons, https://commons.wikimedia.org/wiki/File:Termite_hill_in_forest.jpg.)

have profound local effects on the spatial patterns in soil properties. Termites, ants, worms, and other soil fauna can mix and chemically alter vast volumes of soil material (Figure 11.7). For example, an assessment of surface soil (0–10 cm) P concentrations in a rain forest in Costa Rica suggested that ant (*Atta cephalotes*, a leaf cutter ant) mounds significantly alter extractable P pools in consistent ways (Figure 11.8). Because ants can drive substantial physical turnover of tropical soils, the effects of these organisms on tropical soil P are of significant consequence (Perfecto and Vandermeer 1993). Phosphorus can also regulate the activity of soil animals; for example, McGlynn et al. (2007) tested the effects of soil and litter nutrient stoichiometry on the invertebrate litter fauna of a Costa Rican tropical rain forest by assessing the animal densities from 15 sites across a P gradient. There was a significant variation in the density of the invertebrate litter fauna, which was strongly tied to soil and litter P concentrations: an increase in P concentrations related to an equally large increase in animal density. Interestingly, in a study of 87 species of rain forest ants, Kaspari et al. (2015) found that the P content of the ants was strongly correlated with the organisms' thermal tolerance. It has been hypothesized that ants active at high temperatures invest in P-rich machinery to buffer their metabolism against thermal extremes, and the P content of the ants varied threefold, with a temperature sensitivity that was lower and a thermal range that was higher in P-rich ant species. An improved understanding of how plants, soils, and animals interact in tropical forests to regulate P cycling is greatly needed.

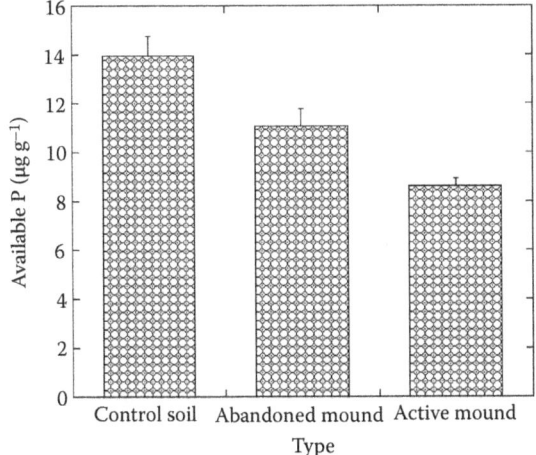

**FIGURE 11.8** Extractable soil P concentrations for highly weathered soils (0–10 cm depth) from La Selva Biological Station in Costa Rica. Soils were collected from active *Atta cephalotes* ant mounds (Active mound), from mounds that were no longer active (Abandoned mound), and from soils adjacent to both active and abandoned mounds (Control soil). Data show that ant activity significantly reduced topsoil P availability and that this reduction lasts even after the mounds are abandoned. Because *Atta cephalotes* are believed to turnover all the soil of this forest on only a 200–300 year timescale (Perfecto and Vendermeer 1993), this activity could represent an important control over the P cycle.

### 11.4.3 NUTRIENT LIMITATION

In relatively static ecosystems, the traditional view of tropical nutrient limitation to ecosystem processes, such as net primary production (NPP) and soil respiration, suggests that younger less well-developed soils would be limited by N, and older soils limited by P. Looking at the results from a chronosequence in Hawaii, which maintains ecosystems where it is possible to vary a single state factor while holding the rest relatively constant (Vitousek 2004), the transition from N limitation of tree growth on young soils to N and P colimitation on intermediate-aged soils to P limitation on old soils is clear (Vitousek and Farrington 1997). An analogous assessment of soil C cycling along the same chronosequence suggested that soil microbial growth and respiration were more affected by N at the youngest site and P at the oldest site, in line with the results observed aboveground (Reed et al. 2011b). Thus, the scarcity of P in many lowland forests and the tendency toward N limitation in tropical montane systems that typically grow on less well-developed soil make sense (Tanner et al. 1998; Adamek et al. 2009). A number of cross-forest comparisons also suggest a positive relationship between soil P and tropical forest aboveground and belowground biomass, decomposition, soil respiration, and soil fauna (Wardle et al. 2004; Paoli et al. 2005; Espeleta and Clark 2007; McGlynn et al. 2007; Cleveland et al. 2011; Baribault et al. 2012), as do results from some fertilization studies where additions of P stimulate these and other lowland tropical forest ecosystem pools and fluxes (Cuevas and Medina 1988; Reed et al. 2007; Wieder et al. 2009;

Alvarez-Clare et al. 2013). And these patterns fit into the concepts of retrogression (Peltzer et al. 2010), where in the absence of large rejuvenating disturbances over timescales of thousands to millions of years, ecosystem properties such as NPP, decomposition, and nutrient cycling rates undergo substantial declines (Wardle et al. 2004; Paoli et al. 2005; Espeleta and Clark 2007; McGlynn et al. 2007; Cleveland et al. 2011; Baribault et al. 2012). Phosphorus is suggested as a strong factor in structuring tropical productivity, soil C cycling, and plant community composition. For example, a study conducted in a seasonally dry tropical forest in Panama assessed the responses of 550 tree species to eight environmental factors across a number of climatic and geological gradients (Condit et al. 2013). These results showed that soil P was one of the two strongest predictors of community composition, together affecting the distribution of more than half of the species.

However, while multiple lines of evidence suggest that soil P cycling has a role in regulating aspects of tropical ecosystem structure and function, particularly in lowland forests growing on highly weathered soil (Figure 11.9), a number of studies in tropical lowland forests suggest that there is much more complexity need for considerations of how P and other nutrients regulate ecosystem functions (Mirmanto et al. 1999; Newbery et al. 2002; Kaspari et al. 2008; Wright et al. 2011; Alvarez-Clare et al. 2013; Powers and Peréz-Aviles 2013; Turner et al. 2013b; Mayor et al. 2014; Turner and Wright 2014). In other words, nutrient limitation in tropical ecosystems is not as simple as wholesale limitation by P, and changes in the availability of different nutrients may elicit distinct responses. For example, an investment in the acquisition of a nutrient may be greater when that nutrient is limiting to growth. A fertilization study in Hawaii found that P additions consistently reduced phosphatase activity, mycorrhizal colonization, and P uptake capacity across sites. In contrast, excess N

(a)                              (b)

**FIGURE 11.9** **(See color insert.)** Phosphorus availability on highly weathered soils can be a major constraint to agriculture in the tropics. (a) Adding P to nutrient-poor tropical soils can dramatically enhance crop production, and the picture shows the effects of P fertilization on plant growth in Brazil's cerrado. Maize plants in the background were fertilized with P, and they are much larger than the unfertilized maize plants in the foreground. (Courtesy of D.M.G. de Sousa.) (b) Some crops are much more commonly grown in the tropics than anywhere else on earth. Shown here is cassava farming in Nigeria where the International Institute for Tropical Agriculture. (Courtesy of Tunga Media, Niger State, Nigeria, http://tungamediang.com.)

was allocated to the construction of extracellular phosphatases to acquire P. Along these lines, a $^{32}PO_4$ fertilization experiment in wet tropical forest found that P fertilization saturated the soils, which overwhelmed biological P demand, thereby reducing biological controls. In contrast, N fertilization increased the available P through reduced P sorption to the soil surface (Olander and Vitousek 2004). A long-term fertilization study in Panama also found a significant decline in phosphatase activity in response to fertilization (Turner and Wright 2014). The increase in phosphatase production and reduced P sorption in response to N fertilization implies that even P-limited systems might increase productivity in response to increased N availability and highlights the complex effects of nutrient availability on ecosystem processes (Treseder and Vitousek 2001; Vitousek et al. 2010).

A fertilization experiment by Alvarez-Clare et al. (2013) further illustrates the complexity of nutrient limitation in tropical forested ecosystems. This study used a N × P fertilization experiment in 30 × 30 m plots in a lowland rain forest in Costa Rica to show that after nearly 3 years, at the community scale, the growth of smaller trees (5–10 cm diameter at breast height) was doubled in plots receiving P, and seedling survival increased from 59 to 78%, with no effects of N addition on any characteristics measured. But P had no effect on litterfall productivity, root growth, or growth of intermediate (10–30 cm) or large trees (>30 cm). The effect of P also varied by plant species: *Pentaclethra macroloba*, the most abundant species at the site, did not increase growth rates with fertilization, but *Socratea exorrhiza*, the most abundant palm, had more than two times higher stem growth rates with P additions. Thus, while the research suggests soil P availability is indeed a significant driver of plant processes, it also highlights the importance of considering different aspects of the plant community (e.g., size class and species identity) when considering an overall framework of nutrient limitation.

Another complexity when evaluating nutrient limitation in terrestrial ecosystems is the form and the quantity of fertilizer utilized (Sayer et al. 2012). For example, microbes produce enzymes according to economic rules, whereby the microbial enzyme production increases when complex nutrients are abundant and simple nutrients are scarce; however, this response is also mediated by the availability of C and N (Allison and Vitousek 2005). A large-scale litter manipulation experiment conducted in a wet tropical forest in Costa Rica found a significant positive effect of litter addition on leaf litter production 5 months after application. The magnitude of the effect was significantly positively correlated with the total P added in litter form (Wood et al. 2006); however, a similar long-term litter manipulation experiment in Panama found no effect of litter addition on forest productivity (Sayer et al. 2012). This study did, however, find a significant effect of litter addition on the concentration of organic P in the soil (Vincent et al. 2010). It is also possible that the volume of aboveground biomass added in response to P fertilization could depend upon the concentration and the timing of P application.

Micronutrients (e.g., molybdenum) in addition to N and P may also exert important controls on tropical forest ecosystems. A long-term fertilization experiment in a tropical forest in Panama has consistently found that micronutrients (B, Ca, Cu, Fe, Mg, Mn, Mo, S, Zn), as well as N and P, can exert varied effects on ecosystem pools and processes (Kaspari et al. 2008; Wright et al. 2011; Santiago et al. 2012; Turner et al. 2013b; Mayor et al. 2014; Turner and Wright 2014; Wurzburger and Wright 2015). Although P was the only nutrient to increase soil microbial biomass or affect measured enzyme activities (Turner and Wright 2014), this fertilization study found no clear effect of P limitation on plant growth, root growth, or decomposition (Kaspari et al. 2008; Wright et al. 2011). This diversity of responses to P fertilization, both within and across tropical ecosystems, makes sense when considered in the context of multinutrient limitation (as well as limitation by other resources; Kaspari et al. 2008; Townsend et al. 2011) and reflects the heterogeneity of nutrient availability in soils across the biome and the tropics as a whole (Quesada et al. 2010; Cleveland et al. 2011). These results underscore that soil age alone is not sufficient to explain all the observed patterns in N versus P (or other nutrient) limitation and the importance of multiple processes for determining the balance of nutrient inputs, losses, and biological demand (Vitousek et al. 2010). Nevertheless, even with the advances made in our understanding of soil nutrient controls in tropical ecosystems over the last years, these results still represent a focus on only a handful of sites relative to the high within- and across-ecosystem diversities of the tropics. Clearly, more work is needed to understand how soil P and other nutrients help regulate the structure and the function of earth's tropical ecosystems. Finally, it is important to recognize that there are strong interactions among these nutrient cycles, such that P availability can affect N, for example, via $N_2$ fixation (Reed et al. 2011c; Wurzburger et al. 2012; Batterman et al. 2013; Reed et al. 2013; Nasto et al. 2014), and other nutrients can affect P (e.g., N stimulation of phosphatase activity; Marklein and Houlton 2012).

## 11.5   THE TEMPORALLY DYNAMIC NATURE OF TROPICAL PHOSPHORUS CYCLING ON HIGHLY WEATHERED SOIL

In addition to the high spatial heterogeneity of soil P, available soil P also exhibits a high temporal variation in tropical forested ecosystems, with the potential to substantially vary on timescales ranging from hours (Vandecar et al. 2011; Turner et al. 2003) to days (Wood et al. 2016) to months (McGrath et al. 2001). Temporal variation in biologically available P often correlates well with soil moisture availability, which can significantly fluctuate in tropical forested ecosystems, due to both short punctuated rainfall events and broad seasonal changes in rainfall (Singh et al. 1989; Lodge et al. 1994; Campo et al. 1998; McGrath et al. 2000; Wood and Lawrence 2008). These changes in soil moisture alter a suite of variables that are likely to affect the availability of P in tropical soils. For example, large rainfall events can drive abrupt spikes in soil moisture that stimulate rapid changes in soil chemistry (e.g., redox potential, pH, desorption of nutrients from minerals; Kieft 1987; Chacon et al. 2006; Liptzin et al. 2011; Vandecar et al. 2011; Hall et al. 2013; Wood et al. 2013), which, in turn, stimulates both microbial and plant activities (Kieft 1987; Singh et al. 1989; Tiessen et al. 1994; Cleveland et al. 2004). Decomposition rates are also correlated with soil moisture availability, and a large portion of available soil P is likely replenished via the decomposition of OM by microbes (Singh et al. 1989; Tiessen et al. 1994, Cleveland et al. 2013). Thus, increased soil moisture can result in P release via OM decomposition, and this P is made part of the available P pool. The time the P spends in the available pool depends upon immobilization, sorption, and leaching controls.

In addition to the effects of soil moisture availability, diurnal and seasonal changes in temperature can influence microbial decomposition rates, thereby releasing P bound in OM and increasing the amount of available P in the soil. While an increase in microbial mineralization rates increases the amount of available P, microbes also allocate resources to uptake and growth, which would instead immobilize labile P, thereby reducing its availability in the soil as microbial activity increases (Lodge et al. 1994; Cleveland et al. 2004). Furthermore, environmental controls on above-ground processes, such as plant photosynthesis, can stimulate root activity (Tang et al. 2005; Vandecar et al. 2009; Yuste et al. 2010). This increase in root activity can draw down labile P in the soil to meet the demands of the plant processes (Vandecar et al. 2009). Furthermore, sites with higher root density may facilitate greater labile soil P uptake and increase competition for P with the microbial community (Pregitzer et al. 1998; Vandecar et al. 2009, 2011).

As a whole, which of these processes exert primary control on labile P in the soils is not well understood, and there is evidence that the dominant controls may substantially differ among tropical forest sites along varied timescales. In more humid tropical forests, available soil P tends to be higher in the wet season (Cleveland et al. 2002; McGrath et al. 2001; Turner et al. 2013b; Wood and Silver 2012). In contrast, dry tropical forest sites have observed higher available soil P during dry season months (Singh et al. 1989; Campo et al. 1998; Srivastava 1997). The timing of high soil P availability, regardless of the soil moisture content, tends to correspond with times when the microbial biomass is also high (Singh et al. 1989; Cleveland et al. 2004; Turner and Wright 2014). Many theories have been put forth to explain these patterns: Soil P availability may be higher during the dry season months in seasonally dry tropical forests because (1) the demand by the vegetation is low, (2) P is stored in microbial biomass and then released upon the rewetting of soils (Singh et al. 1989), and (3) there is reduced loss of P due to low leaching in the dry season months (Campo et al. 1998). In humid forest soils, higher P availability in the wet season has been hypothesized to be due to (1) reducing conditions in the soil that release Fe-bound P (Chacon et al. 2006), (2) cloudier skies during wet season months that lower plant demand for P and thus uptake (3) higher P mineralization rates in the wet season (Campo et al. 1998; Wood and Lawrence 2008).

This raises the question of why we observe opposite effects of moisture on P in wet versus dry forests. There is increasing evidence to suggest that prior conditions play an important role in determining the fate of P in soil. Soil incubation experiments have found that the addition of water to wet soils results in rapid immobilization of available soil P (Campo et al. 1998; DeLonge et al. 2013). In contrast, the addition of water to dry soils results in the immediate release of available P (via microbial turnover or mineralization; Campo et al. 1998; Turner and Haygarth 2001). The decomposer community could be contributing to observed changes; for example, fungi, microfungi, and bacteria vary in their response to moisture (Lodge 1993; Cornejo et al. 1994; Lodge et al. 1994). The periodicity and amplitude of redox oscillations could also play a role in determining the fate of P in tropical soils, with more stable conditions driving a decline in microbial diversity, changes in mineral reactivity, enzyme activity, as well as higher N losses from the soil system (DeAngelis et al. 2005; Pett-Ridge and Firestone 2005; Thompson et al. 2006; Hall and Silver 2013;

Hall et al. 2015). Taken together, these data suggest that P availability is varying on extremely fast timescales, and this dynamism has the potential to affect a myriad of aspects of tropical function and response to change.

## 11.6  TROPICAL SOIL PHOSPHORUS CYCLING AND GLOBAL CHANGE

Tropical forests exchange more $CO_2$ with the atmosphere than any other class of ecosystem (Foley et al. 2003; Beer et al. 2010), as well as accounting for over two-thirds of the earth's living terrestrial plant biomass (Pan et al. 2013) and represent-ing nearly one-third of all soil C stocks (Jobbagy and Jackson 2000; Tarnocai et al. 2009). Given this expansive amount of C, investigations of tropical forest response to environmental drivers will be critical for our understanding of future climate and biogeochemical cycling at the global scale. Tropical ecosystems also house the majority of the world's population, which makes them exceptionally important when considering change in these systems, including at local, regional, and pantropi-cal scales (Lewis et al. 2015). However, the vulnerability of tropical ecosystems to climate-related change is poorly understood and a topic of much debate (Lloyd and Farquhar 2000; Lewis et al. 2009; Clark et al. 2013; Cox et al. 2013; Good et al. 2013; Randerson 2013). At the same time, a limited ability to characterize tropi-cal responses to global change may represent our largest challenge in appropriately predicting the future climate of our planet (Bonan and Levis 2010; Huntingford et al. 2013; Piao et al. 2013). It is in this context that nutrient cycling may be central to improving our understanding of and capacity to predict tropical ecosystem response to global change.

Human activities have significantly affected each of the soil-forming state factors, with large implications for soil development and function (Amundson and Jenny 1991). The tropics are predicted to experience striking increases in atmospheric $CO_2$ concentrations, unprecedented changes in temperature, considerable modifications to the amount and the timing of rainfall (Diffenbaugh and Scherer 2011; Mora et al. 2013), and significant changes to N and P deposition (Matson et al. 2002; Mahowald et al. 2005; Peñuelas et al. 2012; Peñuelas et al. 2013; Carnicer et al. 2015; Houlton and Morford 2015). Each of these anthropogenic perturbations has the potential to markedly affect tropical soil P cycling, which, as discussed earlier, could affect numerous aspects of tropical ecosystem structure and function. However, the man-ner in which an ecosystem's P cycle will respond to anthropogenic perturbation will depend upon both the nature of the perturbation and the characteristics of the soil.

Soil nutrient availability may be at the heart of determining how tropical ecosys-tems will respond to global change (Silver 1998a; Cernusak et al. 2013; Yang et al. 2014; Cavaleri et al. 2015; Reed et al. 2015). Increases in nutrient availability (e.g., via fertilization) consistently increase the productivity of terrestrial plants of both man-aged and unmanaged ecosystems (Sanchez and Buol 1975; Sanchez 1976; Elser et al. 2007; Vitousek et al. 2009). Soil nutrient availability has also been shown to posi-tively affect the rates of soil $CO_2$ efflux and microbial biomass growth (Reed et al. 2011b; Wood and Silver 2012). Indeed, a synthesis of data from 92 forests in different climate zones underscored the crucial role nutrient availability plays in determining net ecosystem productivity, suggesting that nutrient availability exerted a stronger control over the net ecosystem productivity than that of C input (i.e., gross primary production; Fernández-Martínez et al. 2014). This result conflicts with assumptions of nearly all global coupled C cycle–climate models, which assume that C inputs via photosynthesis are the dominant driver of biomass production and C sequestration. In addition, results from past free-air $CO_2$ enrichment (FACE) studies that explored plant and soil responses to increased $CO_2$ showed that nutrient limitation strongly constrained plant responses, and the data suggested that nutrient limitation became progressively more extreme over time for plants experiencing elevated $CO_2$ (Luo et al. 2004; Norby et al. 2010; Johnson 2006). But none of these $CO_2$ experiments occurred in tropical ecosystems, and none focused on P (but see Huang et al. 2014; Crous et al. 2015; Hoosbeek 2015; Norby et al. 2016). Due to the lack of a significant gaseous phase, the P cycle may be less able to respond to increased plant demand in the face of elevated $CO_2$ (Reed et al. 2015), which would mean that P regulation of C cycling responses to global change could become increasingly pronounced (Yang et al. 2014). A recent study comparing modeling, satellite, and FACE experimental data indeed suggests that $CO_2$ fertilization effects on plants are potentially much lower than what models would predict, and nutrient limitation was cited as a key reason for this response (Smith et al. 2015).

Our understanding of how warming and altered precipitation patterns will affect tropical forests via effects on soil P cycling is also quite poor (Silver 1998b; Wood et al. 2012; Meir et al. 2015). Nevertheless, kinetic controls dictate that changes in temperature could result in strong effects on decomposition and nutrient mineralization rates, as well as changes in soil weathering. Indeed, warming has been shown to increase nutrient mineralization rates in other biomes (Robinson et al. 1997), and increased mineralization could increase nutrient losses via leaching, thereby lowering nutrient availability and constraining C cycling over the longer term (Lodge et al. 1994). If warming initially increases and then drives down nutrient availability, processes such as microbial respiration and decomposition could become slow due to nutrient limitation (Hobbie and Vitousek 2000; Reed et al. 2011b). Reduced rainfall in tropical regions could significantly affect P cycling in tropical forest ecosystems. For example, experimental reduction of soil moisture in a wet tropical forest in Puerto Rico led to significantly reduced soil P availability, likely due to increased soil aeration and sorption of P to Fe oxides in the soil (Wood and Silver 2012). Changes in the patterns of rainfall could generate complex effects on soil P cycling. As discussed earlier, prior soil conditions matter, whereby the rewetting of dry soils compared with the wetting of soils that are already wet led to different fates of P in the soil. Furthermore, tropical microbes have the potential to adapt to preexposure to dry conditions (Bouskill et al. 2013), further adapting as wetting and drying cycles intensify (Evans and Wallenstein 2014).

In addition to $CO_2$ and climatic changes, human-induced changes to nutrient inputs, for example, N deposition, have the potential to both directly and indirectly affect tropical soil P cycling. These changes can stem not only from the differences in the absolute input of any single element (e.g., increased N or P deposition) but also from an imbalance in the relative proportions of these inputs (Peñuelas et al. 2012, 2013). However, our understanding of these effects and their controls remain notably poor, and studies of coupled biogeochemical cycles in the context of change and stoichiometric perspectives offer an exciting line of tropical research.

As the human population continues to grow, the demand for agricultural productivity to meet the increasing demand for food and biofuel production will likely lead to the intensification of agricultural practices, particularly in tropical regions (Coomes et al. 2000; Tilman et al. 2002; Laurance et al. 2014; Smith et al. 2014; Lal 2015). Intensification of agriculture occurs in three ways: (1) expansion of the land area used, (2) increase in the frequency that a given area is used, and (3) intensification of the management controls (Trenbath et al. 1990). Changes to agricultural practices have been found to affect soil P availability. For example, changes to timing whereby the number of crop cycles are increased and the period the forest is allowed to recover (fallow) is reduced, has been found to reduce P availability in tropical soils (Arnason et al. 1982; Fujisaka 1991). These shorter fallow periods are thought to facilitate the loss of P from the system via soil mining and accelerated erosion (Fujisaka 1991; Mertz 2002) and could ultimately reduce the efficacy of this land management technique (e.g., Jakovac et al. 2015). In systems that utilize shifting agriculture, the timing of slashing and burning is also critical and also affects soil P cycling. Biomass is cut, left to dry, and burned just prior to the start of the wet season. The timing of slashing is important for allowing sufficient time for the biomass to dry, and the timing of burning is timed to minimize the loss of P in ash that can be blown off of the land surface and to maximize the plant growth and uptake of nutrients. In these agricultural systems, and many others, farmers rely on historical experience to determine the timing of slashing, burning, and planting (Ewel 1986). Changes in precipitation patterns could have significant effects on the efficacy of these practices as historical knowledge may no longer clearly inform land management decision-making. In addition to the intensification of cultivation practices, the increasing demand for agricultural lands has also led to the expansion of agriculture into areas previously considered lower quality.

Secondary forests (forests regenerating following land use conversion) in the tropics currently comprise more land area than that of their old growth counterparts. The legacy of land use drives profound changes in the community composition and the soil nutrient cycles of these secondary forests, and the trajectory of their recovery is influenced by P cycling. For example, Davidson et al. (2007) showed that after agricultural abandonment, secondary tropical forests on highly weathered soils may transition from N limitation near to the time of abandonment to P limitation as the forests reach later stages of succession. Over the course of succession, N became relatively more available and P relatively less available. Of course, this trajectory would be expected to differ depending on the type and the intensity of prior land use. For instance, the intensification of shifting cultivation currently occurring across the tropics was recently found to significantly reduce secondary forest resilience (Jakovac et al. 2015), which may be exacerbated by a slow recovery of P as land use intensifies.

## 11.7 CONCLUSIONS

Phosphorus availability is a critical component determining terrestrial ecosystem structure and function, particularly for ecosystems growing on highly weathered soils (i.e., Ultisols, Oxisols). Due to the effects of extensive weathering, these soils maintain unique soil P traits, and the specialized manner in which the soils cycle P has cascading effects on numerous aspects of the ecosystems themselves. For example, the availability of soil P has been linked with plant community composition, NPP, and a wide range of fundamental soil processes (e.g., OM decomposition, soil respiration, $N_2$ fixation, C, N, and P-related enzyme activity). Moreover, agriculture in the tropics is an increasingly important component of providing food for our growing human population, and effectively managing tropical soils in this context is at the heart of agricultural success. Even in unmanaged tropical ecosystems, due to the massive amounts of C, water, and energy exchanged between the tropical forest biosphere and the atmosphere, it is critical to understand the role P plays in determining the tropical responses to global change factors, such as increasing atmospheric $CO_2$ and climate change. Nevertheless, our understanding of tropical soil P cycling remains notably incomplete, particularly in the context of global change, although the implications of these multielement connections are large. As highlighted in this chapter, the scientific community has made great strides in our understanding of the P cycle on highly weathered tropical soils, but our ability to measure, scale, and forecast ecosystem functions in tropical ecosystems lags behind that of many other biomes. Further, the exceptional diversity and biogeochemical variation at multiple scales in the tropics complicates this task. Taken together, the data suggest that a soil perspective is a solid foundation on which to build an improved understanding of the tropical P cycle.

## ACKNOWLEDGMENTS

S. C. R. and T. E. W. are exceptionally grateful to Cory Cleveland, Alan Townsend, Peter Vitousek, Deborah Lawrence, DJ Lodge, Whendee Silver, Stephen Porder, Ben Turner, Ariel Lugo, and Xiaojuan Yang for research and perspectives that have shaped their understanding of the phosphorus cycle. Finally, S. C. R. and T. E. W. owe a debt of gratitude to Jessica Metcalf and Nataly Ascarrunz for improvements on previous drafts of the chapter, and to the following agencies for supporting their tropical phosphorus research: National Science Foundation, Andrew W. Mellon Foundation, National Center for Ecological Analysis and Synthesis (NCEAS), U.S. Geological Survey Powell Center for Analysis and Synthesis, U.S.D.A. Forest Service International Institute of Tropical Forestry, Department of Energy Terrestrial Ecosystem Science Program, and U.S. Geological Survey Ecosystems Mission Area. Any use of trade, firm, or product names is for descriptive purposes only and does not imply endorsement by the U.S. Government.

# REFERENCES

Adamek, M., Corre, M. D., and Hölscher, D. (2009). Early effect of elevated nitrogen input on above-ground net primary production of a lower montane rain forest, Panama. *Journal of Tropical Ecology, 25*(06), 637–647.

Aerts, R. (1996). Nutrient resorption from senescing leaves of perennials: Are there general patterns? *Journal of Ecology, 84*, 597–608.

Aldrich-Wolfe, L. (2007). Distinct mycorrhizal communities on new and established hosts in a transitional tropical plant community. *Ecology, 88*, 559–566.

Allison, S. D., and Vitousek, P. M. (2005). Responses of extracellular enzymes to simple and complex nutrient inputs. *Soil Biology and Biochemistry, 37*, 937–944.

Alvarez-Clare, S., Mack, M. C., and Brooks, M. (2013). A direct test of nitrogen and phosphorus limitation to net primary productivity in a lowland tropical wet forest. *Ecology, 94*, 1540–1551.

Amundson, R., and Jenny, H. (1991). The place of humans in the state factor theory of ecosystems and their soils. *Soil Science, 151*(1), 99–109.

Arnason, T., Lambert, J. D. H., Gale, J., Cal, J., and Vernon, H. (1982). Decline of soil fertility due to intensification of land use by shifting agriculturists in Belize, Central America. *Agro-ecosystems, 8*, 27–37.

Artaxo, P., Martins, J. V., Yamasoe, M. A., Procópio, A. S., Pauliquevis, T. M., Andreae, M. O., Guyon, P., Gatti, L. V., and Leal, A. M. C. (2002). Physical and chemical properties of aerosols in the wet and dry seasons in Rondônia, Amazonia. *Journal of Geophysical Research: Atmospheres, 107*(D20).

Artaxo, P., Storms, H., Bruynseels, F., Van Grieken, R., and Maenhaut, W. (1988). Composition and sources of aerosols from the Amazon Basin. *Journal of Geophysical Research: Atmospheres, 93*(D2), 1605–1615.

Asner, G. P. (1997). The decoupling of terrestrial carbon and nitrogen cycles. *BioScience, 47*(4), 226–234.

Baldwin, D. S., and Mitchell, A. M. (2000). The effects of drying and re-flooding on the sediment and soil nutrient dynamics of lowland river–floodplain systems: A synthesis. *Regul. Rivers: Res. Mgmt., 16*, 457–467. doi: 10.1002/1099-1646(200009/10)16:5 <457::AID-RRR597>3.0.CO;2-B.

Barber, S. (1984). *Soil Nutrient Bioavailability.* New York: Wiley.

Baribault, T. W., Kobe, R. K., and Finley, A. O. (2012). Tropical tree growth is correlated with soil phosphorus, potassium, and calcium, though not for legumes. *Ecological Monographs, 405*, 189–203.

Batterman, S. A., Wurzburger, N., and Hedin, L. O. (2013). Nitrogen and phosphorus interact to control tropical symbiotic $N_2$ fixation: A test in *Inga punctata. Journal of Ecology, 101*(6), 1400–1408.

Beer, C., Reichstein, M., Tomelleri, E., Ciais, P., Jung, M., Carvalhais, N., Rödenbeck, C. et al. (2010). Terrestrial gross carbon dioxide uptake: Global distribution and covariation with climate. *Science, 329*, 834–838.

Ben-Ami, Y., Koren, I., Rudich, Y., Artaxo, P., Martin, S. T., and Andreae, M. O. (2010). Transport of North African dust from the Bodélé depression to the Amazon Basin: A case study. *Atmospheric Chemistry and Physics, 10*(16), 7533–7544.

Berg, A., and Joern, B. (2006). Sorption dynamics of organic and inorganic phosphorus compounds in soil. *Journal of Environmental Quality, 35*(5), 1855–1862.

Bern, C. R., Townsend, A. R., and Farmer, G. L. (2005). Unexpected dominance of parent-material strontium in a tropical forest on highly weathered soils. *Ecology, 86*, 626–632.

Binkley, D., and Giardina, C. (1998). Why do tree species affect soils? The warp and woof of tree-soil interactions. *Biogeochemistry, 42*(1–2), 89–106.

Bonan, G., and Levis, S. (2010). Quantifying carbon-nitrogen feedbacks in the Community Land Model (CLM4). *Geophysical Research Letters*, *37*, L07401.

Bouskill, N. J., Lim, H. C., Borglin, S., Salve, R., Wood, T. E., Silver, W. L., and Brodie, E. L. (2013). Pre-exposure to drought increases the resistance of tropical forest soil bacterial communities to extended drought. *The ISME Journal*, *7*(2), 384–394.

Brown, S., and Lugo, A. E. (1982). The storage and production of organic matter in tropical forests and their role in the global carbon cycle. *Biotropica*, *14*, 161–187.

Bünemann, E. K., Steinebrunner, F., Smithson, P. C., Frossard, E., and Oberson, A. (2004). Phosphorus dynamics in a highly weathered soil as revealed by isotopic labeling techniques. *Soil Science Society of America Journal*, *68*(5), 1645–1655.

Campo, J., Jaramillo, V. J., and Maass, J. M. (1998). Pulses of soil phosphorus availability in a Mexican tropical dry forest: Effects of seasonality and level of wetting. *Oecologia*, *115*(1–2), 167–172.

Carnicer, J., Sardans, J., Stefanescu, C., Ubach, A., Bartrons, M., Asensio, D., and Peñuelas, J. (2015). Global biodiversity, stoichiometry and ecosystem function responses to human-induced C–N–P imbalances. *Journal of Plant Physiology*, *172*, 82–91.

Cavaleri, M., Reed, S., Smith, W., and Wood, T. (2015). Urgent need for warming experiments in tropical forests. *Global Change Biology*, *21*, 2111–2121.

Cernusak, L. A., Winter, K., Dalling, J. W., Holtum, J. A., Jaramillo, C., Körner, C., Leakey, A. D. et al. (2013). Tropical forest responses to increasing atmospheric $CO_2$: Current knowledge and opportunities for future research. *Functional Plant Biology*, *40*(6), 531–551.

Chacon, N., Silver, W. L., Dubinsky, E. A., and Cusack, D. F. (2006). Iron reduction and soil phosphorus solubilization in humid tropical forests soils: The roles of labile carbon pools and an electron shuttle compound. *Biogeochemistry*, *78*(1), 67–84.

Chadwick, O., Derry, L., Vitousek, P. M., Huebert, B., and Hedin, L. (1999). Changing sources of nutrients during four million years of ecosystem development. *Nature*, *397*, 491–497.

Clark, D., Clark, D., and Oberbauer, S. (2013). Field-quantified responses of tropical rainforest aboveground productivity to increasing $CO_2$ and climatic stress, 1997–2009. *Journal of Geophysical Research-Biogeosciences*, *118*, 1–12.

Clark, D. B., Palmer, M. W., and Clark, D. A. (1999). Edaphic factors and the landscape scale distributions of tropical rain forest trees. *Ecology*, *80*(8), 2662–2675.

Cleveland, C. C., Houlton, B. Z., Smith, W. K., Marklein, A. R., Reed, S. C., Parton, W., Running, S. W. et al. (2013). Patterns of new versus recycled primary production in the terrestrial biosphere. *Proceedings of the National Academy of Sciences*, *110*(31), 12733–12737.

Cleveland, C. C., and Liptzin, D. (2007). C: N: P stoichiometry in soil: Is there a "Redfield ratio" for the microbial biomass? *Biogeochemistry*, *85*(3), 235–252.

Cleveland, C. C., and Townsend, A. R. (2006). Nutrient additions to a tropical rain forest drive substantial soil carbon dioxide losses to the atmosphere. *Proceedings of the National Academy of Sciences, 103*, 10316–10321.

Cleveland, C. C., Townsend, A. R., Constance, B. C., Ley, R. E., and Schmidt, S. K. (2004). Soil microbial dynamics in Costa Rica: Seasonal and biogeochemical constraints. *Biotropica*, *36*(2), 184–195.

Cleveland, C. C., Townsend, A. R., and Schmidt, S. K. (2002). Phosphorus limitation of microbial processes in moist tropical forests: Evidence from short-term laboratory incubations and field studies. *Ecosystems*, *5*(7), 680–691.

Cleveland, C. C., Townsend, A. R., Taylor, P., Alvarez-Clare, S., Bustamante, M., Chuyong, G., Dobrowski, S. Z. et al. (2011). Relationships among net primary productivity, nutrients and climate in tropical rain forest: A pan-tropical analysis. *Ecology Letters*, *14*(9), 939–947.

Condit, R., Engelbrecht, B. M. J., Pino, D., Pérez, R., and Turner, B. L. (2013). Species distributions in response to individual soil nutrients and seasonal drought across a community of tropical trees. *Proceedings of the National Academy of Sciences, 110*(13), 5064–5068.

Coomes, O. T., Grimard, F., and Burt, G. J. (2000). Tropical forests and shifting cultivation: Secondary forest fallow dynamics among traditional farmers of the Peruvian Amazon. *Ecological Economics, 32*, 109–124.

Cordell, D., Drangert, J., and White, S. (2009). The story of phosphorus: Global food security and food for thought. *Global Environmental Change, 19*, 292–305.

Cornejo, F. H., Varela, A., and Wright, S. J. (1994). Tropical forest litter decomposition under seasonal drought: Nutrient release, fungi and bacteria. *Oikos*, 183–190.

Cox, P., Pearson, D., Booth, B., Friedlingstein, P., Huntingford, C., Jones, C., and Luke, C. M. (2013). Sensitivity of tropical carbon to climate change constrained by carbon dioxide variability. *Nature, 494*, 341–344.

Crous, K. Y., Ósvaldsson, A., and Ellsworth, D. S. (2015). Is phosphorus limiting in a mature Eucalyptus woodland? Phosphorus fertilisation stimulates stem growth. *Plant and Soil, 391*(1–2), 293–305.

Cuevas, E., and Medina, E. (1986). Nutrient dynamics within Amazonian forest ecosystems: I. Nutrient flux in fine litter fall and efficiency of nutrient utilization. *Oecologia 68*, 466–472.

Cuevas, E., and Medina, E. (1988). Nutrient dynamics with Amazonian forests: II. Fine root growth, nutrient availability, and leaf litter decomposition. *Oecologia, 76*, 222–235.

Das, R., Evan, A., and Lawrence, D. (2013). Contributions of long-distance dust transport to atmospheric P inputs in the Yucatan Peninsula. *Global Biogeochemical Cycles, 27*, 167–175.

Davidson, E. A., de Carvalho, C. J. R., Figueira, A. M., Ishida, F. Y., Ometto, J. P. H., Nardoto, G. B., Sabá, R. T. et al. (2007). Recuperation of nitrogen cycling in Amazonian forests following agricultural abandonment. *Nature, 447*(7147), 995–998.

DeAngelis, K. M., Ji, P., Firestone, M. K., and Lindow, S. E. (2005). Two novel bacterial biosensors for detection of nitrate availability in the rhizosphere. *Applied and Environmental Microbiology, 71*(12), 8537–8547.

DeLonge, M., Vandecar, K. L., D'Odorico, P., and Lawrence, D. (2013). The impact of changing moisture conditions on short-term P availability in weathered soils. *Plant and Soil, 365*(1–2), 201–209.

Diffenbaugh, N. S., and Scherer, M. (2011). Observational and model evidence of global emergence of permanent, unprecedented heat in the 20th and 21st centuries. *Climatic Change, 107*(3–4), 615–624.

Dubinsky, E. A., Silver, W. L., and Firestone, M. K. (2010). Tropical forest soil microbial communities couple iron and carbon biogeochemistry. *Ecology, 91*(9), 2604–2612.

Edixhoven, J. D., Gupta, J., and Savenije, H. H. G. (2013). Recent revisions of phosphate rock reserves and resources: Reassuring or misleading? An in-depth literature review of global estimates of phosphate rock reserves and resources. *Earth Systems Dynamics 4*, 1005–1034.

Ellsworth, D., Crous, K., Lambers, H., and Cooke, J. (2015). Phosphorus recycling in photorespiration maintains high photosynthetic capacity in woody species. *Plant, Cell and Environment, 38*, 1142–1156.

Elser, J., Bracken, M. E., Cleland, E. E., Gruner, D. S., Harpole, W. S., Hillebrand, H., Ngai, J. T., Seabloom, E. W., Shurin, J. B., and Smith, J. E. (2007). Global analysis of nitrogen and phosphorus limitation of primary producers in freshwater, marine and terrestrial ecosystems. *Ecology Letters, 10*(12), 1135–1142.

Espeleta, J., and Clark, D. A. (2007). Multi-scale variation in fine-root biomass in a tropical rain forest: A seven-year study. *Ecological monographs, 77*(3), 377–404.

Evans, S. E., and Wallenstein, M. D. (2014). Climate change alters ecological strategies of soil bacteria. *Ecology Letters, 17*(2), 155–164.

Ewel, J. J. (1986). Designing agricultural ecosystems for the humid tropics. *Annual Review of Ecology and Systematics*, 245–271.

Fernández-Martínez, M., Vicca, S., Janssens, I. A., Sardans, J., Luyssaert, S., Campioli, M., Chapin III, F.S. et al. (2014). Nutrient availability as the key regulator of global forest carbon balance. *Nature Climate Change, 4*(6), 471–476.

Finzi, A. C., Van Breemen, N., and Canham, C. D. (1998). Canopy tree-soil interactions within temperate forests: Species effects on soil carbon and nitrogen. *Ecological applications, 8*(2), 440–446.

Foley, J. A., Costa, M. H., Delire, C., Ramankutty, N., and Snyder, P. (2003). Green surprise: How terrestrial ecosystems could affect earth's climate. *Frontiers in Ecology and Environment, 1*, 38–44.

Fox, R. L. (1982). Some highly weathered soils of Puerto Rico: 3. Chemical properties. *Geoderma, 27*(1), 139–176.

Fujisaka, S. (1991). A diagnostic survey of shifting cultivation in northern Laos: Targeting research to improve sustainability and productivity. *Agroforestry Systems, 13*, 95–109.

Gambrell, R. P., and Patrick, W. H. (1978). Chemical and microbiological properties of anaerobic soils and sediments. In Hook, D. D., and Crawford, R. M. M. (Eds.), *Plant Life in Anaerobic Environments* (pp. 233–247). Ann Arbor, MI: Ann Arbor Science Publishing. Inc.

Gersper, P. L., and Holowaychuk, N. (1971). Some effects of stem flow from forest canopy trees on chemical properties of soils. *Ecology, 52*, 691–702.

Giardina, C. P., Sanford, R. L., and Døckersmith, I. C. (2000). Changes in soil phosphorus and nitrogen during slash-and-burn clearing of a dry tropical forest. *Soil Science Society of America Journal, 64*, 399–405.

Gillman, G. (1984). Using variable charge characteristics to understand the exchangeable cation status of oxic soils. *Australian Journal of Agricultural Research 22*, 71–80.

Good, P., Jones, C., Lowe, J., Betts, R., and Gedney, N. (2013). Comparing tropical forest projections from two generations of Hadley Centre Earth System Models, HadGEM2-ES and HadCM3LC. *Journal of Climate, 26*, 495–511.

Guppy, C. N., Menzies, N. W., Moody, P. W., and Blamey, F. P. C. (2005). Competitive sorption reactions between phosphorus and organic matter in soil: A review. *Soil Research, 43*(2), 189–202.

Hall, S. J., McDowell, W. H., and Silver, W. L. (2013). When wet gets wetter: Decoupling of moisture, redox biogeochemistry, and greenhouse gas fluxes in a humid tropical forest soil. *Ecosystems, 16*(4), 576–589.

Hall, S. J., and Silver, W. L. (2013). Iron oxidation stimulates organic matter decomposition in humid tropical forest soils. *Global Change Biology, 19*(9), 2804–2813.

Hall, S. J., Silver, W. L., Timokhin, V. I., and Hammel, K. E. (2015). Lignin decomposition is sustained under fluctuating redox conditions in humid tropical forest soils. *Global Change Biology, 21*(7), 2818–2828.

Hanrahan, G., Salmassi, T. M., Khachikian, C. S., and Foster, K. L. (2005). Reduced inorganic phosphorus in the natural environment: Significance, speciation and determination. *Talanta, 66*, 435–444.

Harrison, A. (1987). *Soil Organic Phosphorus: A Review of World Literature*. Wallingford: CAB International.

Hedin, L. O. (1995). Patterns of nutrient loss from unpolluted, old-growth temperate forests: Evaluation of biogeochemical theory. *Ecology, 76*, 493–509.

Herrera, R., Merida, T., Stark, N., and Jordan, C. F. (1978). Direct phosphorus transfer from leaf litter to roots. *Naturwissenschaften 65*, 208–209.

Hilgard, E. W. (1906). *Soils*. London: Macmillan.

Hobbie, S. E., Reich, P. B., Oleksyn, J., Ogdahl, M., Zytkowiak, R., Hale, C., and Karolewski, P. (2006). Tree species effects on decomposition and forest floor dynamics in a common garden. *Ecology*, *87*(9), 2288–2297.

Hobbie, S. E., and Vitousek, P. M. (2000). Nutrient limitation of decomposition in Hawaiian forests. *Ecology*, *81*(7), 1867–1877.

Holdridge, L. (1967). *Life Zone Ecology*. San Jose: Centro Científico Tropical.

Hong, H. N., Rumpel, C., Henry des Tureaux, T., Bardoux, G., Billou, D., Tran Duc, T., and Jouquet, P. (2011). How do earthworms influence organic matter quantity and quality in tropical soils? *Soil Biology and Biochemistry*, *43*(2), 223–230.

Hoosbeek, M. R. (2015). Elevated $CO_2$ increased phosphorous loss from decomposing litter and soil organic matter at two FACE experiments with trees. *Biogeochemistry*, *127*, 1–9.

Houlton, B., and Morford, S. (2015). A new synthesis for terrestrial nitrogen inputs. *SOIL*, *1*, 381–397.

Hsu, P. H. (1989). Aluminum hydroxides and oxyhydroxides. *Minerals in Soil Environments*, *2*, 331–378.

Huang, W., Zhou, G., Liu, J., Duan, H., Liu, X., Fang, X., and Zhang, D. (2014). Shifts in soil phosphorus fractions under elevated $CO_2$ and N addition in model forest ecosystems in subtropical China. *Plant Ecology*, *215*(11), 1373–1384.

Huntingford, C., Zelazowski, P., and Galbraith, D. (2013). Simulated resilience of tropical rainforests to $CO_2$-induced climate change. *Nature Geoscience*, *6*, 268–273.

Iyamuremye, F., Dick, R., and Baham, J. (1996). Organic amendments and phosphorus dynamics: I. Phosphorus chemistry and sorption. *Soil Science*, *161*(7), 426–435.

Jakovac, C. C., Peña-Claros, M., Kuyper, T. W., and Bongers, F. (2015). Loss of secondary-forest resilience by land-use intensification in the Amazon. *Journal of Ecology*, *103*(1), 67–77.

Jansa, J., Finlay, R., Wallander, H., Smith, F., and Smith, S. (2011). Role of mycorrhizal symbioses in phosphorus cycling. In Bunemann, E., Oberson, A., and Frossard, E. (Eds.), *Phosphorus in Action: Biological Processes in Soil Phosphorus Cycling* (Vol. Soil Biology). Amsterdam: Springer.

Jenny, H. (1941). *Factors of Soil Formation*. New York: McGraw Hill.

Jenny, H. (1981). The soil resource-origin and behavior: Ecological studies 37. *Soil Science*, *132*(5), 380.

Jobbagy, E. G., and Jackson, R. B. (2000). The vertical distribution of soil organic carbon and its relation to climate and vegetation. *Ecological Applications*, *10*(2), 423–436.

John, R., Dalling, J. W., Harms, K. E., Yavitt, J. B., Stallard, R. F., Mirabello, M., Hubbell, S. P. et al. (2007). Soil nutrients influence spatial distributions of tropical tree species. *Proceedings of the National Academy of Sciences*, *104*(3), 864–869.

Johnson, D. W. (2006). Progressive N limitation in forests: Review and implications for long-term responses to elevated $CO_2$. *Ecology*, *87*(1), 64–75.

Johnson, A. H., Frizano, J., and Vann, D. R. (2003). Biogeochemical implications of labile phosphorus in forest soils determined by the Hedley fractionation procedure. *Oecologia*, *135*(4), 487–499.

Kaspari, M., Clay, N. A., Lucas, J. A., Revzen, S., Kay, A. D., and Yanoviak, S. P. (2015). Thermal adaptation and phosphorus shape thermal performance in an assemblage of rainforest ants. *Ecology*, *97*, 1038–1047.

Kaspari, M., Wright, S. J., Yavitt, J. B., Harms, K. E., Garcia, M., and Santana, M. (2008). Multiple nutrients limit litterfall and decomposition in a tropical forest. *Ecology Letters*, *11*, 35–43.

Keller, A. B., Reed, S. C., Townsend, A. R., and Cleveland, C. C. (2013). Effects of canopy tree species on belowground biogeochemistry in a lowland wet tropical forest. *Soil Biology and Biochemistry*, *58*, 61–69.

Kieft, T. L. (1987). Microbial biomass response to a rapid increase in water potential when dry soil is wetted. *Soil Biology and Biochemistry*, *19*(2), 119–126.

Kitayama, K., and Aiba, S. I. (2002). Ecosystem structure and productivity of tropical rain forests along altitudinal gradients with contrasting soil phosphorus pools on Mount Kinabalu, Borneo. *Journal of Ecology*, *90*(1), 37–51.

Kusel, K. C., Wagner, T., Trinkwalter, A. S., Gobner, A. S., Baumler, R., and Drake, H. L. (2002). Microbial reduction of Fe(III) and turnover of acetate in Hawaiian soils. *Microbiolial Ecology 40*, 73–81.

Lal, R. (2009). Soil degradation as a reason for inadequate human nutrition. *Food Security*, *1*, 45–57.

Lal, R. (2015). Sustainable intensification for adaptation and mitigation of climate change and advancement of food security in Africa. In *Sustainable Intensification to Advance Food Security and Enhance Climate Resilience in Africa* (pp. 3–17). Amsterdam: Springer.

Lambers, H., Raven, J. A., Shaver, G. R., and Smith, S. E. (2008). Plant nutrient-acquisition strategies change with soil age. *Trends in Ecology and Evolution*, *23*(2), 95–103.

Laurance, W. F., Sayer, J., and Cassman, K. G. (2014). Agricultural expansion and its impacts on tropical nature. *Trends in Ecology and Evolution*, *29*, 107–116.

Lawrence, D., and Schlesinger, W. H. (2001). Changes in soil phosphorus during 200 years of shifting cultivation in Indonesia. *Ecology*, *82*, 2769–2780.

Lewis, S. L., Edwards, D. P., and Galbraith, D. (2015). Increasing human dominance of tropical forests. *Science*, *349*, 827–832.

Lewis, S., Lloyd, J., Sitch, S., Mitchard, E., and Laurance, W. (2009). Changing ecology of tropical forests: Evidence and drivers. *Annual Review of Ecology Evolution and Systematics*, *40*, 529–549.

Liptzin, D., and Silver, W. L. (2009). Effects of carbon additions on iron reduction and phosphorus availability in a humid tropical forest soil. *Soil Biology and Biochemistry*, *41*(8), 1696–1702.

Liptzin, D., Silver, W. L., and Detto, M. (2011). Temporal dynamics in soil oxygen and greenhouse gases in two humid tropical forests. *Ecosystems*, *14*(2), 171–182.

Lloyd, J., and Farquhar, G. D. (2000). Do slow-growing species and nutrient-stressed plants consistently respond less to elevated $CO_2$? A clarification of some issues raised by Poorter (1998). *Global Change Biology*, *6*, 871–876.

Lodge, D. J. (1993). *Nutrient Cycling by Fungi in Wet Tropical Forests*. Retrieved from British Mycological Society Symposium Series (Vol. 19, pp. 37–57). Cambridge University Press.

Lodge, D. J., Cantrell, S. A., and González, G. (2014). Effects of canopy opening and debris deposition on fungal connectivity, phosphorus movement between litter cohorts and mass loss. *Forest Ecology and Management*, *332*, 11–21.

Lodge, D. J., McDowell, W. H., and McSwiney, C. P. (1994). The importance of nutrient pulses in tropical forests. *Trends in Ecology and Evolution*, *9*(10), 384–387.

López-Hernández, D. (1977). *La química del fósforo en suelos ácidos*. Caracas: Universidad Central de Venezuela.

Lougheed, T. (2011). Facing up to phosphorus. *Environment*, 20–25.

Lovley, D. R. (1991). Dissimilatory Fe(III) and Mn(IV) reduction. *Microbiology Review*, *55*, 259–287.

Lovley, D. R., Fraga, J. L., Blunt-Harris, E. L., Hayes, L. A., Phillips, E. J. P., and Coates, J. D. (1998). Humic substances as a mediator for microbially catalyzed metal reduction. *Acta Hydrochimica Hydrobiology*, *26*, 152–157.

Luo, Y., Su, B. O., Currie, W. S., Dukes, J. S., Finzi, A., Hartwig, U., Pataki, D. E. et al. (2004). Progressive nitrogen limitation of ecosystem responses to rising atmospheric carbon dioxide. *Bioscience*, *54*(8), 731–739.

Mahowald, N. M., Artaxo, P., Baker, A. R., Jickells, T. D., Okin, G. S., Randerson, J. T., and Townsend, A. R. (2005). Impacts of biomass burning emissions and land use change on Amazonian atmospheric phosphorus cycling and deposition. *Global Biogeochemical Cycles*, 19(4).

Marichal, R., Mathieu, J., Couteaux, M. M., Mora, P., Roy, J., and Lavelle, P. (2011). Earthworm and microbe response to litter and soils of tropical forest plantations with contrasting C: N: P stoichiometric ratios. *Soil Biology and Biochemistry*, 43, 1528–1535.

Marklein, A. R., and Houlton, B. Z. (2012). Nitrogen inputs accelerate phosphorus cycling rates across a wide variety of terrestrial ecosystems. *New Phytologist*, 193(3), 696–704.

Matson, P., Lohse, K. A., and Hall, S. J. (2002). The globalization of nitrogen deposition: Consequences for terrestrial ecosystems. *Ambio*, 31, 113–119.

Mayor, J. R., Wright, S. J., and Turner, B. L. (2014). Species-specific responses of foliar nutrients to longterm nitrogen and phosphorus additions in a lowland tropical forest. *Journal of Ecology*, 102, 36–44.

McGill, W. B., and Cole, C. V. (1981). Comparative aspects of cycling of organic C, N, S, and P through soil organic matter. *Geoderma*, 26, 267–286.

McGlynn, T. P., Salinas, D. J., Dunn, R. R., Wood, T. E., Lawrence, D., and Clark, D. A. (2007). Phosphorus limits tropical rain forest litter fauna. *Biotropica*, 39(1), 50–53.

McGrath, D. A., Comerford, N. B., and Duryea, M. L. (2000). Litter dynamics and monthly fluctuations in soil phosphorus availability in an Amazonian agroforest. *Forest Ecology and Management*, 131(1), 167–181.

McGrath, D. A., Duryea, M. L., and Cropper, W. P. (2001). Soil phosphorus availability and fine root proliferation in Amazonian agroforests 6 years following forest conversion. *Agriculture, Ecosystems & Environment*, 83(3), 271–284.

McGroddy, M., Daufresne, T., and Hedin, L. (2004). Scaling of C:N:P stoichiometry in forests worldwide: Implications of terrestrial Redfield-type ratios. *Ecology*, 85, 2390–2401.

Meir, P., Wood, T. E., Galbraith, D. R., Brando, P. M., Da Costa, A. C., Rowland, L., and Ferreira, L. V. (2015). Threshold responses to soil moisture deficit by trees and soil in tropical rain forests: Insights from field experiments. *BioScience*, 65(9), 882–892.

Mertz, O. (2002). The relationship between length of fallow and crop yields in shifting cultivation: A rethinking. *Agroforestry Systems*, 55, 149–159.

Mirmanto, E., Proctor, J., Green, J., Nagy, L., and Suriantata. (1999). Effects of nitrogen and phosphorus fertilization in a lowland evergreen rain forest. *Philosophical Transactions of the Royal Society of London B*, 354, 1825–1829.

Mora, C., Frazier, A. G., Longman, R. J., Dacks, R. S., Walton, M. M., Tong, E. J., Sanchez, J. J. et al. (2013). The projected timing of climate departure from recent variability. *Nature*, 502(7470), 183–187.

Myers, N., Mittermeier, R. A., Mittermeier, C. G., Da Fonseca, G. A., and Kent, J. (2000). Biodiversity hotspots for conservation priorities. *Nature*, 403, 853–858.

Nasto, M. K., Alvarez-Clare, S., Lekberg, Y., Sullivan, B. W., Townsend, A. R., and Cleveland, C. C. (2014). Interactions among nitrogen fixation and soil phosphorus acquisition strategies in lowland tropical rain forests. *Ecology Letters*, 17(10), 1282–1289.

Natural Resources Conservation Service (Soil Science Division). (2005). *World Soil Resources* (September 2005 version). U.S. Department of Agriculture, Washington, DC.

Newbery, D. M., Chuyong, G. B., Green, J. J., Songwe, N. C., Tchuenteu, F., and Zimmermann, L. (2002). Does low phosphorus supply limit seedling establishment and tree growth in groves of ectomycorrhizal trees in a central African rainforest? *New Phytologist*, 156, 297–311.

Norby, R. J., De Kauwe, M. G., Domingues, T. F., Duursma, R. A., Ellsworth, D. S., Goll, D. S., Lapola, D. M. et al. (2016). Model–data synthesis for the next generation of forest free-air $CO_2$ enrichment (FACE) experiments. *New Phytologist*, 209(1), 17–28.

Norby, R. J., Warren, J. M., Iversen, C. M., Medlyn, B. E., and McMurtrie, R. E. (2010). $CO_2$ enhancement of forest productivity constrained by limited nitrogen availability. *Proceedings of the National Academy of Sciences, 107*(45), 19368–19373.

Oberson, A., Friesen, D., Morel, C., and Tiessen, H. (1997). Determination of phosphorus released by chloroform fumigation from microbial biomass in high P sorbing tropical soils. *Soil Biology and Biochemistry, 29*, 1579–1583.

Okin, G. S., Mahowald, N., Chadwick, O. A., and Artaxo, P. (2004). Impact of desert dust on the biogeochemistry of phosphorus in terrestrial ecosystems. *Global Biogeochemical Cycles, 18*.

Olander, L. P., and Vitousek, P. M. (2004). Biological and geochemical sinks for phosphorus in soil from a wet tropical forest. *Ecosystems, 7*, 404–419.

Olsson, P. A., Jakobsen, I., and Wallander, H. (2003). Foraging and resource allocation strategies of mycorrhizal fungi in a patchy environment. *Mycorrhizal Ecology* (pp. 93–115). Amsterdam: Springer.

Palm, C., Sanchez, P., Ahamed, S., and Awiti, A. (2007). Soils: A contemporary perspective. *Annual Review of Environment and Resources, 32*, 99–129.

Pan, Y., Birdsey, R., Phillips, O., and Jackson, R. (2013). The structure, distribution, and biomass of the world's forests. *Annual Review of Ecology, Evolution, and Systematics, 44*, 593–622.

Paoli, G. D., Curran, L. M., and Zak, D. R. (2005). Phosphorus efficiency of Bornean rain forest productivity: Evidence against the unimodal efficiency hypothesis. *Ecology, 86*, 1548–1561.

Parfitt, R. L. (1978). Anion adsorption by soils and soil materials. *Advances in Agronomy, 30*, 50.

Parfitt, R. L., Atkinson, R. J., and Smart, R. S. C. (1975). The mechanism of phosphate fixation by iron oxides. *Soil Science Society of America Journal, 39*(5), 837–841.

Peltzer, D. A., Wardle, D. A., Allison, V. J., Baisden, W. T., Bardgett, R. D., Chadwick, O. A., Condron, L. M. et al. (2010). Understanding ecosystem retrogression. *Ecological Monographs, 80*, 509–529.

Peñuelas, J., Poulter, B., Sardans, J., Ciais, P., van der Velde, M., Bopp, L., Boucher, O. et al. (2013). Human-induced nitrogen–phosphorus imbalances alter natural and managed ecosystems across the globe. *Nature Communications, 4*.

Peñuelas, J., Sardans, J., Rivas-Ubach, A., and Janssens, I. A. (2012). The human-induced imbalance between C, N and P in Earth's life system. *Global Change Biology, 18*(1), 3–6.

Perfecto, I., and Vandermeer, J. (1993). Distribution and turnover rate of a population of *Atta cephalotes* in a tropical rain forest in Costa Rica. *Biotropica, 25*, 316–321.

Pett-Ridge, J. (2009). Contributions of dust to phosphorus cycling in tropical forests of the Luquillo Mountains, Puerto Rico. *Biogeochemistry, 94*(1), 63–80.

Pett-Ridge, J., and Firestone, M. K. (2005). Redox fluctuation structures microbial communities in a wet tropical soil. *Applied and Environmental Microbiology, 71*(11), 6998–7007.

Piao, S., Sitch, S., and Ciais, P. (2013). Evaluation of terrestrial carbon cycle models for their response to climate variability and to $CO_2$ trends. *Global Change Biology, 19*, 2117–2132.

Ponnamperuma, F. N., Tianco, E. M., and Loy, T. (1967). Redox equilibria in flooded soils: I. The iron hydroxide systems. *Soil Science 103*, 374–382.

Porder, S., Vitousek, P. M., Chadwick, O. A., Chamberlain, C. P., and Hilley, G. E. (2007). Uplift, erosion, and phosphorus limitation in terrestrial ecosystems. *Ecosystems, 10*(1), 159–171.

Powers, J. S., and Peréz-Aviles, D. (2013). Edaphic factors are a more important control on surface fine roots than stand age in secondary tropical dry forests. *Biotropica, 45*(1), 1–9.

Pregitzer, K. S., Laskowski, M. J., Burton, A. J., Lessard, V. C., and Zak, D. R. (1998). Variation in sugar maple root respiration with root diameter and soil depth. *Tree Physiology*, *18*(10), 665–670.

Quesada, C. A., Lloyd, J., Schwarz, M., Patiño, S., Baker, T. R., Czimczik, C., Fyllas, N. M. et al. (2010). Variations in chemical and physical properties of Amazon forest soils in relation to their genesis. *Biogeosciences*, *7*, 1515–1541.

Randerson, J. (2013). Climate science: Global warming and tropical carbon. *Nature*, *494*, 319–320.

Reed, S. C., Cleveland, C. C., and Townsend, A. R. (2007). Controls over leaf litter and soil nitrogen fixation in two lowland tropical rain forests. *Biotropica*, *39*, 585–592.

Reed, S. C., Cleveland, C. C., and Townsend, A. R. (2008). Tree species control rates of free-living nitrogen fixation in a tropical rain forest. *Ecology*, *89*(10), 2924–2934.

Reed, S., Townsend, A., Taylor, P., and Cleveland, C. (2011a). Phosphorus cycling in tropical forests growing on highly weathered soils. *Phosphorus in Action* (pp. 339–369). Amsterdam: Springer.

Reed, S., Vitousek, P., and Cleveland, C. (2011b). Are patterns in nutrient limitation belowground consistent with those aboveground: Results from a 4 million year chronosequence. *Biogeochemistry*, *106*, 323–336.

Reed, S., Cleveland, C., and Townsend, A. (2011c). Functional ecology of free-living nitrogen fixation: A contemporary perspective. *Annual Review of Ecology, Evolution and Systematics*, *42*, 489–512.

Reed, S. C., Cleveland, C. C., and Townsend, A. R. (2013). Relationships among phosphorus, molybdenum and free-living nitrogen fixation in tropical rain forests: Results from observational and experimental analyses. *Biogeochemistry*, *114*(1–3), 135–147.

Reed, S. C., Townsend, A. R., Davidson, E. A., and Cleveland, C. C. (2012). Stoichiometric patterns in foliar nutrient resorption across multiple scales. *New Phytologist*, *196*(1), 173–180l.

Reed, S., Yang, X., and Thornton, P. (2015). Incorporating phosphorus cycling into global modeling efforts: A worthwhile, tractable endeavor. *New Phytologist Tansley Insights*, *208*, 324–329.

Richardson, A. E., and Simpson, R. J. (2011). Soil microorganisms mediating phosphorus availability update on microbial phosphorus. *Plant Physiology*, *156*, 989–996.

Richter, D. D., Allen, H. L., Li, J., Markewitz, D., and Raikes, J. (2006). Bioavailability of slowly cycling soil phosphorus: Major restructuring of soil P fractions over four decades in an aggrading forest. *Oecologia*, *150*(2), 259–271.

Richter, D. D., and Babbar, L. (1991). Soil Diversity in the Tropics. *Advances in Ecological Research*, *21*, 315–389.

Richter, D. D., Saplaco, S. R., and Nowak, P. (1985). Watershed management problems in humid tropical uplands. *Nature and Resources (UNESCO)*, *21*, 10–21.

Robinson, C. H., Michelsen, A., Lee, J. A., Whitehead, S. J., Callaghan, T. V., Press, M. C., and Jonasson, S. (1997). Elevated atmospheric $CO_2$ affects decomposition of *Festuca vivipara* (L.) Sm. litter and roots in experiments simulating environmental change in two contrasting arctic ecosystems. *Global Change Biology*, *3*(1), 37–49.

Sanchez, P. (1976). *Properties and Management of Soils in the Tropics*. New York: John Wiley & Sons.

Sanchez, P. A. (2002). Soil Fertility and Hunger in Africa. *Science*, *295*, 2019–2020.

Sanchez, P. A., and Buol, S. W. (1975). Soils of the tropics and the world food crisis. *Science*, *188*, 598–603.

Sanchez, P. A., and Swaminathan, M. S. (2005). Hunger in Africa: The link between unhealthy people and unhealthy soils. *Lancet*, *365*, 442–444.

Santiago, L. S., Wright, S. J., Harms, K. E., Yavitt, J. B., Korine, C., M. N. Garcia, and Turner, B. L. (2012). Tropical tree seedling growth responses to nitrogen, phosphorus and potassium addition. *Journal of Ecology*, *100*, 309–316.

Sayer, E. J., Sutcliffe, L. M., Ross, R. I., and Tanner, E. V. (2010). Arthropod abundance and diversity in a lowland tropical forest floor in Panama: The role of habitat space vs. nutrient concentrations. *Biotropica, 42*, 194–200.

Sayer, E. J., Wright, S. J., Tanner, E. V., Yavitt, J. B., Harms, K. E., Powers, J. S., and Turner, B. L. (2012). Variable responses of lowland tropical forest nutrient status to fertilization and litter manipulation. *Ecosystems, 15*, 387–400.

Scatena, F., and Lugo, A. (1995). Geomorphology, disturbance, and the soil and vegetation of two subtropical wet steepland watersheds of Puerto Rico. *Geomorphology 13*, 199–213.

Schipper, W. (2014). Phosphorus: Too big to fail. *European Journal of Inorganic Chemistry 10*, 1567–1571.

Schwertmann, U., and Taylor, R. M. (1977). Iron oxides. In Dixon, J. B., and Weed, S. B. (Eds.), *Minerals in Soil Environments*. Madison, WI: Soil Science Society of America.

Sharpley, A. N., Rekolainen, S., Tunney, H., Carton, O. T., Brookes, P. C., and Johnston, A. E. (1997). Phosphorus in agriculture and its environmental implications. *Phosphorus Loss from Soil to Water*. Wexford: CAB International.

Shin, H. et al. (2005). Complex regulation of Arabidopsis AGR1/PIN2-mediated root gravitropic response and basipetal auxin transport by cantharidin-sensitive protein phosphatases. *The Plant Journal 42*, 188–200.

Silver, W. L. (1994). Is nutrient availability related to plant nutrient use in humid tropical forests? *Oecologia, 98*(3–4), 336–343.

Silver, W. L. (1998a). The potential effects of elevated $CO_2$ and climate change on tropical forest soils and biogeochemical cycling. *Climatic Change, 39*, 337–361.

Silver, W. L. (1998b). The potential effects of elevated $CO_2$ and climate change on tropical forest soils and biogeochemical cycling. In *Potential Impacts of Climate Change on Tropical Forest Ecosystems* (pp. 197–221). Netherlands: Springer.

Silver, W. H., Lugo, A. E., and Keller, M. (1999). Soil oxygen availability and biogeochemistry along rainfall and topographic gradients in upland wet tropical forest soils. *Biogeochemistry, 44*, 301–328.

Sims, J. T., and Sharpley, A. N. (2005). *Phosphorus: Agriculture and the Environment*. Madison, WI: American Society of Agronomy.

Singh, B., and Gilkes, R. (1991). Phosphorus sorption in relation to soil properties for the major soil types of south-western Australia. *Soil Research, 29*(5), 603–618.

Singh, J. S., Raghubanshi, A. S., Singh, R. S., and Srivastava, S. C. (1989). Microbial biomass acts as a source of plant nutrients in dry tropical forest and savanna. *Nature, 338*(6215), 499–500.

Smith, W. K., Cleveland, C. C., Reed, S. C., and Running, S. W. (2014). Agricultural conversion without external water and nutrient inputs reduces terrestrial vegetation productivity. *Geophysical Research Letters, 41*(2), 449–455.

Smith, W. K., Reed, S. C., Cleveland, C. C., Ballantyne, A. P., Anderegg, W. R., Wieder, W. R., Liu, Y. Y., and Running, S. W. (2015). Large divergence of satellite and Earth system model estimates of global terrestrial $CO_2$ fertilization. *Nature Climate Change, 6*, 306–310.

Smyth, T., and Sanchez, P. (1982). Phosphate rock dissolution and availability in Cerrado soils as affected by phosphorus sorption capacity. *Soil Science Society of America Journal, 46*(2), 339–345.

Soil Survey Staff. (1998). Keys to soil taxonomy (326 pp.). Washington, DC: Natural Resources Conservation Service.

Sollins, P., Robertson, G., and Uehara, G. (1988). Nutrient mobility in variable- and permanent-charge soils. *Biogeochemistry 6*, 181–199.

Srivastava, S. C. (1997). Microbial contribution to extractable N and P after air-drying of dry tropical soils. *Biol Fertil Soils, 26*, 31–34.

Stark, N. M., and Jordan, C. F. (1978). Nutrient retention by the root mat of an Amazonian rain forest. *Ecology, 58*, 434–437.

Steidinger, B. S., Turner, B. L., Corrales, A., and Dalling, J. W. (2015). Variability in potential to exploit different soil organic phosphorus compounds among tropical montane tree species. *Functional Ecology*, *29*(1), 121–130.

Swap, R., Garstang, M., Greco, S., Talbot, R., and Kallberg, P. (1992). Saharan dust in the Amazon Basin. *Tellus*, *44*, 133–149.

Tang, J., Baldocchi, D. D., and Xu, L. (2005). Tree photosynthesis modulates soil respiration on a diurnal time scale. *Global Change Biology*, *11*(8), 1298–1304.

Tanner, E. V. J., Vitousek, P. M., and Cuevas, E. (1998). Experimental investigation of nutrient limitation of forest growth on wet tropical mountains. *Ecology*, *79*, 10–22.

Tarnocai, C., Canadell, J., Schuur, E., Kuhry, P., Mazhitova, G., and Zimov, S. (2009). Soil organic carbon pools in the northern circumpolar permafrost region. *Global Biogeochemical Cycles*, *23*(2).

Theodorou, M., and Plaxton, W. (1996). Metabolic adaptations of plant respiration to nutritional phosphate deprivation. *Plant Physiology*, *101*, 339–344.

Thompson, A., Chadwick, O. A., Rancourt, D. G., and Chorover, J. (2006). Iron-oxide crystallinity increases during soil redox oscillations. *Geochimica et Cosmochimica Acta*, *70*(7), 1710–1727.

Tiessen, H. (2001). *Phosphorus Availability in the Environment*. New York: John Wiley & Sons, Ltd.

Tiessen, H. (2005). Phosphorus dynamics in tropical soils. In, Sims, J., and Sharpley, A. (Eds.) *Phosphorus: Agriculture and the Environment* (pp. 253–262). Madison, WI: American Society Agronomy.

Tiessen, H., Stewart, J. W. B., and Cole, C. V. (1984). Pathways of phosphorus transformations in soils of differing pedogenesis. *Soil Science Society of America Journal*, *48*(4), 853–858.

Tiessen, H., Stewart, J. W. B., and Oberson, A. (1994). Innovative soil phosphorus availability indices: Assessing organic phosphorus. *Soil Testing: Prospects for Improving Nutrient Recommendations*, 143–162.

Tilman, D., Cassman, K. G., Matson, P. A., Naylor, R., and Polasky, S. (2002). Agricultural sustainability and intensive production practices. *Nature*, *418*, 671–677.

Townsend, A. R., Asner, G. P., and Cleveland, C. C. (2008). The biogeochemical heterogeneity of tropical forests. *Trends in Ecology and Evolution*, *23*(8), 424–431.

Townsend, A. R., Cleveland, C. C., Houlton, B. Z., Alden, C. B., and White, J. W. (2011). Multi-element regulation of the tropical forest carbon cycle. *Frontiers in Ecology and the Environment*, *9*(1), 9–17.

Trenbath, B. R., Conway, G. R., and Craig, I. A. (1990). Threats to sustainability in intensified agricultural systems: Analysis and implications for management. *Agroecology* (pp. 337–365). New York: Springer.

Treseder, K. K., and Vitousek, P. M. (2001). Effects of soil nutrient availability on investment in acquisition of N and P in Hawaiian rain forests. *Ecology*, *82*(4), 946–954.

Tsui, C.-C., Chen, Z.-S., and Hsieh, C.-F. (2004). Relationships between soil properties and slope position in a lowland rain forest of southern Taiwan. *Geoderma*, *123*(1), 131–142.

Turner, B. L., Driessen, J. P., Haygarth, P. M., and Mckelvie, I. D. (2003). Potential contribution of lysed bacterial cells to phosphorus solubilisation in two rewetted Australian pasture soils. *Soil Biology and Biochemistry*, *35*(1), 187–189.

Turner, B. L., and Haygarth, P. M. (2001). Biogeochemistry: Phosphorus solubilization in rewetted soils. *Nature*, *411*(6835), 258.

Turner, B. L., Lambers, H., Condron, L. M., Cramer, M. D., Leake, J. R., Richardson, A. E., and Smith, S. E. (2013a). Soil microbial biomass and the fate of phosphorus during long-term ecosystem development. *Plant and Soil*, *367*(1–2), 225–234.

Turner, B. L., and Wright, S. J. (2014). The response of microbial biomass and hydrolytic enzymes to a decade of nitrogen, phosphorus, and potassium addition in a lowland tropical rain forest. *Biogeochemistry* *117*(1), 115–130.

Turner, B. L., Yavitt, J. B., Harms, K. E., Garcia, M. N., Romero, T. E., and Wright, S. J. (2013b). Seasonal changes and treatment effects on soil inorganic nutrients following a decade of fertilization in a lowland tropical forest. *Soil Science Society of America Journal*, *77*(4), 1357–1369.

Uehara, G., and Gillman, G. (1981). *The Mineralogy, Chemistry, and Physics of Tropical Soils with Variable Charge Clays*. Boulder: Westview Press.

Uhde-Stone, C., Gilbert, G., Johnson, J. M., Litjens, R., Zinn, K. E., Temple, S. J., Vance, C. P., and Allan, D. L. (2003). Acclimation of white lupin to phosphorus deficiency involves enhanced expression of genes related to organic acid metabolism. *Plant and Soil*, *248*(1–2), 99–116.

Ulrich, A. E., Stauffacher, M., Krutli, P., Schnug, E., and Frossard, E. (2013). Tackling the phosphorus challenge: Time for reflection on three key limitations. *Environmental Development*, *8*, 137–144.

Uriarte, M., Turner, B. L., Thompson, J., and Zimmerman, J. K. (2015) Linking spatial patterns of leaf litterfall and soil nutrients in a tropical forest: A neighborhood approach. *Ecological Applications*, *25*(7), 2022–2034.

U.S. Department of Energy. (2012). *Research Priorities for Tropical Ecosystems Under Climate Change Workshop Report*. Available at http://science.energy.gov/ber/.

Vance, C. P., Uhde-Stone, C., and Allan, D. L. (2003). Phosphorus acquisition and use: Critical adaptations by plants for securing a nonrenewable resource. *New Phytologist*, *157*(3), 423–447.

Vandecar, K. L., Lawrence, D., and Clark, D. (2011). Phosphorus sorption dynamics of anion exchange resin membranes in tropical rain forest soils. *Soil Science Society of America Journal*, *75*(4), 1520–1529.

Vandecar, K. L., Lawrence, D., Wood, T., Oberbauer, S. F., Das, R., Tully, K., and Schwendenmann, L. (2009). Biotic and abiotic controls on diurnal fluctuations in labile soil phosphorus of a wet tropical forest. *Ecology*, *90*(9), 2547–2555.

Vincent, A. G., Turner, B. L., and Tanner, E. V. J. (2010). Soil organic phosphorus dynamics following perturbation of litter cycling in a tropical moist forest. *European Journal of Soil Science*, *61*, 48–57.

Vira, B., Wildburger, C., and Mansourian, S. (2015). *Forests, Trees, and Landscapes for Food Security and Nutrition. A Global Assessment Report*. Vienna: Open Book Publishers.

Vitousek, P. M. (1984). Litterfall, nutrient cycling, and nutrient limitation in tropical forests. *Ecology*, *65*, 285–298.

Vitousek, P. (2004). *Nutrient Cycling and Limitation: Hawai'i as a Model System*. Princeton, NJ: Princeton University Press.

Vitousek, P., Chadwick, O., Matson, P., Allison, S., Derry, L., Kettley, L., Luers, A. et al. (2003). Erosion and the rejuvenation of weathering-derived nutrient supply in an old tropical landscape. *Ecosystems*, *6*(8), 762–772.

Vitousek, P. M., and Farrington, H. (1997). Nutrient limitation and soil development: Experimental test of a biogeochemical theory. *Biogeochemistry*, *37*, 63–75.

Vitousek, P. M., Ladefoged, T. N., Kirch, P. V., Hartshorn, A. S., Graves, M. W., Hotchkiss, S. C., and Chadwick, O. A. (2004). Soils, agriculture, and society in precontact Hawaii. *Science*, *304*, 1665–1669.

Vitousek, P. M., Naylor, R., Crews, T., David, M. B., Drinkwater, L. E., Holland, E., Johnes, P. J. et al. (2009). Nutrient imbalances in agricultural development. *Science*, *324*(5934), 1519.

Vitousek, P. M., Porder, S., Houlton, B. Z., and Chadwick, O. A. (2010). Terrestrial phosphorus limitation: Mechanisms, implications, and nitrogen–phosphorus interactions. *Ecological Applications*, *20*, 5–15.

Walker, T. (1965). *The Significance of Phosphorus in Pedogenesis*. Lincoln: Lincoln College.

Walker, T. W., and Syers, J. K. (1976). The fate of phosporus during pedogenesis. *Geoderma*, *15*, 1–19.

Walker, A. P., Beckerman, A. P., Gu, L., Kattge, J., Cernusak, L. A., Domingues, T. F., Woodward, F. I. et al. (2014). The relationship of leaf photosynthetic traits–Vcmax and Jmax–to leaf nitrogen, leaf phosphorus, and specific leaf area: A meta-analysis and modeling study. *Ecology and Evolution*, *4*(16), 3218–3235.

Wardle, D. A., Walker, L. R., and Bardgett, R. D. (2004). Ecosystem properties and forest decline in contrasting long-term chronosequences. *Science*, *305*, 509–513.

Waring, B. G., Álvarez-Cansino, L., Barry, K. E., Becklund, K. K., Dale, S., Gei, M. G., Keller, A. B. et al. (2015). Pervasive and strong effects of plants on soil chemistry: A meta-analysis of individual plant 'Zinke' effects. *Proc. R. Soc. B*, *282*.

Wassen, M., Venterink, H., Lapshina, E., and Tanneberger, F. (2005). Endangered plants persist under phosphorus limitation. *Nature* *437*, 547–550.

Webb, L. J., Tracey, J. G., and Williams, W. T. (1972). Regeneration and pattern in the subtropical rainforest. *Journal of Ecology 60*, 675–695.

Werner, F. A., and Homeier, J. (2015). Is tropical montane forest heterogeneity promoted by a resource-driven feedback cycle? Evidence from nutrient relations, herbivory and litter decomposition along a topographical gradient. *Functional Ecology*, *29*(3), 430–440.

White, G., and Zelazny, L. (1986). Charge properties of soil colloids. In Sparks, D. (Ed.), *Soil Physical Chemistry* (pp. 39–81). Boca Raton, FL: CRC.

White, A. F., Blum, A. E., Schulz, M. S., Vivit, D. V., Stonestrom, D. A., Larsen, M., and Eberl, D. (1998). Chemical weathering in a tropical watershed, Luquillo Mountains, Puerto Rico: I. Long-term versus short-term weathering fluxes. *Geochimica et Cosmochimica Acta*, *62*, 209–226.

Wieder, W., Cleveland, C., and Townsend, A. (2009). Controls over leaf litter decomposition in wet tropical forests. *Ecology 90*, 3333–3341.

Withers, P. J. A., van Dijk, K. C., Neset, T.-S. S., Nesme, T., Oenema, O., Rubæk, G. H., Schoumans, O. F., Smit, B., and Pellerin, S. (2015). Stewardship to tackle global phosphorus inefficiency: The case of Europe. *Ambio*, *44*, 193–206.

Wood, T. E., Cavaleri, M. A., and Reed, S. C. (2012). Tropical forest carbon balance in a warmer world: A critical review spanning microbial–to ecosystem–scale processes. *Biological Reviews*, *87*(4), 912–927.

Wood, T. E., Detto, M., and Silver, W. L. (2013). Sensitivity of soil respiration to variability in soil moisture and temperature in a humid tropical forest. *PloS One*, *8*(12), e80965.

Wood, T. E., and Lawrence, D. (2008). No short-term change in soil properties following four-fold litter addition in a Costa Rican rain forest. *Plant and Soil*, *307*(1–2), 113–122.

Wood, T. E., Lawrence, D., and Clark, D. A. (2006). Determinants of leaf litter nutrient cycling in a tropical rain forest: Soil fertility versus topography. *Ecosystems*, *9*(5), 700–710.

Wood, T. E., Lawrence, D., and Wells, J. A. (2011). Inter-specific variation in foliar nutrients and resorption of nine canopy-tree species in a secondary neotropical rain forest. *Biotropica*, *43*(5), 544–551.

Wood, T. E., Matthews, D., Vandecar, K., and Lawrence, D. (2016). Short-term variability in labile soil phosphorus is positively related to soil moisture in a humid tropical forest in Puerto Rico. *Biogeochemistry*, *127*(1), 35–43.

Wood, T. E., and Silver, W. L. (2012). Strong spatial variability in trace gasdynamics following experimental drought in a humid tropical forest. *Global Biogeochemical Cycles*, *26*(3).

Wright, S. J., Yavitt, J. B., Wurzburger, N., Turner, B. L., Tanner, E. V. J., Sayer, E. J., Santiago, L. S. et al. (2011). Potassium, phosphorus, or nitrogen limit root allocation, tree growth, or litter production in a lowland tropical forest. *Ecology*, *92*(8), 1616–1625.

Wurzburger, N., Bellenger, J. P., Kraepiel, A. M., and Hedin, L. O. (2012). Molybdenum and phosphorus interact to constrain asymbiotic nitrogen fixation in tropical forests. *PloS One*, *7*(3), e33710.

Wurzburger, N., and Wright, S. J. (2015). Fine root responses to fertilization reveal multiple nutrient limitation in a lowland tropical forest. *Ecology*, 2137–2146.

Yang, X., and Post, W. M. (2011). Phosphorus transformations as a function of pedogenesis: A synthesis of soil phosphorus data using Hedley fractionation method. *Biogeosciences*, 8(10), 2907–2916.

Yang, X., Thornton, P., Ricciuto, D., and Post, W. (2014). The role of phosphorus dynamics in tropical forests—A modeling study using CLM-CNP. *Biogeosciences*, 11(6), 1667–1681.

Yuan, Z., and Chen, H. Y. (2009). Global-scale patterns of nutrient resorption associated with latitude, temperature and precipitation. *Global Ecology and Biogeography*, 18(1), 11–18.

Yuste, J. C., Ma, S., and Baldocchi, D. D. (2010). Plant-soil interactions and acclimation to temperature of microbial-mediated soil respiration may affect predictions of soil $CO_2$ efflux. *Biogeochemistry*, 98(1–3), 127–138.

Zemunik, G., Turner, B. L., Lambers, H., and Laliberté, E. (2015). Diversity of plant nutrient-acquisition strategies increases during long-term ecosystem development. *Nature Plants*, 1, 15050. Available at http://www.nature.com/articles/nplants201550 -supplementary-information.

Zinke, P. J. (1962). The pattern of influence of individual forest trees on soil properties. *Ecology*, 43, 130–133.

# 12 The Use of Phosphorus Radioisotopes to Investigate Soil Phosphorus Dynamics and Cycling in Soil–Plant Systems

*Long Nguyen, Felipe Zapata, and Joseph Adu-Gyamfi*

## CONTENTS

## 12.1 INTRODUCTION

Phosphorus (P) is one of three major elements (nitrogen, phosphorus, and potassium) that are required for plants and livestock (Syers et al. 2008; Tiessen 2008). It is essential for photosynthesis in plants, for energy transformations, and for the activity of some hormones in both plants and animals (Syers et al. 2008; Marschner 2012). Plants obtain P from the soil in which they grow, and grazing animals such as sheep and cattle obtain their P from ingested plant materials (Nguyen and Goh 1992; Haynes and Williams 1993; Simpson et al. 2014).

Phosphorus deficiency can affect crop and animal production, biological nitrogen fixation by leguminous crops in arable and arable farming systems, fertilizer nitrogen (N) use efficiency, and crop water use efficiency (Nguyen et al. 1989; Hatfield et al. 2001; Høgh-Jensen et al. 2002; Turner 2004; Sutton et al. 2011; Kröbel et al. 2012). Since soil-available P for plant growth is normally low in many high P-fixing (sorption) soils, particularly in highly weathered and tropical acid soils (e.g., Acrosols, Andosols, Ferralsols, and Nitisols) of Asia, Africa, and Latin America (Lal 1990; Tiessen 1995; Formoso 1999; Syers et al. 2008), P as fertilizers (such as superphosphate and phosphate rocks), organic wastes, and animal manure need to be applied to correct the P deficiency.

The present world population of 7.3 billion is expected to reach 8 billion by the year 2024 (http://www.worldometers.info/world-population/). Most of this increase in population will occur in developing countries, where the majority depends upon agriculture for their livelihoods (Lal 2000). Such an increase will place great pressure on the need to enhance crop and livestock production in the many parts of the world and the cost of P fertilizers (Syers et al. 2008). This is particularly so in the least developed countries where costly P fertilizers are normally out of reach of many farmers, and PRs, from which P fertilizers are manufactured, are finite, nonrenewable resources (Cordell et al. 2009; Van Kauwenbergh 2010; Van Vuuren et al. 2010; Cordell and White 2014). In addition to the rising costs of P fertilizers and the finite P resources, the efficiency of P use by plants from soil and fertilizer sources is often poor despite many soils containing a relatively large amount of TP that is only sparingly available to plants. In many situations in developed countries, the inefficient use of P inputs resulting from excessive or inappropriate applications of P from inorganic (P fertilizers) and organic P (manure and compost) sources over the years can lead to the accumulation of P in top 7.5 cm soils (Nguyen and Goh 1992;

Kleinman et al. 2002, 2007, 2011; Motavalli and Miles 2002) and potentially increase pollution risks, accelerate eutrophication, and create toxic algal blooms in streams, rivers, and lakes (Sharpley and Rekolainen 1997; Carpenter et al. 1998; Hesketh and Brookes 2000; McDowell and Sharpley 2001; Koopmans et al. 2002; Hart et al. 2004; McDowell et al. 2004; Elsner and Bennet 2011).

It is commonly known that only 10–25% of applied P fertilizers in the year of their application are taken up by plants (Syers et al. 2008). Most of the applied P is accumulated in soils depending on the ability of soils to retain the released phosphate from P fertilizers by soil particles as sorbed P or precipitated P and by immobilized P as soil organic P by soil microbial biomass (Syers et al. 2008). Some of this accumulated soil P will potentially be lost as dissolved P through leaching down the soil profile or losses of eroded soils and particulate P resulting from soil erosion and runoff (Hesketh and Brookes 2000; McDowell and Sharpley 2001; Koopmans et al. 2002; McDowell et al. 2004; Lucci et al. 2010). An understanding of the physical–chemical (sorption–desorption–precipitation)–biological (rate taken up by plants and soil microbial biomass) processes that influence the fate of phosphate fertilizer added to soils and the rate at which phosphate transfers from one soil P pool to another (e.g., inorganic and organic P pools) is important to enhance the efficient use of phosphate fertilizers for both economic and environmental reasons and to prolong the life of limited P rock resources (Richardson et al. 2011; Selles et al. 2011).

The purpose of this chapter is to highlight the role of P radioisotopes ($^{32}$P and $^{33}$P) to investigate the transformations and the dynamics of applied P in soils and in soil–plant systems, to trace the fate of P fertilizers in these systems, to assess P fertilizer recovery by the crops as influenced by soil types and soil-water management practices, and to evaluate the agronomic effectiveness of different P fertilizers such as reactive PRs. Information obtained from the use of P radioisotopes can help unravel the main soil and plant factors that influence soil P fluxes and the availability of P from soils and external P inputs (e.g., P fertilizers) to plants so that farm management practices (e.g., soil cultivation, mulch, and liming) can be put in place to enhance the P use efficiency (PUE) and minimize the potential P losses to the environment.

The review will briefly focus on soil P pools, soil P dynamics, release of inorganic P from inorganic (e.g., superphosphate and PRs) and organic P (e.g., biosolids, compost, and animal manure) fertilizers, and uptake of P by plants and soil microorganisms. The use of P radioisotopes ($^{32}$P and $^{33}$P) in each of the areas mentioned earlier will then be elaborated. These P radioisotopes with short half-lives (14.26–25.34 days for $^{32}$P and $^{33}$P, respectively; Nguyen et al. 2011) can only be deployed in short-term (<2–3 to 3–6 months, respectively) studies. The half-life is defined as the time required for half of the radioactive atoms to disintegrate (i.e., undergo radioactive decay). The use of these short-lived P radioisotopes is not appropriate to investigate the long-term fate of P fertilizers in field plots and agricultural catchments because of the radiation protection and safety issues involved. Advanced development has been made by various research workers (Frossard et al. 2011; Tamburini et al. 2014) on the use of the natural abundance of oxygen (O) isotopes ($^{18}$O/$^{16}$O) that are bound to phosphate in soils to investigate soil P transformation and P transfers in agricultural ecosystems. This aspect will only be briefly discussed here since this review is mainly focused on the use of $^{32}$P and $^{33}$P in soil P dynamics. Likewise, this review will not focus on the use of $^{31}$P-nuclear magnetic resonance (NMR) and X-ray absorption near edge structure spectroscopy to characterize various P compounds in organic manure, compost, and biosolids as soil amendments to provide P for plant growth since these techniques have been covered in a recent review (Nguyen et al. 2011).

## 12.2 AN OVERVIEW OF SOIL PHOSPHORUS POOLS AND PHOSPHORUS DYNAMICS IN SOIL–PLANT SYSTEMS

### 12.2.1 INTRODUCTION

Phosphorus (P) is an essential element for plant growth because it is a structural component of nucleic acids, coenzymes, phosphoproteins, phospholipids, and ATP, an energy-bearing molecule found in all living cells. Due to this essential nature of P, primary production is greatly dependent on plant P availability in soil (Richardson et al. 2005; Bünemann and Condron 2007; Syers et al. 2008).

The concentration of TP in agricultural topsoil (normally 30 cm depth) varies widely from 100–3000 mg P ha$^{-1}$ (Nguyen and Goh 1990, 1992; Richardson et al. 2005; Bünemann and Condron 2007), and less than 1% of TP is immediately available to plants at any one time (about 0.01–10 µmol P m$^{-3}$ in soil solution; Pierzynski 1991).

Many factors influence the soil TP content such as parent material, degree of weathering, and climatic conditions. In addition, soil TP levels are affected by soil erosion, crop removal, and P fertilization. In native natural ecosystems, the parent material mainly determines the TP content in the soil, while in agricultural soils, P fertilizer additions largely determine the soil TP status and the distribution of P forms (Nguyen et al. 1989; Nguyen and Goh 1990). This section provides an overview of the soil P pools and dynamics that influence P cycling in soil–plant systems, the plant-available soil P pool, and plant the P uptake for crop productivity.

### 12.2.2 SOIL PHOSPHORUS POOLS

Soil TP is classified into two broad groups, inorganic ($P_i$) and organic ($P_o$). Soil $P_o$ is estimated as the difference between soil TP and $P_i$ (Condron et al. 1990; Saunders and Williams 1995; Agbenin et al. 1999). In most agricultural soils, $P_i$ constitutes 50–75% of the TP, while $P_o$ ranges from 30% to 65% (Harrison 1987). In organic soils, up to 90% of soil TP can be in organic forms, while in mineral topsoils, values vary from 15% to 80% (Magid et al. 1996). In acid noncalcareous soils, hydrous sesquioxides and amorphous and crystalline aluminum (Al) and iron (Fe) compounds are the dominant $P_i$ forms. In alkaline and calcareous soils, calcium (Ca) compounds such as monocalcium phosphate ($Ca(H_2PO_4)_2$) and dicalcium phosphate ($CaPO_4$) dominate. Some apatite minerals originating from the parent material may also be present as $P_i$ in both acid and calcareous soils.

### 12.2.2.1 Soil Inorganic Phosphorus

The main $P_i$ forms in a soil solution are $H_3PO_4$, $H_2PO_4^-$, and $HPO_4^{2-}$. Plants and soil microorganisms can take up these $P_i$ forms, principally $H_2PO_4^-$ and, to a lesser extent $HPO_4^{2-}$ (Syers et al. 2008). These ions undergo physical–chemical–biological–microbiological

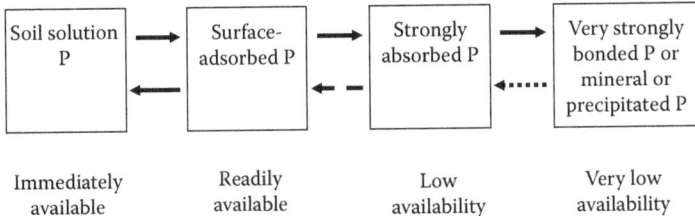

**FIGURE 12.1** Soil inorganic phosphorus pools and their availability for plant uptake. (Adapted from Syers, J. K., Johnston, A. E., and Curtin, D., *FAO Fertilizer and Plant Nutrition Bulletin* No. 18, Food and Agriculture Organization, Rome, 2008. With permission.)

reactions in soils, some occurring within a few seconds, others over several years, and are distributed in a large number of pools (Fardeau 1996; Bünemann and Condron 2007; Syers et al. 2008). These reactions include (1) sorption (adsorption and then absorption) into soil aggregates containing Al and Fe oxides and clay aluminosilicates, (2) precipitation as calcium phosphates in soils above pH 6 and particularly those containing free calcium carbonate, and (3) microbial immobilization and plant uptake. The movement of phosphate into plants influences $P_i$ concentrations in soil solution and promotes dissolution and desorption reactions. According to Syers et al. (2008), sorbed $P_i$ is not irreversibly fixed in soils and can be plant available over a long term. They proposed that soil $P_i$ could be considered to exist in four pools, with different levels of plant availability (Figure 12.1). The amount of $P_i$ in each of the four pools is related to differences in the bonding energy for P between and within soil constituents. In addition, it is assumed that there is a reversible transfer of P between the soil solution pool, the readily plant-available P pool, and the less readily plant-available pool. In other words, there is equilibrium between the P in these pools.

Since the concentration of $P_i$ forms in a soil solution is very low, usually less than $0.01–10$ $\mu$mol P m$^{-3}$ (Pierzynski 1991), plants depend on replenishing the soil solution with phosphates from soil P forms (desorption–dissolution) or from the application of P fertilizers, manure, biosolids, or crop residues (Goh and Condron 1989; Nguyen et al. 1995; Bertrand et al. 2006).

### 12.2.2.2 Soil Organic Phosphorus

Soil organic P ($P_o$) is derived largely from plant and animal residues and microbial tissues. Studies using $^{31}$P NMR techniques have revealed that soil $P_o$ is largely esters (compounds with C–O–P linkages) of orthophosphoric acid ($H_3PO_4$), phosphonates, and anhydrides, based on the nature of the P bond (Newman and Tate 1980; Condron et al. 1985, 2005). Orthophosphate monoesters are the dominant form of $P_o$ (Dalal 1977) with inositol phosphate (phytic acid) as the dominant form of orthophosphate monoesters. Inositol hexaphosphate is stable in soil but can also form sparingly soluble salts with Al, Fe, and Ca, and it is adsorbed by clay minerals and hydrous oxides. Orthophosphate diesters typically represent about less than 10% of the $P_o$ in agricultural soils. They include nucleic acids, phospholipids, teichoic acid, and aromatic compounds. Among the different forms of P identified by $^{31}$P NMR, orthophosphate diesters, including phospholipids and nucleic acids, are most readily converted to plant-available P under favorable soil conditions, compared to monoester phosphates (Tate and Newman 1982; Hinedi et al. 1989; Dai et al. 1996).

### 12.2.2.3  Soil Phosphorus Dynamics

The amounts, forms, and associated dynamics of soil $P_i$ and $P_o$ are affected by a combination of biological, chemical, and physical factors, which in turn are influenced by soil type (e.g., soil pH), climatic conditions (e.g., temperature, moisture), land use, and management practices. The effects of P fertilizer applications on different P forms depend on the type and the rate of P applied (Condron and Goh 1989; Zhang and MacKenzie 1997; Motavalli and Miles 2002) and also on the soil P status (Perrott et al. 1992). For example, Nguyen and Goh (1992) showed that long-term P fertilizer applications to a grazed pasture in New Zealand increased both $P_i$ and $P_o$ to a soil depth of 22.5 cm. The accumulation of $P_i$ was most pronounced when superphosphate fertilizer was applied annually at 376 kg ha$^{-1}$ for 35 years, which was in excess of the pasture plant P requirements (18–20 kg P ha$^{-1}$). Other researchers reported either both $P_o$ and $P_i$ or only $P_o$ increases due to long-term continued applications of P fertilizers and/or animal manure under a crop production system (e.g., Richards et al. 1993; Zhang and MacKenzie 1997; Motavalli and Miles 2002). For example, Motavalli and Miles (2002) reported that either manufactured fertilizer or manure significantly increased the soil $P_i$ and $P_o$ levels in soils that had been cropped for 111 years. However, Koopmans et al. (2003), using $^{31}P$ NMR, reported that after 11 years of continued P fertilizer and manure applications, $P_i$ accumulation occurred in soils treated with NPK fertilizer or animal manures, while $P_o$ increased only in the soil treated with pig slurry.

Several studies have reported the influence of cultivation on soil $P_i$ and $P_o$ pools. For example, long-term cultivation significantly reduced soil $P_o$ levels, while no-tillage or minimum tillage systems increased the $P_o$ accumulation near the soil surface (e.g., Hedley et al. 1982; Langdale et al. 1984; Weil et al. 1988; Beck and Sanchez 1994; Motavalli and Miles 2002). Crop removal (Tiessen et al. 1992) or afforestation of grassland soils (Condron et al. 1996; Chen et al. 2000) also decreased soil $P_o$.

The release of P from $P_o$ depends on $P_o$ mineralization. The turnover of soil $P_o$ is primarily determined by the rates of immobilization and mineralization (Condron et al. 2005). The net immobilization of P occurs during pasture development with continued application of P fertilizers (Nguyen and Goh 1992; Magid et al. 1996). The rate of soil $P_o$ accumulation declines when equilibrium is established between $P_o$ inputs and mineralization (Condron and Goh 1989). Since soil microbial activity depends on SOM, the C:P ratio is regarded as critical in balancing the processes of P mineralization and immobilization. It has been proposed that the critical C:P values for decomposing plant residues are less than 200:1 for net mineralization and greater than 300:1 for net immobilization (Sharpley 1985). For the substrate being decomposed, the critical C:P ratio calculated from microbial growth yield ranged from 50 to 70 (White 1980).

Inorganic P may be released from $P_o$ associated with SOM components or during the decomposition of plant residues. It is widely believed that $P_o$ mineralization is controlled by a combination of biological and biochemical factors which determine the solubility (accessibility) of $P_o$ and $P_o$ hydrolysis by the phosphatase enzyme

(Adams and Pate 1992; Magid et al. 1996; Gressel and McColl 1997; Frossard et al. 2000; Joner et al. 2000). Soil physical factors such as soil aggregation and associated effects of drying and wetting (Perrott et al. 1999; Chepkowny et al. 2001) together with the exudation of organic anions by plant roots and mycorrhizae (Fox and Comerford 1990; Hayes et al. 2000; Chen et al. 2002) may also affect the solubility and the susceptibility of $P_o$ to mineralization.

### 12.2.3  SOIL PHOSPHORUS FOR PLANT UPTAKE

Plant P uptake involves a two-step process: the supply of available P by soil and the uptake of that available P by plants. Other factors affecting P availability include climate (temperature, moisture), soil pH, P fertilizer placement, soil biota, and plant species. Some plants have developed various mechanisms to enhance the acquisition of P from soils such as modified root architecture (Lambers and Shane 2007; Shu et al. 2007), mycorrhizal association (Bolan 1991; Miller 2000), and production of phosphatase enzymes to catalyze the hydrolysis and release of $P_i$ from $P_o$ forms (Richardson et al. 2005; George et al. 2006). Various soil test methods were developed to estimate the P availability and the requirements of crops to maintain optimal soil P levels (Dyer 1984; Bolland et al. 1994). Although rapid soil tests provide a means for determining optimal P fertilizer recommendations (Bolland et al. 1994), they do not provide information on the processes affecting the P availability and dynamics in soils as stated in a previous section (Section 12.2.2). Quantitative estimates of P fluxes are important to ensure that there adequate P for plant uptake while at the same time not exceeding levels that could contribute to P losses.

### 12.2.4  CONCLUSIONS

The overview indicates that chemical, physical, and biological processes influence the size of the plant-available soil P, the soil P dynamics, and the overall fate of phosphate fertilizer added to soils. An understanding of these processes, the measurement of the size of the various fractions or pools of phosphates in soils, and the rate at which phosphate transfers from one pool to another is important to enhance the efficient use of P resources (e.g., P fertilizers and manure). Phosphorus radioisotope ($^{32}P$ and $^{33}P$) tracers which will be reviewed in the next section (Section 12.3) may be used in conjunction with routine P soil analyses to provide a greater understanding of the interactions of P from different soil P pools that can affect the availability of $P_i$ for plant uptake in soil–plant systems and the management factors that improve the PUE and enhance both the sustainability of agriculture and environmental quality.

## 12.3  BASIC PRINCIPLES OF PHOSPHORUS RADIOISOTOPES

### 12.3.1  PHOSPHORUS RADIOISOTOPES

Isotopes are defined as atoms of the same atomic number ($Z$) but of different atomic weight ($A$). The atom is the basic structural unit of matter and has a set number of positively charged protons and neutral neutron in the nucleus (Nguyen et al. 2011).

The number of protons ($Z$) and neutrons ($N$) present in the nucleus is referred to as the mass number (or also termed *atomic mass* or *atomic weight*, $A = Z + N$), while the number of protons ($Z$) is termed *atomic number*. An atom of a given element may have two or more isotopes. Isotopes may exist in both stable and unstable (radioactive) forms, depending on the stability of the nucleus in an atom. In the case of the P atom, it has twenty-three isotopes from $^{24}$P to $^{46}$P and only one of these isotopes is stable ($^{31}$P). The longest-lived radioactive isotopes are $^{33}$P with a half-life of 25.34 days and $^{32}$P with a half-life of 14.26 days. All other isotopes have very short half-lives of less than 2.5 minutes (IAEA 1976, 2001a). Thus, only two radioactive isotopes ($^{32}$P and $^{33}$P) are discussed here since their half-lives are sufficiently long for soil and soil–plant studies (IAEA 1976, 2001a,b; Zaharah and Zapata 2003). Both these isotopes and $^{31}$P stable isotope have the same number of protons ($Z = 15$) but different atomic weight ($A = 31$, 32, and 33, respectively) and different number of neutrons ($N$, defined as the difference between $A$ and $Z$, of 16, 17, and 18, respectively). Both $^{32}$P and $^{33}$P have unstable nuclei that spontaneously emit radiation with beta particles. The radiations emitted can be monitored by means of specific radiation detection devices, for instance, Geiger–Müller counters or liquid scintillation counters for beta emitters. The use of $^{32}$P and $^{33}$P requires the consideration of radiation protection and safety procedure for both personnel and equipment and chemical storage/disposal according to international standards and the National Radiation Protection Authorities (IAEA 1996, 1999, 2001a,b, 2004, 2005, 2006; IAEA and ILO 1999).

The international unit of radioactivity is the becquerel (Bq), which is equal to one nuclear disintegration per second (dps). The old unit commonly used was the curie, which is equivalent to $3.7 \times 10^{10}$ dps or $3.7 \times 10^{10}$ Bq (Axmann and Zapata 1990). The radioactivity at time $t$ is expressed as follows:

$$A = A_o e^{-(0.693t/T_{1/2})}.$$

where $e$ is the base of the natural logarithm, $A_o$ is the initial radioactivity, $t$ is the time elapsed since the initial activity was measured, and $T_{1/2}$ is the half-life of the phosphorus radioisotopes used. The half-lives of $^{32}$P and $^{33}$P are short (14.26 and 25.34 days, respectively), and the radiation energy of $^{32}$P and $^{33}$P is 1.71 and 0.25 million eV, respectively. One million electron volts is equal to $1.6 \times 10^{-13}$ J and is equivalent to the kinetic energy acquired by an electron accelerated through a potential difference of 1 million V.

The specific radioactivity (SR), also referred to as *specific activity* (SA), is used to describe the amount of radioactivity per unit of material added (e.g., $^{32}$P or $^{33}$P added to soils as labeled fertilizer products or plant materials) as well as the amount of radiotracer per unit of the element being traced. For instance, when using $^{32}$P in a tracer experiment, the SR of $^{32}$P in the soil or in the plant material is becquerel $^{32}$P per 1 g of soil or per 1 g of plant dry matter (DM), and the SR of $^{32}$P in the soil P or in the plant P is becquerel $^{32}$P per mg P (becquerel $^{32}$P mg$^{-1}$ P).

When $^{32}$P or $^{33}$P is applied without the addition of stable (nonradioactive) $^{31}$P, it is described as a carrier-free $^{32}$P (or carrier-free $^{33}$P) isotope. When $^{32}$P or $^{33}$P is accompanied by a $^{31}$P stable isotope, it is termed *carrier $^{32}$P* or *carrier $^{33}$P*. In this case, the SA of $^{32}$P or $^{33}$P is the ratio of $^{32}$P or $^{33}$P radioactive atoms to carrier $^{31}$P atoms (IAEA 1976).

### 12.3.2 MEASUREMENTS OF PHOSPHORUS RADIOACTIVITY

Liquid scintillation counting is the most common method of counting beta emitters such as $^{32}$P or $^{33}$P. Using this technique, the sample to be counted is placed in counting vials containing a scintillation cocktail. This cocktail is a mixture of organic solvents such as toluene or dioxane and organic scintillators such as 1,4-bis-2-(5-phenyloxazolyl)benzene (POPOP). The major part of the liquid scintillation cocktails, the so-called emulsifying cocktails, contain a combination of surfactants (detergents) to be able to hold aqueous samples (IAEA 1976, 2001b). Liquid scintillation counting instruments often termed *liquid scintillation counters* (LSC) measure the photons that are generated when the beta particles of $^{32}$P or $^{33}$P radioactive samples interact with the scintillator. The counting rate or pulses recorded by the LSC depends not only on the total activity present in the radioactive samples but also on the efficiency of the energy transfer from the radioactive samples to the solvent and scintillator molecules.

Any process that reduces the light output from the scintillator is termed *quenching*. Various types of quenching such as self-quenching, color quenching, chemical quenching, and chemiluminescence have been described (Faires and Boswell 1981). It is beyond the scope of this review to provide further details concerning this aspect. However, it is important to take into account this quenching effect (termed *quenching correction*) when assessing the radioactivity (counts) of the samples (IAEA 2001b; L'Annunziata 2003).

### 12.3.3 PRINCIPLES INVOLVED IN THE USE OF PHOSPHORUS RADIOISOTOPES AS TRACERS

Phosphorus radioactive isotopes are used as tracers to assess soil P availability, study soil P dynamics in the presence and the absence of applied P, and evaluate the P recovery of labeled P fertilizers by plants as affected by soil type and soil–water–fertilizer management practices (e.g., timing, method, type). The two major principles involved in the use of $^{32}$P or $^{33}$P radioisotopes as tracers in soil dynamics and soil–plant interactions are isotope dilution technique and inverse dilution technique. In the isotopic dilution technique, $^{32}$P (or $^{33}$P) carrier-free (i.e., containing no $^{31}$P) solutions or $^{32}$P-labeled orthophosphate solutions or $^{32}$P-labeled superphosphate fertilizers (i.e., containing $^{31}$P) are added to soils as tracers, and the SRs of radioactive isotopes in soils and plants are determined. In contrast, a reverse dilution technique is used where a materials such as PRs, farmyard manure, and compost cannot be homogeneously labeled with $^{32}$P/$^{33}$P solutions because of the changes induced in their physical and chemical characteristics during labeling (IAEA 2001b). In this reverse dilution technique, carrier-free $^{32}$P solution is added to the soil or the soil–plant systems. After an equilibrium period, PRs, crop residues, manure, or compost are then applied. In treatments where PRs or composts are readily released phosphate into soils, the phosphate released substantially dilutes the SA of $^{32}P$ in soils and plants. The reduction in the SA thus reflects the release of P from these sources (Kucey and Bole 1984; Zapata et al. 1986; Kato et al. 1995; Zapata and Axmann 1995; Fardeau et al. 1996).

### 12.3.3.1  Isotopic Dilution Technique of Radioactive Phosphorus (P) to Quantify Available Soil Phosphorus

Several studies have applied the principle of isotopic dilution technique to measure the quantity of available P in soils. Depending on the conditions under which this isotopic dilution technique is conducted, the quantity value is referred to as the *E* value (or exchangeable P), the *L* value, or the *A* value. The *E* value is obtained from laboratory studies in the absence of growing plants (McAuliffe et al. 1947; Wiklander 1950). The *L* value (or labile P) and the *A* value are obtained from studies using plants under glasshouse (Larsen 1952) and field (Fried and Dean 1952; Fried 1964) conditions, respectively.

Comparative studies of the amount of soil-available P (quantity factor) in different soils that have been treated with and without fertilizer P sources have been made by various researchers (Fried 1964; Fardeau and Jappe 1976). Good correlations between the *E* and *L* values were reported by Fardeau and Jappe (1976), and both the E and L values were found to be better in assessing soil P availability than the *A* values (Ipinmidun 1973; Reddy et al. 1982).

#### 12.3.3.1.1  E Value or Isotopically Exchangeable Soil Phosphorus

The *E* value is an estimation of the amount of P that is immediately available to plants by measuring the isotopically exchangeable P pool in a soil, i.e., the solid-phase P is in equilibrium with the P in the solution at a given time (McAuliffe et al. 1947; Russell et al. 1957; Russell 1959). Unlike the *A* and *L* value methods, the *E* value method involves the thorough mixing of $^{32}P$ isotope with the soil under laboratory conditions. The change in the SR of the $^{32}P$ is used to determine the isotopically exchangeable P (IAEA 2001b). This method assumes that the $^{32}P$ and the $^{31}P$ come to equilibrium, which is known not to happen, but this is not a severe limitation of the method.

The *E* value for a $^{32}P$ carrier solution is calculated as follows:

$$E = \left( \frac{r_i}{r_f} \times {}^{31}P_f \right) - {}^{31}P_i$$

where $r_i$ and $r_f$ are the initial and final activities (KBq/mL) of $^{32}P$ in the soil solution, respectively, and $^{31}P_i$ and $^{31}P_f$ are the initial and final solution P concentrations, respectively, expressed as microgram P per gram soil.

The *E* value may not be similar to the L value (Section 12.3.3.1.2) since the *E* value does not reflect the same conditions as those of the *L* value in which the soil–plant interactions on soil P availability are taken into account. The *E* value has been reported to be less successful in soils with high P retention (Amer et al. 1969) and also in soils with very low P levels such as soils in the tropics (Wolf et al. 1986; Salcedo et al. 1991). Buehler et al. (2003) suggested that the discrepancies between the *E* and *L* values could be due to (1) the specific sorption of the labeled phosphate on soil surfaces, (2) the difficulty of accurately determining very small amounts of inorganic P ($P_i$), and (3) the assumed isotopic equilibrium had not been reached between the added $P_i$ and the soil phosphate at a given time. In addition, the interference of seed P was also cited as a possible factor (Amer et al. 1969; Truong and Pichot 1976).

### 12.3.3.1.2  L Value or Labile Phosphorus

The $L$ value is the amount of available soil P determined under glasshouse conditions where the added $^{32}P$ is thoroughly mixed with the soil (Larsen 1952). The $L$ value is based on the assumption of an equilibrium being attained between the added $^{32}P$ and the exchangeable P in the soil (Larsen 1952, 1967). Such equilibrium may not be established, particularly in soils where P fertilizer has been recently applied. If this condition is not met, the $L$ value measured long after labeling may overestimate the exchangeable P value (Di et al. 1997). For tropical acid soils, Buehler et al. (2003) reported that the $E$ and $L$ values were suitable for ranking treatments but not precise enough to identify plant species or cultivars in terms of their ability to take up P from slowly or nonexchangeable P pools or to quantify the rate of P mineralization in SOM.

The $L$ value (milligram P/100 g soil) is calculated as follows:

$$L = \text{amount of P fertilizer added} \times \left( \frac{S_A \text{ fertilizer}}{S_A \text{ plant}} - 1 \right)$$

where $S_A$ (kilobecquerel per milligram P) is the specific activity of $^{32}P$ in fertilizer and in the plant DM.

### 12.3.3.1.3  A Value or Available Phosphorus

The $A$ value concept was developed by Fried and Dean (1952). The basic assumption of the $A$ value is that when a plant is presented with two sources of P (soil and fertilizer), it will absorb P in direct proportion to the respective amount available from each source. The measurement relies on determination of the SA of a P fertilizer added to the soil and the consequent SA of the plants grown in that soil.

The A value is determined as follows:

$$A = \frac{B(\text{Fraction of plant P derived from soil})}{\text{Fraction of plant P derived from fertilizer}}$$

or

$$A = B \times \left( \frac{S_A \text{ of fertilizer}}{S_A \text{ of plant}} - 1 \right)$$

where $B$ is the amount of P applied (milligram of fertilizer P/100 g soil) and $S_A$ is the specific activities (kilobecquerel per milligram P) in the fertilizer and in the plant DM.

The *A* value is particularly useful in comparing the P status of different soils, the efficacy of P sources that can be labeled, the residual P from previous P fertilizer applications, and the methods and the time of P applications. The *A* value approach has been found to be very useful to assess the P availability of PRs because it is not possible to label PRs with $^{32}$P in the same manner as with manufactured fertilizers (e.g., superphosphate and TSP) without making significant changes to their properties. Although the *A* values have been used to determine the agronomic effectiveness of PRs (Zapata and Axmann 1991, 1995; Shrivastava et al. 2007), they have been found to vary with many factors such as rates and methods of fertilizer application (Reddy et al. 1982).

### 12.3.3.2   The Isotopic Exchange Kinetics Method for Determining Soil Phosphorus Availability

The isotopic exchange kinetics (IEK) method has been used by various workers (e.g., Fardeau 1993, 1996; Fardeau et al. 1993, 1996; Frossard and Sinaj 1997; Casanova et al. 2002; Owusu-Bennoah et al. 2002; Xiong et al. 2002; Bah et al. 2003; Shrivastava et al. 2007) to assess the P dynamics in soil with or without the addition of P fertilizers and to measure the soil P–fixation capacity (Frossard et al. 1993) and the P availability in the presence or the absence of P fertilizers. This method is based on the transfer of radioactive $^{32}$PO$_4$ (phosphate) ions from the soil solution to the soil solid phase (Fardeau et al. 1985; Fardeau 1993). The IEK takes into account not only the quantity (Q) factor of soil-available P (measured by *E*, *L*, and *A* methods) but also the intensity (I) and capacity (C) factors. Many reports (e.g., Fardeau et al. 1985; Fardeau 1993, 1996; IAEA 2002) have demonstrated that the IEK is valuable to assess the P dynamics in soil with or without the addition of PRs and WSP fertilizers and that there is no single soil chemical P test that can be universally used to estimate plant-available soil.

In the IEK method, the quantity factor is defined as the isotopically exchangeable P $(E_t)$ measured in a soil suspension (water) at a time *t*. The $E_t$ represents the plant-available P reserve (Fardeau and Jappe 1976; Frosssard et al. 1994; Morel and Plenchette 1994). The capacity factor is described by the rate at which P disappears with time from the soil suspension (Frossard et al. 1993), while the intensity factor is represented by the P concentration in the soil suspension. In the IEK approach, soil P is considered as a continuum of P pools and not as a number of distinct pools of plant-available P. The pools release P via one another into the soil solution over time (Frossard et al. 2000, 2011). Many investigators have defined P exchangeability operationally into that exchangeable within discrete time frames. For example, Fardeau et al. (1985) showed that the data derived from the short term (100 min) kinetics of isotopic exchange can be used to derive P exchangeable over at least 3 months by using a model.

According to the IEK method, when $^{32}$P (or $^{33}$P) ions are added to a soil system that is in a steady-state equilibrium, the radioactivity in the solution decreases with time according to the following equation:

$$r_{(t)}/R = (r_{(1)}/R) \times [t + (r_{(1)}/R)^{1/n}]^{-n} + r_{(\infty)}/R$$

where *R* is the total radioactivity introduced; $r_{(1)}$ and $r_{(\infty)}$ are the radioactivities remaining in the solution after 1 min and at infinity, respectively, and *n* is the parameter describing the rate of disappearance of the radioactive tracer from the solution after the first minute of exchange. The $n$ is calculated as the factor (slope) of the linear regression between $\log(r_{(t)}/R)$ and $\log(t)$. The ratio of $r(\infty)/R$ is the proportion of $^{32}P$ in the soil solution when the isotopic equilibrium or the maximum possible dilution is reached (Fardeau et al. 1996).

For isotopic exchange of less than 60 min, the radioactivity in a solution can be simplified as follows because $(r_{(1)}/R)^{1/n}$ and $r(\infty)/R$ can be ignored:

$$r_{(t)}/R = (r_{(1)}/R) \times t^{-n}$$

The quantity, $E_{(t)}$ (milligram P per kilogram soil), of isotopically exchangeable P at time $t$ can be calculated by assuming that $^{31}P$ and $^{32}P$ ions have the same fate in the system, and at time $t$, the SA of the phosphate ions in the soil solution is similar to that of the isotopically exchangeable phosphate ions in the whole system:

$$r_{(t)}/(a \times C_p) = R/E_{(t)}$$

where $a$ represents the soil/solution ratio and $C_p$ is the WSP (milligram P per liter). The term $a \times C_p$ is equivalent to the WSP in the soil expressed as milligram P per kilogram soil.

$E_{(t)}$ can be obtained as follows:

$$E_{(t)} = E_{1min} t^n$$

For $t = 1$ min, $E_{1min} = a \times C_p/(r_{(1)}/R)$.

$r_{(1)}/R$ is the ratio of the radioactivity remaining in the solution after 1 min to the total radioactivity introduced and is considered as an estimate of the P-buffering capacity of the soil (Tran et al. 1988; Salcedo et al. 1991; Frossard et al. 1992, 1994b).

The two major features of the IEK method are that it can be used to generate multiple estimates of P availability from only one IEK measurement and there is no requirement for the system to reach equilibrium (Morel et al. 1994). However, the amount of added tracer must be smaller in comparison with the total pool of P that receives the tracer (Fardeau et al. 1996). In addition, for soils in the tropics with very little P present, the accurate determination of this small amount of P may be a problem (Buehler et al. 2003), although improved methods for measuring very small concentrations of P have been developed (Ohno and Zibilski 1991; Fardeau 1993).

The results of the IEK experiments suggest that it is not possible to divide the soil total $P_i$ into an available $P_i$ pool and a nonavailable $P_i$ pool since most of the $P_i$ can become available at some point in time. Although some soil $P_i$ is rapidly exchangeable with $P_i$ in the solution and is therefore rapidly available, most of inorganic $P_i$ is slowly exchangeable (Fardeau 1993). These results are in agreement with those obtained from the long-term sorption experiments (Barrow 1974, 1983, 1991; Barrow and Shaw 1975a,b) that show that $P_i$ ions located on the solid phase of the soil are distributed along a continuum of solubility, some being in rapid equilibrium with $P_i$ in the solution and some being in very slow equilibrium with $P_i$ in the soil solution.

The soil P–fixation capacity can also be determined from the $^{32}$P IEK measurements (Frossard et al. 1993). The evaluation made after about 1 month of incubation of soil P fertilizer can be used to predict the agronomic effectiveness of P fertilizers (Morel and Fardeau 1991). Furthermore, the method can be used to identify various kinetic pools of soil P and their changes with time (Fardeau 1993; Fardeau et al. 1995).

### 12.3.3.3 Direct and Indirect (Reverse) Isotopic Dilution Technique of Radioactive Phosphorus to Investigate Phosphorus Utilization in Soil–Plant Systems

#### 12.3.3.3.1 Direct Isotopic Dilution Technique of Radioactive Phosphorus

In the direct isotopic dilution technique, the $^{32}$P tracer (e.g., $^{32}$P-labeled superphosphate) is assumed to be uniformly mixed in the particular pool of material (soil P) and that the rates with which the $^{32}$P participate in the various processes are not substantially different from the unlabeled material ($^{31}$P). The P fertilizer utilization or efficiency (PFU) can be calculated from the total $^{32}$P uptake by plants divided by the total $^{32}$P applied in $^{32}$P-labeled fertilizer:

$$PFU = (S_{Rp} \times P_{py})/(S_{Rf} \times P_{fa})$$

where $S_{Rp}$ is the SA of $^{32}$P in the plants, $P_{py}$ (gram P) is the total plant P uptake, $S_{Rf}$ is the SA of the fertilizer applied, and $P_{fa}$ is the total amount of $^{32}$P in plants. The SA of the fertilizer applied ($S_{Rf}$) is equal to the amount (becquerel) of $^{32}$P in the applied P fertilizer ($^{32}$P$_f$) divided by the amount (gram) of P applied ($P_f$) in fertilizer:

$$S_{Rf} = {}^{32}P_f/P_f$$

The SA of P in the plants ($S_{Rp}$) is equal to the amount (becquerel) of plant $^{32}$P ($^{32}$P$_p$) divided by the total plant P uptake ($P_p$):

$$S_{Rp} = {}^{32}P_p/P_p$$

The proportions (%) of P in the plant derived from the $^{32}$P or $^{33}$P-labeled fertilizer material ($P_{dff}$) and from the soil ($P_{dfs}$) are as follows:

$$\%P_{dff} = (S_A \text{ in plants}/S_A \text{ in labeled fertilizer}) \times 100$$

$$\%P_{dfs} = 100 - \%P_{dff}$$

#### 12.3.3.3.2 Reverse or Indirect Isotope Dilution Method

The reverse dilution method involves labeling the studied soil with a carrier-free $^{32}$P–labeled solution (e.g., $KH_2PO_4$ or $NaH_2PO_4$ solution with low P concentrations of 10–50 mg P/L and containing $4–7 \times 10^6$ Bq $^{32}$P/kg soil or $7–18 \times 10^6$ Bq $^{32}$P/m$^2$; Zapata 1990) and using the plant to measure the changes in the SA of the $^{32}$P supplied by the labeled soil. In this reverse dilution method, the dilution of $^{32}$P in the labeled soil by $^{31}$P released from the unlabeled P source (e.g., PRs, manure, or compost) is monitored by measuring the changes in the SR of $^{32}$P in plants growing in the soil. A control treatment (without application of the unlabeled source) is also required. In the reverse dilution, the SR of the control is always higher than that of the fertilized treatments as the SR in the fertilizer treatments is reduced by the P released from the unlabeled P sources. In contrast, in the direct dilution technique, the control treatment (C), which receives no isotope, has an SR of 0.

The $^{32}$P in the plant can only be derived from the $^{32}$P-labeled soil-available P pool, and the PFU by plants is calculated as follows:

$$PFU = P_{Ppff} \times P_{py}/P_{fa}$$

where $P_{Ppff}$ is the proportion of the total plant P derived from the fertilizer, $P_{py}$ (gram P) is the total plant P uptake, and $P_{fa}$ is the total amount of $^{32}$P applied.

The proportion of the total plant P derived from the fertilizer ($P_{Ppff}$) is:

$$P_{Ppff} = Ppff/Pp = (Pp - Ppfa)/Pp$$

where $Ppff$ is the plant P concentration derived from the applied fertilizer, $Pp$ is the P concentration in plants, and $Ppfa$ is the plant P concentration derived from the available soil P pool.

### 12.3.3.3.2.1 Specific Activities of Plant Phosphorus in the Control Treatment

In the control-treated plants (plants receiving no fertilizers), the values of $^{32}$P and P concentration ($^{32}P_{pc}$ and $P_{pc}$) are assumed to provide an ongoing estimate of the SR of the available soil P ($S_{Ra}$) as follows:

$$^{32}P_{pc}/P_{pc} = S_{Rpc} = S_{ra}$$

where $S_{Rpc}$ is the SA of P in the plants taken from the control treatment.

### 12.3.3.3.2.2 Specific Activities of Phosphorus in the Fertilizer-Treated Plants and the Calculated Proportion of the Total Plant P Derived from the Fertilizer

For the fertilizer-treated plants receiving nonlabeled P fertilizer of $P_f$ concentration (containing no $^{32}$P and hence no SA; $S_{Rf} = 0$), the SA of the fertilizer-treated plants ($S_{Rp}$) is as follows:

$$S_{Rp} = {^{32}P_p}/P_p$$

$$P_p = P_{pff} + P_{pfa}$$

where $^{32}Pp$ is the activity (becquerel) of $^{32}$P in plants, $P_p$ is the P concentration in plants, $P_{pff}$ is the plant P concentration derived from the applied fertilizer, and $P_{pfa}$ is the P concentration derived from the available soil P pool.

The $^{32}P$ in the plant could only come from the labeled soil-available P pool and must have been accompanied by the corresponding amount of P from that pool. The $P_{pfa}$ is given as follows:

$$P_{pfa} = {}^{32}P_p/S_{Ra}$$

Thus, the $P_{Ppff}$ can be rewritten as follows:

$$P_{Ppff} = (P_p - {}^{32}P_p/S_{Ra})$$

$$P_p = 1 - S_{Rp}/S_{Rpc}$$

### 12.3.3.4 The Use of Isotopic Dilution Technique to Investigate Phosphorus Mineralization

The isotopic dilution technique has also been used to quantify the amount of P mineralized from SOM and from microbial P as reported in the review conducted by Frossard et al. (2011). This technique involves the labeling of isotopically exchangeable soil $P_i$ with $^{32}P_i$ or $^{33}P_i$ in sterile and nonsterile soils. Organic P mineralization is then calculated by subtracting the amount of isotopically exchangeable $P_i$ related solely to the physicochemical reactions obtained from a short-term batch experiment using sterile soils and extrapolated for a period of 7–14 days from the amount of isotopically exchangeable $P_i$ measured after a 7–14-day incubation experiment (nonsterile soils), in which physicochemical and biological reactions were allowed to take place. The experiment should be done under the following conditions: (1) a steady-state equilibrium for $P_i$, (2) no recent or freshly inputs of rapidly degradable OM, and (3) the incubation period would be 7–14 days to avoid remineralization of organic P compounds that would have been labeled with $^{32}P_i$ or $^{33}P_i$.

The amount of P immobilized in soil microbial biomass during the incubation is calculated from the difference between the amount of $^{32}P_i/^{33}P_i$ and that of $^{31}P$ in microbial biomass, while the net P mineralization is the difference between the gross P mineralization and the microbial immobilization. Using the isotopic technique, Bünemann et al. (2007) reported the gross and net mineralization rates of 0.9–1.2 mg P kg$^{-1}$ soil day$^{-1}$ and 0.5–0.9 mg P kg$^{-1}$ soil day$^{-1}$, respectively. This gross mineralization rate was equivalent to 8–11% of the amount of exchangeable $P_i$ within a 24-hour incubation period. Similar findings (5–9% of the amount of exchangeable $P_i$) were reported by Oehl et al. (2004).

Phosphorus radioisotopes have also been found to be useful in incubation studies (70–154 days) to elucidate the role of soil microbial biomass in soil P cycling (Oehl et al. 2001a; Bünemann et al. 2007; Achat et al. 2009; Oberson et al. 2011). The turnover rate of microbial P depends on the cropping systems and the addition of glucose and nitrogen. In the study conducted by Achat et al. (2009), the gross mineralization rate of the total organic P and the increase in $P_i$ concentration in the soil solution were essentially related to the mineralization of microbial P. Although long-term incubations can be difficult to manage (e.g., maintaining constant soil moisture and temperature conditions), they are useful in investigating the P fluxes and the rate of organic P mineralization from dead OM (Frossard et al. 2011).

### 12.3.3.5 The Use of Phosphorus Radioisotopes to Investigate Phosphorus Dynamics in Soils Receiving Different Phosphorus Sources and Phosphorus Acquisition by Plants as Influenced by Mycorrhizae and Farm Management Practices

Phosphorus radioisotopes ($^{32}$P or $^{33}$P) have been used under both laboratory and glasshouse experiments to investigate P dynamics in soils treated with PRs (Zapata 1990; Zapata and Axmann 1995; Zaharah and Zapata 2003) and green manure (Daroub et al. 2000; Bah et al. 2006). Information obtained together with agronomic data on the effectiveness of different types of PRs relative to water-soluble phosphate fertilizers (e.g., TSP) in a range of soils with different soil P statuses (e.g., Zapata 1995, 2002; Rajan et al. 1996; IAEA 2000, 2002; Owusu-Bennoah et al. 2002; Rajan and Chien 2003; Zapata and Roy 2004; Shrivastava et al. 2007) have been compiled into a PR database, and a website decision-support system for the direct application of PR was developed in a joint venture with the International Fertilizer Development Center (http://www-iswam.iaea.org/dapr/srv/en/about) to assist land managers and policy makers on the economic use of PRs for sustainable agricultural production.

Phosphorus radioisotopes have also been used to measure soil P acquisition by mycorrhizae mycorrhizal fungi (Joner and Jakobsen 1995; Ravnskov and Jakobsen 1995; Schweiger and Jakobsen 1999; Wang et al. 2002; Jansa et al. 2005; Cameron et al. 2007; Jansa et al. 2011), to investigate the main processes occurring in the root–soil interface (rhizosphere) that determine P acquisition by different crop genotypes (IAEA 2013) and to evaluate the effects of different tillage practices on crop residues (Daroub et al. 2000).

### 12.3.4 Limitations in the Use of Phosphorus Radioisotopes as Tracers

Phosphorus radioisotopes ($^{32}$P and $^{33}$P) are not appropriate for long-term studies of soil $P_i$ and $P_o$ dynamics and the fate of P fertilizers in agricultural catchments because of their short half-lives (14.26–25.34 days) and a large-scale radiation protection and safety issues required (Nguyen et al. 2011). Frossard et al. (2011) in their review have identified limitations and practical solutions relating to the measurement of isotopically exchangeable phosphate in highly weathered tropical soils with low $P_i$ concentrations in the soil solution. Although these soils can have high P retention, there is no significant isotopic discrimination between added $^{32}P_i$ or $^{33}P_i$ and $^{31}P_i$ in the soil solution during isotopic exchange kinetic experiments. The most important consideration is the determination of low $P_i$ concentrations in the soil solution using the analytical approaches proposed by Frossard et al. (2011).

Since the fluxes and the exchange of different P pools can occur at different timescales (Section 12.3.3.2) and these can ultimately influence the plant-available soil P and the long-term fate of applied P sources in soils, it is important to consider the use of not only phosphorus radioisotopes but also oxygen stable isotope ratios in various soil P pools as a tracer to understand the P cycling in soils. Various studies (Tamburini et al. 2010; Angert et al. 2011; Jaisi and Blake 2014; Stout et al. 2014; Roberts et al.

2015; Wu et al. 2015) have indicated that the isotopic ratio (R) of oxygen-18 ($^{18}O$) to oxygen-16 ($^{16}O$) in phosphate ($\delta^{18}O$–P) of various soil P pools and in plants can be used as a promising tracer to investigate soil P transformations including phosphorus (P) recycling by soil microorganisms, plant P uptake, P sorption–desorption, P precipitation, and release by oxides and minerals, thus increasing our understanding on the P cycling and the transfer of P in terrestrial ecosystems.

The differences in the isotopic ratio (R) of the heavy ($^{18}O$) to light ($^{16}O$) isotopes between the sample (R sample) and the standard (R standard), are reported as the delta ($\delta$) notation (expressed as parts per thousand or per mile (‰)) deviation from the standard as follows:

$$\delta^{18}O \text{ sample} = [(^{18}O/^{16}O \text{ sample})/(^{18}O/^{16}O \text{ standard}) - 1] \times 1000$$

$$\delta^{18}O \text{ sample} = [(R \text{ sample}/R \text{ standard}) - 1] \times 1000$$

R standard is the $^{18}O/^{16}O$ in the international standard Vienna Standard Mean Ocean Water (Nguyen et al. 2011).

Gross et al. (2015) have reported that $\delta^{18}O$–P can be used to calculate the fraction of the soil P that is turned over by the soil microbial biomass and provide additional data on soil P dynamics. Similarly Angert et al. (2011) have shown that $\delta^{18}O$–P in resin-extractable soil P can be a marker for the rate of biological P transformation. In recent comprehensive reviews, both Frossard et al. (2011) and Tamburini et al. (2014) have emphasized the potential importance of the $\delta^{18}O$–P technique to provide important information on biological processes and to trace the origin and the fate of P in soil–plant systems. However, they have emphasized the need for further development to consolidate this $\delta^{18}O$–P technique with regards to the information on the kinetics of equilibrium between the $\delta^{18}O$–P in different soil P compartments, the proper inventory of the $\delta^{18}O$–P of different $P_i$ and $P_o$ pools in the soil, the $\delta^{18}O$–P in different P sources and waters, and the effects of different parameters such as atmospheric humidity, temperature, and supply of P, $CO_2$, and water on the $\delta^{18}O$–P of metabolic P and structural P in plants.

## 12.4 SUMMARY AND CONCLUSIONS

Phosphorus radioisotopes ($^{32}P$ and $^{33}P$) have been used to measure P processes such as P dynamics, P kinetics, mineralization of soil $P_o$, mycorrhizal acquisition of $P_i$ and to investigate the effects of farming practices such as P sources, green manure addition, and soil cultivation on soil $P_i$ and $P_o$. Since $^{32}P$ and $^{33}P$ are radioactive, their use in field experiments are limited in terms of quantifying on-farm P fluxes from different P sources as influenced by farm management practices. Although there exist limitations relating to the use of short-lived $^{32}P$ and $^{33}P$ in P cycling studies as comprehensively reviewed by Frossard et al. (2011), these radioisotopes can be used in laboratory glasshouse experiments to investigate P transformation processes, and the information obtained can be used in modeling studies and routine soil chemical analyses to estimate long-term plant-available soil P under the influence of the amount (freshly and previously applied P fertilizers) and the type (WSP fertilizers,

PRs, and green manure) of P sources, soil–water–nutrient management practices, crop varieties/genotypes, and changes in environmental conditions (e.g., drying-flooding; Nguyen et al. 1997; Nguyen 2000).

## REFERENCES

Achat, D. L., Bakker, M. R., and Morel, C. 2009. Process-based assessment of P availability in a low P-sorbing forest soil using isotopic dilution methods. *Soil Science Society of America Journal* 73: 2131–2142.

Adams, M. A., and Pate, J. S. 1992. Availability of organic and inorganic forms of phosphorus to lupins. *Plant and Soil* 145: 107–113.

Agbenin, J. O., Iwuafor, E. N. O., and Ayuba, B. 1999. A critical assessment of methods for determining organic phosphorus in savana soils. *Biology and Fertility of Soils* 28: 177–181.

Amer, F., Mahdi, S., and Alradi, A. 1969. Limitations in isotopic measurements of labile phosphate in soils. *Journal of Soil Science* 20: 91–100.

Angert, A., Weiner, T., Mazeh, S., Tamburini, F., Frossard, E., Bernasconi, S. M., and Sternberg, M. 2011. Seasonal variability of soil phosphate stable oxygen isotopes in rainfall manipulation experiments. *Geochimica et Cosmochimica Acta* 75: 4216–4227.

Axmann, H., and Zapata, F. 1990. Stable and radioactive isotopes. In Hardarson, G. (ed.), *Use of Nuclear Techniques in Studies of Soil-Plant Relationships*. Training Course Series 2, pp. 9–34. International Atomic Energy Agency, Vienna.

Bah, A. R., Zaharah, A. R., Hussin, A., Husni, M. H. A., and Halimi, M. S. 2003. Phosphorus status of amended soil as assessed by conventional and isotopic methods. *Communications in Soil Science and Plant Analysis* 34: 2659–2681.

Bah, A. R., Zaharah, A. R., and Hussin, A. 2006. Phosphorus uptake from green manures and phosphate fertilizers applied in an acid tropical soil. *Communications in Soil Science and Plant Analysis* 37: 2077–2093.

Barrow, N. J. 1974. The slow reactions between soil and anions: 1. Effects of time, temperature and water content of a soil on the decrease in effectiveness of phosphate for plant growth. *Soil Science* 118: 380–386.

Barrow, N. J. 1983. A mechanistic model for describing the sorption and desorption of phosphate by soil. *Journal of Soil Science* 34: 733–750.

Barrow, N. J. 1991. Testing a mechanistic model: XI. The effects of time and of level of application on isotopically exchangeable phosphate. *Journal of Soil Science* 42: 277–288.

Barrow, N. J., and Shaw, T. C. 1975a. The slow reactions between soil and anions. 5. Effects of period of prior contact on the desorption of phosphate from soils. *Soil Science* 119: 311–320.

Barrow, N. J., and Shaw, T. C. 1975b. The slow reactions between soil and anions: 3. The effects of time and temperature on the decrease in isotopically exchangeable phosphate. *Soil Science* 119: 190–197.

Beck, M. A., and Sanchez, M. P. A. 1994. Soil phosphorus fraction dynamics during 18 years of cultivation on a Typic Paleudult. *Soil Science Society of America Journal* 58: 1424–1431.

Bertrand, I., McLaughlin, M. J., Holloway, R. E., Armstrong, R. D., and McBeath, T. 2006. Changes in P bioavailability induced by liquid and powder sources of P, N and Zn fertilisers in alkaline soils. *Nutrient Cycling in Agroecosystems* 74: 27–40.

Bolan, N. S. 1991. A critical review on the role of mycorrhizal fungi in the uptake of phosphorus by plants. *Plant and Soil* 134: 189–207.

Bolland, M. D. A., Kumar, V., and Gilkes, R. G. 1994. A comparison of five soil phosphorus tests for five crop species for soil previously fertilized with super phosphate and rock phosphate. *Fertilizer Research* 37: 125–132.

Buehler, S., Oberson, A., Sinaj, S., Friesen, D. K., and Frossard, E. 2003. Isotope methods for assessing plant available phosphorus in acid tropical soils. *European Journal of Soil Science* 54: 605–616.

Bünemann, E., and Condron, L. 2007. Phosphorus and sulphur cycling in terrestrial ecosystems. In Marschner, P., and Rengel, Z. (eds.), *Nutrient Cycling in Terrestrial Ecosystems*, pp. 65–94. Springer-Verlag, New York.

Bünemann, E. K., Marschner, P., McNeill, A. M., and McLauchlan, M. J. 2007. Measuring rates of gross and net mineralisation of organic phosphorus in soils. *Soil Biology and Biochemistry* 39: 900–913.

Cameron, D. D., Johnson, I., Leake, J. R., and Read, D. J. 2007. Mycorrhizal acquisition of inorganic phosphorus by the green-leaved terrestrial orchid Goodyera repens. *Annals of Botany* 99: 831–834.

Carpenter, S. R., Caraco, N. F., Correll, D. L., Howarth, R. W., Sharpley, A. N., and Smith, V. H. 1998. Nonpoint pollution of surface waters with phosphorus and nitrogen. *Ecological Applications* 8: 559–568.

Casanova, E., Salas, A. M., and Toro, M. 2002. Evaluating the effectiveness of phosphate fertilizers in some Venezuelan soils. *Nutrient Cycling in Agroecosystems* 63: 13–20.

Chen, C. R., Condron, L. M., Davis, M. R., and Sherlock, R. R. 2000. Effects of afforestation on phosphorus dynamics and biological properties in a New Zealand grassland soil. *Plant and Soil* 220: 151–163.

Chen, C. R., Condron, L. M., Davis, M. R., and Sherlock, R. R. 2002. Phosphorus dynamics in the rhizosphere of perennial ryegrass (*Lolium perene* L.) and radiata pine (*Pinus radiata* D. Don). *Soil Biology and Biochemistry* 34: 487–499.

Chepkowny, C. K., Haynes, R. J., Swift, R. S., and Harrison, R. 2001. Mineralization of soil organic P induced by drying and rewetting as a source of plant-available P in limed and unlimed samples of an acid soil. *Plant and Soil* 234: 83–90.

Condron, L. M., and Goh, K. M. 1989. Effects of long-term phosphatic fertilizer applications on amounts and forms of phosphorus in soils under irrigated pasture in New Zealand. *Journal of Soil Science* 40: 383–395.

Condron, L. M., Davis, M. R., Newman, R. H., and Cornforth, I. S. 1996. Influence of conifers on the forms of phosphorus in selected New Zealand grassland soils. *Biology and Fertility of Soils* 21: 37–42.

Condron, L. M., Frossard, E., Tiessen, H., Newman, R. H., and Stewart, J. W. B. 1990. Chemical nature of organic phosphorus in cultivated and uncultivated soils under different environmental conditions. *Journal of Soil Science* 41: 41–50.

Condron, L. M., Goh, K. M., and Newman, R. H. 1985. Nature and distribution of soil phosphorus as revealed by a sequential extraction method followed by $^{31}$P nuclear magnetic resonance analysis. *Journal of Soil Science* 36: 199–207.

Condron, L. M., Turner, B. L., and Cade-Menun, B. J. 2005. Chemistry and dynamics of soil organic phosphorus. In Sims, J. T., and Sharpley, A. N. (eds.), *Phosphorus: Agriculture and the Environment*. Monograph No. 46, pp. 87–121. American Society of Agronomy, Madison, WI.

Cordell, D., and White, S. 2014. Life's bottleneck: Sustaining the world's phosphorus for a food secure future. *Annual Review of Environment and Resources* 39: 161–188.

Cordell, D., Drangert, J., and White, S. 2009. The story of phosphorus: Global food security and food for thought. *Global Environmental Change* 19: 292–305.

Dai, K. H., David, M. B., Vance, G. F., and Kryszowaska, A. J. 1996. Characterization of phosphorus in a spruce-fir Spodosol by phosphorus-31 nuclear magnetic resonance spectroscopy. *Soil Science Society of America Journal* 60: 1943–1950.

Dalal, R. C. 1977. Soil organic phosphorus. *Advanced in Agronomy* 29: 85–117.

Daroub, S. H., Pierce, F. J., and Ellis, B. G. 2000. Phosphorus fractions and fate of phosphorus-33 in soils under ploughing and no-tillage. *Soil Science Society of America Journal* 64: 170–176.

Dean, L. A., Nelson, W. L., MacKenzie, A. J., Armiger, W. H., and Hill, W. L. 1947. Application of radioactive tracer technique to studies of phosphatic fertilizer by crops. *Soil Science Society of America Proceedings* 12: 107–112.

Di, H. J., Condron, L. M., and Frossard, E. 1997. Isotope techniques to study phosphorus cycling in agricultural and forest soils: A review. *Biology and Fertility of Soils* 27: 1–12.

Dyer, B. 1984. On the analytical determination of probably available "mineral" plant food in soils. *Journal of the Chemical Society* 65: 115–167.

Eisner, J., and Bennett, E. 2011. A broken biogeochemical cycle. *Nature* 489: 29–31.

Faires, R. A., and Boswell, G. G. J. (eds.). 1981. *Radioisotope Laboratory Techniques*. Butterworth-Heinemann Ltd, Oxford.

Fardeau, J. C. 1993. Le phosphore assimilable des sols: Sa représentation par un modèle fonctionnel a plusieurs compartiments. Available soil phosphate: Its representation by a functional multiple compartment model. *Agronomie* 13: 317–331.

Fardeau, J. C. 1996. Dynamics of phosphate in soils: An isotopic outlook. *Fert Res* 45: 91–100.

Fardeau, J. C., and Jappe, J. 1976. Nouvelle méthode de détermination du phosphore assimilable par les plantes: Extrapolation des cinétiques de dilution isotopique. *Comptes Rendus des Séances de l'Académie Sci.* Series D282, pp. 1137–1140.

Fardeau, J. C., Guiraud, G., and Marol, C. 1995. Bioavailable soil P as a key to sustainable agriculture: Functional model determined by isotopes tracers. *Proceedings of the International Symposium on Nuclear and Related Techniques in Soil-Plant Studies in Sustainable Agriculture and Environmental Preservation, IAEA-SM-334/3, 131–144. International Atomic Energy Agency, Vienna.*

Fardeau, J. C., Guiraud, G., and Marol, C. 1996. The role of isotopic techniques on the evaluation of the agronomic effectiveness of P fertilizers. *Fertilizer Research* 45: 101–109.

Fardeau, J. C., Morel, C. H., and Boniface, R. 1993. Cinetiques de transfert des ions phosphates du sol vers la solution. *Agronomie* 11: 787–797.

Fardeau, J. C., Morel, C., and Jappe, J. 1985. Cinétique d'échange des ions phosphate dans les systèmes sol:solution. Vérification expérimentale de l'équation théorique. *Comptes Rendus des Séances de l'Académie Sci.* Series D300, pp. 371–376.

Formoso, M. L. L. 1999. *Workshop on Tropical Soils*. Brazilian Academy of Sciences, Rio de Janeiro.

Fox, T. R., and Comerford, N. B. 1990. Low molecular weight organic acids in selected forest soils of the south-eastern USA. *Soil Science Society of America Journal* 54: 1139–1144.

Fried, M. 1954. Quantitative evaluation of processed and natural phosphates. *Journal of Agricultural and Food Chemistry* 2: 241–244.

Fried, M. 1964. "E", "L" and "A" value, *Transactions of the Eighth International Congress Of Soil Science* 4: 29–39.

Fried, M., and Dean, L. A. 1952. A concept concerning the measurement of available soil nutrients. *Soil Science* 73: 263–271.

Frossard, E., and Sinaj, S. 1977. The isotopic exchange technique: A method to describe the availability of inorganic nutrients: Applications to K, $PO_4$, $SO_4$ and Zn. *Isotope Environment Health Studies* 33: 61–77.

Frossard, E., Achat, D. L., Bernasconi, S. M., Bünemann, E. K., Fardeau, J.C., Jansa, J., Morel, C., Rabeharisoa, L., and Randriam, L. 2011. The use of tracers to investigate phosphate cycling in soil–plant systems. In Bünemann, E. K., Oberson, A., and Frossard, E. (eds.), *Phosphorus in Action: Biological Processes in Soil Phosphorus Cycling*. Springer Science & Business Media, Berlin.

Frossard, E., Condron, L. M., Oberson, A., Sinaj, S., and Fardeau, J. C. 2000. Processes governing phosphorus availability in temperate soils. *Journal of Environmental Quality* 29: 15–23.

Frossard, E., Fardeau, J. C., Brossard, M., and Morel, J. L. 1994. Soil isotopically exchangeable phosphorus: A comparison between E and L values. *Soil Science Society of America Journal* 58: 846–851.

Frossard, E., Fardeau, J. C., Ognalaga, M., and Morel, J. L. 1992. Influences of agricultural practices, soil properties and parent material on the phosphate buffering capacity of cultivated soils developed in temperate climates. *European Journal of Agronomy* 1: 45–50.

Frossard, E., Feller, C., Tiessen, H., Stewart, J. W. B., Fardeau, J. C., and Morel, J. L. 1993. Can an isotopic method allow for the determination of the phosphate-fixing capacity of soils? *Communications in Soil Science and Plant Analysis* 24: 367–377.

Frossard, E., López-Hernández, D., and Brossard, M. 1996. Can isotopic exchange kinetics give valuable information on the rate of mineralization of organic phosphorus in soils? *Soil Biology and Biochemistry* 28: 857–864.

George, T. S., Turner, B. L., Gregory, P. J., Cade-Menum, B. J., and Richardson, A. E. 2006. Depletion of organic phosphorus from Oxisols in relation to phosphate activities in the rhizosphere. *European Journal of Soil Science* 57: 47–57.

Goh, K. M., and Condron, L. M. 1989. Plant availability of phosphorus accumulated from long-term applications of superphosphate and effluent to irrigated pastures. *New Zealand Journal of Agricultural Science* 32: 45–51.

Gressel, N., and McColl, J. G. 1997. Phosphorus mineralization and organic matter decomposition: A critical review. In Cadish, G., and Giller, K. E. (eds.), *Driven by Nature: Plant Litter Quality and Decomposition*, pp. 297–309. CAB International, Wallingford.

Gross, A., Nishri, A., and Angert, A. 2015. What processes control the oxygen isotopes of soil bio-available phosphate? *Geochimica et Cosmochimica Acta* 159: 100–111.

Harrison, A. F. 1987. *Soil Organic Phosphorus—A Review of World Literature*. CAB International, Wallingford.

Hart, M. R., Quin, B. F., and Nguyen, M. L. 2004. Phosphorus runoff from agricultural land and direct fertilizer effects: A Review. *Journal of Environmental Quality* 33: 1954–1972.

Hatfield, J. L., Sauer, T. J., and Prueger, J. H. 2001. Managing soils to achieve greater water use efficiency: A review. *Agronomy Journal* 93: 271–280.

Hayes, J. E., Simpson, R. J., and Richardson, A. E. 2000. The growth and phosphorus utilisation of plants in sterile media when supplied with inositol hexaphosphate, glucose 1-phosphate or inorganic phosphate. *Plant and Soil* 220: 165–174.

Haynes, R. J., and Williams, P. H. 1993. Nutrient cycling and soil fertility in the grazed pasture. *Advances in Agronomy* 49: 119–99.

Hedley, M. J., Stewart, J. W. B., and Chauhan, B. S. 1982. Changes in inorganic and organic soil phosphorus fractions induced by cultivation and by laboratory incubations. *Soil Science Society of America Journal* 46: 970–976.

Hesketh, N., and Brookes, P. C. 2000. Development of an indicator for risk of phosphorus leaching. *Journal of Environmental Quality* 29: 105–110.

Hinedi, Z. R., Chang, A. C., and Yesinowski, J. P. 1989. Phosphorus-31 magic angle spinning nuclear magnetic resonance of wastewater sludges and sludge-amended soil. *Soil Science Society of America Journal* 53: 1053–1056.

Høgh-Jensen, H., Schjoerring, J. K., and Soussana, J. 2002. The influence of phosphorus deficiency on growth and nitrogen fixation of white clover plants. *Annals of Botany* 90: 745–53.

IAEA (International Atomic Energy Agency). 1973. *Radiation Protection Procedures*. Safety Series No. 38. IAEA, Vienna.

IAEA. 1976. *Tracer Manual on Crops and Soils.* Technical Report Series No. 171. IAEA, Vienna.

IAEA. 1996. *International Basic Safety Standards for Protection against Ionising Radiation and for the Safety of Radiation Sources.* IAEA Safety Series No. 115. IAEA, Vienna.

IAEA. 1999. *Occupational Radiation Protection.* IAEA Safety Standards Series No. RS-G-1.1. IAEA, Vienna.

IAEA. 2000. *Management and Conservation of Tropical Acid Soils for Sustainable Crop Production.* IAEA TECDOC 1159. IAEA, Vienna.

IAEA. 2001a. *Building Competence in Radiation Protection and Safe Use of Radiation Sources.* IAEA Safety Standards Series No. RS-G-1.4. IAEA, Vienna.

IAEA. 2001b. *Use of Isotope and Radiation Methods in Soil and Water Management and Crop Nutrition.* IAEA Training Manual Series 14. IAEA, Vienna.

IAEA. 2002. *Assessment of Soil Phosphorus Status and Management of Phosphatic Fertilisers to Optimise Crop Production.* IAEA TECDOC 1272. IAEA, Vienna.

IAEA. 2004. *Application of the Concepts of Exclusion, Exemption and Clearance.* IAEA Safety Standards Series No. RS-G-1.7. IAEA, Vienna.

IAEA. 2005. *Categorization of Radioactive Sources.* IAEA Safety Standards Series No. RS-G-1.9. IAEA, Vienna.

IAEA. 2006. *Fundamental Safety Principles.* IAEA Safety Standards Series No. SF-1. IAEA, Vienna.

IAEA. 2010. IAEA Home. Department of Nuclear Safety and Security, Vienna. Available at http://www.ns-iaea.org/.

IAEA. 2013. *Optimizing Productivity of Food Crop Genotypes in Low Nutrient Soils.* IAEA-TECDOC-1721.IAEA, Vienna.

IAEA and International Labour Organization (ILO). 1999. *Occupational Radiation Protection,* Safety Standards Series No. RS-G-1.1. IAEA, Vienna.

Ipinmidun, W. B. 1973. Assessment of residues of phosphate application in some soils of Northern Nigeria: I. Examination of L and E values. *Plant and Soil* 39: 213–225.

Jaisi, D. P., and Blake, R. E 2014. Advances in using oxygen isotope ratios of phosphate to understand phosphorus cycling in the environment. *Advances in Agronomy* 125: 1–53.

Jansa, J., Finlay, R., Wallander, H., Smith, F. .A., and Smith, S. E. 2011. Role of mycorrhizal symbioses in phosphorus cycling. In Bünemann, E., Oberson, A., and Frossard, E. (eds.), *Phosphorus in Action: Biological Processes in Soil Phosphorus Cycling.* Soil Biology Vol. 26. Springer, Heidelberg.

Jansa, J., Mzafar, A., and Frossard, E. 2005. Phosphorus acquisition strategies within arbuscular mycorrhizal fungal community of a single field site. *Plant Soil* 276: 163–176.

Joner, E. J., and Jakobsen, I. 1995. Terranean clover (*Trifolium subterraneum L.*). *Plant Soil* 172: 221–227.

Joner, E. J., van Aarade, I. M., and Voskata, M. 2000. Phosphatase activity of extra-radical arbuscular mycorrhizal hyphae: A review. *Plant and Soil* 226: 199–210.

Kato, N., Zapata, F., and Axmann, H. 1995. Evaluation of agronomic effectiveness of natural and partially acidulated phosphate rocks in several soils using $^{32}P$ isotopic dilution techniques. *Fertilizer Research* 41: 235–242.

Kleinman, P. J. A., Sharpley, A. N., Moyer, B. G., and Elwinger, G. F. 2002. Effect of mineral and manure phosphorus sources onrunoff phosphorus. *Journal of Environmental Management* 31: 2026–2033.

Kleinman, P. J. A., Sharpley, A. N., McDowell, R. W., Flaten, D. N., Buda, A. R., Liang, T., Bergstrom, L., and Zhu, Q. 2011. Managing agricultural phosphorus for water quality protection: Principles for progress. *Plant and Soil* 349: 169–182.

Kleinman, P., Sullivan, D., Wolf, A., Brandt, R., Dou, Z., Elliott, H., Kovar, J. et al. 2007. Selection of a water-extractable phosphorus test for manures and biosolids as an indicator of runoff loss potential. *Journal of Environmental Quality* 36: 1357–1367.

Koopmans, G. F., Chardon, W. J., Dolfing, J., Oenema, O., van der Meer, P., and van Riemsdijk, W. H. 2003. Wet chemical and phosphorus-31 nuclear magnetic resonance analysis of phosphorus speciation in a sandy soil receiving long-term fertilizer or animal manure applications. *Journal of Environmental Quality* 32: 287–295.

Koopmans, G. F., McDowell, R. W., Chardon, W. J., Oenema, O., and Dolfing, J. 2002. Soil phosphorus quantity-intensity relationships to predict increased soil phosphorus loss to overland and subsurface flow. *Chemosphere* 48: 679–687.

Kröbel, R., Campbell, C. A., Zentner, R. P., Lemke, R., Steppuhn, H., Desjardins, R. L., and De Jong, R. 2012. Nitrogen and phosphorus effects on water use efficiency of spring wheat grown in a semi-arid region of the Canadian prairies. *Canadian Journal of Soil Science* 92: 573–587.

Kucey, R, M. N, and Bole, J. B. 1984. Availability of phosphorus from 17 rock phosphates in moderately and weakly acidic soils as determined by P-32 dilution, A-value, and total P uptake methods. *Soil Science* 138: 180–188.

L'Annunziata, M. F. (ed.). 2003. *Handbook of Radioactivity Analysis*, Second ed. Academic Press Ltd, San Diego, CA.

Lal, R. 1990. Tropical soils: Distribution, properties and management. *Resource Management and Optimization* 7: 39–52.

Lal, R. 2000. Soil management in the developing countries. *Soil Science* 165: 57–72.

Lambers, H., and Shane, M. W. 2007. Role of root clusters in phosphorus acquisition and increasing biological diversity in agriculture. In Spiertz, J. H. J., Struik, P. C., and van Laar, H. H. (eds.), *Scale and Complexity in Plant Systems Research: Gene-Plant-Crop Relations*, pp. 237–250. Springer, Dordrecht.

Langdale, G. W., Hargrove, W. L., and Giddens, J. 1984. Residue management in double-crop conservation tillage systems. *Agronomy Journal* 76: 689–694.

Larsen, S. 1952. The use of $^{32}P$ in studies on the uptake of phosphorus by plants. *Plant and Soil* 4: 1–10.

Larsen, S. 1967. Soil phosphorus. *Advances in Agronomy* 19: 151–210.

Lucci, G. M., McDowell, R. W., and Condron, L. M. 2010. Phosphorus and sediment loads/transfer from potential source areas in New Zealand dairy pasture. *Soil Use and Management* 26: 44–52.

Magid, J., Tiessen, H., and Condron, L. M. 1996. Dynamics of organic phosphorus in the soils under natural and agricultural ecosystems. In Piccolo, A. (ed.), *Humic Substances in Terrestrial Ecosystems*, pp. 429–466. Elsevier, Amsterdam.

Marschner, P. (ed.). 2012. *Marschner's Mineral Nutrition of Higher Plants*, Third edn. Academic Press, London.

McAuliffe, C. D., Hall, N. S., and Dean, L. A. 1947. Exchange reactions between phosphates and soils: Hydroxylic surfaces of soil minerals. *Proceedings Soil Science Society of America* 12: 119–123.

McDowell, R. W., and Sharpley, A. N. 2001. Approximating phosphorus release from soils to surface runoff and subsurface drainage. *Journal of Environmental Quality* 30: 508–520.

McDowell, R. W., Biggs, B. J. F., Sharpley, A. N., and Nguyen, L. 2004. Connecting phosphorus loss from agricultural landscapes to surface water quality: A review. *Chemistry and Ecology* 20: 1–40.

Miller, M. H. 2000. Arbuscular mycorrhizae and phosphorus nutrition of maize: A review of Guelph studies. *Canadian Journal of Plant Science* 80: 47–52.

Morel, C., and Fardeau, J. C. 1991. Phosphorus bioavailability of fertilizers: A predictive laboratory method for its evaluation. *Fertilizer Research* 28: 1–9.

Morel, C., and Plenchette, C. 1994. Is the isotopically exchangeable phosphate of a loamy soil the plant-available P? *Plant and Soil* 158: 287–297.

Morel, C., Tiessen, H., Moir, J. O., and Stewart, J. W. B. 1994. Phosphorus transformations and availability under cropping and fertilization assessed by isotopic exchange. *Soil Science Society of America Journal* 58: 1439–1445.

Motavalli, P. P., and Miles, R. J. 2002. Soil phosphorus fractions after 111 years of animal manure and fertilizer applications. *Biology and Fertility of Soils* 36: 35–42.

Newman, R. H., and Tate, K. R. 1980. Soil phosphorus characterisation by [31]P nuclear magnetic resonance. *Communication in Soil Science and Plant Analysis* 11: 835–842.

Nguyen, M. L. 2000. Phosphate incorporation and transformation in surface sediments of a sewage-impacted wetland as influenced by sediment sites, sediment pH and added phosphate concentration. *Ecological Engineering* 14: 139–155.

Nguyen, M. L., and Goh, K. M. 1990. Accumulation of soil sulphur fractions in grazed pastures receiving long-term superphosphate applications. *New Zealand Journal of Agricultural Research* 33: 111–128.

Nguyen, M. L., and Goh, K. M. 1992. Nutrient cycling and losses based on a mass-balance model in grazed pastures receiving long-term superphosphate applications in New Zealand: 1. Phosphorus. *Journal of Agricultural Science, Cambridge* 119: 89–106.

Nguyen, M. L., Cooke, J. G., and McBride, G. B. 1997. Phosphorus retention and release characteristics of sewage-impacted wetland sediments. *Water, Air and Soil Pollution* 100(1–2): 163–179.

Nguyen, M. L., Haynes, R. J., and Goh, K. M. 1995. Nutrient budgets and status in three pairs of conventional alternative mixed cropping farms in Canterbury, New Zealand. *Agriculture, Ecosystems and Environment* 52: 149–162.

Nguyen, M. L., Rickard, D. S., and McBride, S. D. 1989. Pasture production and changes in phosphorus and sulphur status in irrigated pastures receiving long-term applications of superphosphate fertiliser. *New Zealand Journal of Agricultural Research.* 32: 245–262.

Nguyen, L., Zapata, F., Lal, R., and Dercon, G. 2011. Role of nuclear and isotopic techniques in sustainable land management. In Lal, R., and Stewart, B. A. (eds.), *World Soil Resources and Food Security*, Advances in Soil Science, 345–418. CRC Press, Boca Raton, FL.

Oberson, A., Pypers, P., Bünemann, E. K., and Frossard, E. 2011. Management impacts on biological phosphorus cycling in cropped soils. In Bünemann, E., Oberson, A., and Frossard, E. (eds.), *Phosphorus in Action: Biological Processes in Soil Phosphorus Cycling*. Soil Biology vol. 26. Springer, Heidelberg.

Oehl, F., Frossard, E., Fliessbach A., Dubois D., and Oberson, A. 2004. Basal phosphorus mineralisation in soils under different farming systems. *Soil Biology and Biochemistry* 36: 667–675.

Oehl, F., Oberson, A., Probst, M., Fliessbach, A., Roth, H. R., and Frossard, E. 2001b. Kinetics of microbial phosphorus uptake in cultivated soils. *Biology and Fertility of Soils* 34: 31–41.

Oehl, F., Oberson, A., Sinaj, S., and Frossard, E. 2001a. Organic phosphorus mineralisation studies using isotopic dilution techniques. *Soil Science Society of America Journal* 65: 780–787.

Ohno, T., and Zibilski, L. M. 1991. Determination of low concentrations of phosphorus in soil extracts using malachite green. *Soil Science Society of America Journal* 55: 892–895.

Owusu-Bennoah, E., Zapata, F., and Fardeau, J.C. 2002. Comparison of greenhouse and [32]P isotopic laboratory methods for evaluating the agronomic effectiveness of natural and modified rock phosphates in some acid soils of Ghana. *Nutrient Cycling in Agroecosystems* 63: 1–12.

Perrott, K. W., O'Connor, M. B., and Waller, J. E. 1999. Tree stocking effects on soil phosphorus, soil microbial activity and soil phosphate activity at the Tikitere Agroforestry Research Area. *New Zealand Journal of Forestry Science* 29: 116–130.

Perrott, K. W., Sarathchandra, S. U., and Dow, B.W. 1992. Seasonal and fertilizer effects on the organic cycle and microbial biomass in a hill country soil under pasture. *Australian Journal of Soil Research* 3: 383–394.

Pierzynski, G. M. 1991. The chemistry and mineralogy of phosphorus in excessively fertilized soils. *Critical Reviews in Environmental Control* 21: 265–295.

Rajan, S. S. S., and Chien, S. H. (eds.). 2003. Direct application of phosphate rock and related technology: Latest developments and practical experiences. *International Fertilizer Development Center Proceeding of International Meeting, July 16–20. 2001, Kuala Lumpur.*

Rajan, S. S. S., Watkinson, J. H., and Sinclair, A. G. 1996. Phosphate rocks for direct application to soils. *Advances in Agronomy* 57: 77–159.

Ravnskov, S., and Jakobsen, I. 1995. Functional compatibility in arbuscular mycorrhizas measured as hyphal P transport to the plant. *New Phytologist* 129: 611–618.

Reddy, N. V., Saxena, M. C., and Srinivasulu, R. 1982. E-, L-, and A-values for estimation of plant-available soil phosphorus. *Plant and Soil* 69: 3–11.

Richards, J. E., Daigle, J. Y., Leblanc, P., Paulin, R., and Ghanem, I. 1993. Nitrogen availability and nitrate leaching from organo-mineral fertilizers. *Canadian Journal of Soil Science* 73: 197–208.

Richardson, A. E., George, T. S., Hens, M., and Simpson, R. J. 2005. Utilization of soil organic phosphorus by higher plants. In Turner, B. L., Frossard, E., and Baldwin, D. *Utilization of Soil Organic Phosphorus in the Environment*, pp. 165–184. CABI Publishing, Wallingford.

Richardson, A. E., Lynch, J. P., Ryan, P. R., Delhaize, E., Smith, F. A., Smith, S. E., Harvey, P. R. et al. 2011. Plant and microbial strategies to improve the phosphorus efficiency of agriculture. *Plant and Soil* 349: 121–156.

Roberts, K., Defforey, D., Turner, B. L., Condron, L. M., Peek, S., Silva, S., Kendall, C., and Paytan, A. 2015. Oxygen isotopes of phosphate and soil phosphorus cycling across a 6500 year chronosequence under lowland temperate rainforest. *Geoderma* 257–258: 14–21.

Russell, R. S. 1959. The value of measurements of isotopic dilution in the study of soil/plant relationships. *Zeitschrift für Pflanzenernaerung und Bodenkunde* 84: 63–75.

Russell, R. S., Russell, E. W., and Marais, P. G. 1957. Factors affecting the ability of plants to absorb phosphate from soils: I. The relationship between labile phosphate and absorption. *Journal of Soil Science* 8: 248–267.

Salcedo, I. H., Bertino, F., and Sampaio, E. V. S. B. 1991. Reactivity of phosphorus in north eastern Brazilian soils assessed by isotopic dilution. *Soil Science Society of America Journal* 55: 140–145.

Saunders, W. M. H., and Williams, E. G. 1995. Observations on the determination of total organic phosphorus in soils. *Journal of Soil Science* 6: 254–267.

Schweiger, P. F., and Jakobsen, I. 1999. Direct measurement of arbuscular mycorrhizal phosphorus uptake into field-grown winter wheat. *Agronomy Journal* 91: 998–1002.

Selles, F., Campbell, C. A., Zentner, R. P., Curtin, D., James, D. C., and Basnyat, P. 2011. Phosphorus use efficiency and long-term trends in soil available phosphorus in wheat production systems with and without nitrogen fertilizer. *Canadian Journal of Soil Science* 91: 39–52.

Sharpley, A. N. 1985. Phosphorus cycling in unfertilized and fertilized agricultural soils. *Soil Science Society of America Journal* 49: 905–911.

Sharpley, A. N., and Rekolainen, S. 1997. Phosphorus in agriculture and its environmental implications. In Tunney, H., Carton, O. T., Brookes, P. C., and Johnston, A. E. (eds.), *Phosphorus Loss from Soil to Water*, pp. 1–54. CAB International, Wallingford.

Shrivastava, M., Bhujbal, B. M., and D'Souza, S. F. 2007. Agronomic efficiency of Indian rock phosphates in acidic soils employing radiotracer A-value technique. *Communications in Soil Science and Plant Analysis* 38: 461–471.

Shu, L., Shen, J., Rengel, Z., Tang, C., Zhang, F., and Cawthray, G. R. 2007. Formation of cluster roots and citrate exudation by *Lupinus albus* in response to localized application of different phosphorus sources. *Plant Science* 172: 1017–1024.

Simpson, R. J.; Richardson, A. E., Nichols, S. N., and Crush, J. R. 2014. Pasture plants and soil fertility management to improve the efficiency of phosphorus fertiliser use in temperate grassland systems. *Crop and Pasture Science* 65(6): 556–575.

Stout, L. M., Joshi, S. R., Kana, T. M., and Jaisi, D. P. 2014. Microbial activities and phosphorus cycling: An application of oxygen isotope ratios in phosphate. *Geochimica et Cosmochimica Acta* 138: 101–116.

Sutton, M. A., Howard, C. M., Erisman, J. W., Billen, G., Bleeker, A., Grennfel, P., van Grinsven, H., and Grizzetti, B. 2011. *The European Nitrogen Assessment—Sources, Effects and Policy Perspectives*. Cambridge University Press, Cambridge.

Syers, J. K., Johnston, A. E., and Curtin, D. 2008. Efficiency of soil and fertilizer phosphorus: Reconciling changing concepts of soil phosphorus behaviour with agronomic information. *FAO Fertilizer and Plant Nutrition Bulletin* No. 18. Food and Agriculture Organization, Rome.

Tamburini, F., Bernasconi, S. M., Angert, A., Weiner, T., and Frossard, E. 2010. A method for the analysis of the $\delta 18O$ of inorganic phosphate extracted from soils with HCl. *European Journal of Soil Science* 61: 1025–1032.

Tamburini, F., Pfahler, V., von Sperber, C., Frossard, E., and Bernasconi, S. M. 2014. Oxygen isotopes for unraveling phosphorus transformations in the soil-plant system: A review. *Soil Science Society of America Journal* 78: 38–46.

Tate, K. R., and Newman, R. H. 1982. Phosphorus fractions of a climosequence of soils in New Zealand tussock grassland. *Soil Biology and Biochemistry* 14: 191–196.

Tiessen, H. 1995. *Scope 54: Phosphorus in the Global Environment: Transfers, Cycles and Management*. John Wiley & Sons Ltd, Hoboken, NJ.

Tiessen, H. 2008. Phosphorus in the global environment. In White, P. J., and Hammond, J. P., (eds.), *The Ecophysiology of Plant–Phosphorus Interactions*, pp. 1–7. Springer, Dordrecht.

Tiessen, H., Salcedo, I. H., and Sampaio, E. V. S. B. 1992. Nutrient and soil organic matter dynamics under shifting cultivation in semi-arid Northeastern Brazil. *Agriculture, Ecosystems and Environment* 39: 139–151.

Tran, T. S., Fardeau, J. C., and Giroux, M. 1988. Effects of soil properties on plant-available phosphorus determined by the isotopic dilution phosphorus-32 method. *Soil Science Society of America Journal* 52: 1383–1390.

Truong, B., and Pichot, J. 1976. Influence du phosphore des graines de la plante test sur la détermination du phosphore isotopiquement dilable. *Agronomie Tropicale* 31: 379–386.

Turner, N. C. 2004, November. Agronomic options for improving rainfall-use efficiency of crops in dryland farming systems. Water-Saving Agriculture Special Issue. *Journal of Experimental Botany* 55(407): 2413–2425.

Van Kauwenbergh, S. J. 2010. *World Phosphorus Rock Reserves and Resources*. Technical Bulletin IFDC-T-75. International Fertilizer Development Center, Muscle Shoals, AL.

Van Vuuren, D. P., Bouwman, A. F., and Beusen, A. H. W. 2010. Phosphorus demand for the 1970–2100 period: A scenario analysis of resource depletion: Governance, complexity and resilience. *Global Environmental Change* 20: 428–439.

Wang, B., Funakoshi, D. M., Dalpe, Y., and Hamel, C. 2002. Phosphorus-32 absorption and translocation to host plants by arbuscular mycorrhizal fungi at low root-zone temperature. *Mycorrhiza* 12: 93–96.

Weil, R. R., Benetto, P. W., Sikora, L. J., and Bandel, V. A. 1988. Influence of tillage practices on phosphorus distribution and forms in three Ultisols. *Agronomy Journal* 80: 503–509.

White, R. E. 1980. Retention and release of phosphate by soil and soil constituents. In *Soils and Agriculture, Critical Reports on Applied Chemistry*, Vol. 2, pp. 71–114. Blackwell Scientific Publications, Oxford.

Wiklander, L. 1950. Kinetics of phosphate exchange in soils. *Annual Review of Veterinary College*, Sweden 17: 407–424.

Wolf, A. M., Baker, D. E., and Pionke, H. B. 1986. The measurement of labile phosphorus by the isotopic dilution and anion resin methods. *Soil Science* 141: 60–70.

Wu, J., Paudel, P., Sun, M. J., Joshi, S. R., Stout, L. M., Greiner, R., and Jaisi, D. P. 2015. Mechanisms and pathways of phytate degradation: Evidence from oxygen isotope ratios of phosphate, HPLC, and phosphorus-31 NMR spectroscopy. *Soil Science Society of America Journal* 79: 1615–1628.

Xiong, L. M., Zhou, Z. G., Fardeau, J. C., Feng, G. L., Lu, R. K. 2002. Isotopic assessment of soil phosphorus fertility and evaluation of rock phosphates as phosphorus sources for plants in subtropical China. *Nutrient Cycling in Agroecosystems* 63: 91–98.

Zaharah, A. R., and Zapata, F. 2003. The use of P isotope techniques to study soil P dynamics and to evaluate the agronomic effectiveness of phosphate fertilizers; In Rajan, S. S. S., and Chien, S. H. (ed.), *Proceedings of the International Symposium Direct Application of Phosphate Rock and Related Appropriate Technology*, Malaysia, July 2001, pp. 225–235. International Fertilizer Development Center Publication G-1, Muscle Shoals, AL.

Zapata, F. 1990. Isotope techniques in soil fertility and plant nutrition studies. In Hardarson, G. (ed.), *Use of Nuclear Technique in Studies of Soil-Plant Relationships*, pp. 109–127. IAEA, Vienna.

Zapata, F. (ed.). 1995. Evaluation of the agronomic effectiveness of phosphate fertilizers through the use of nuclear and related techniques. Special issue. *Fert. Res.* 41: 167–242.

Zapata, F. (ed.). 2002. Utilisation of phosphate rocks to improve soil status for sustainable crop production in acid soils. Special issue. *Nutrient Cycling in Agroecosystems* 63(1): 1–98.

Zapata, F., and Roy, R. N. (eds.). 2004. Use of phosphate rocks for sustainable agriculture. *FAO Fertilizer and Plant Nutrition Bulletin* No. 13. Food and Agriculture Organization, Rome.

Zapata, F., and Axmann, H. 1991. Agronomic evaluation test of rock phosphate material by radio isotope techniques. *Pedologie* 41: 291–301.

Zapata, F., and Axmann, H. 1995. $^{32}$P isotopic techniques for evaluating the agronomic effectiveness of rock phosphate materials. *Fertilizer Research* 41: 189–195.

Zapata, F., Axmann, H., and Braun, H. 1986. Agronomic evaluation of roch phosphate materials by means of radioisotope techniques. *Proceedings 13th Congress International Soil Science Society*, Vol. 3, pp. 1012–1013. International Society of Soil Science, Hamburg, Germany.

Zhang, T. Q., and Mackenzie, A. F. 1997. Changes of soil phosphorus fractions under long-term corn monoculture. *Soil Science Society of America Journal* 61: 485–493.

# 13 Pathways and Fate of Phosphorus in Agroecosystems

*Rattan Lal*

## CONTENTS

This chapter addresses thematic issues regarding global phosphorus (P) cycle, effects of P on human health, soil P and crop growth, fate of P transported into the aquatic ecosystems, and algal bloom syndrome caused by the eutrophication of surface waters. The objective of this concluding chapter is to summarize the principal issues, identify researchable priorities, and explore science–policy interaction.

## 13.1 THE FINITE RESOURCE OF P AND ITS CYCLING AMONG PRINCIPAL RESERVOIRS

Phosphorus (P) is a finite resource. Despite revised estimates of the proven reserves of rock phosphates (Scholz et al. 2014), P remains a limited resource (International Fertilizer Industry Association 2009; Cordell et al. 2009; Syers et al. 2011). Yet its indiscriminate use in soils of agroecosystems (especially in North America and China) is an important cause of environmental pollution, water eutrophication, and attendant issues of algal bloom and coastal dead zones characterized by anoxia. Therefore, the nature of different reservoirs of P and of fluxes among them must be understood to facilitate the identification of systems of management that maximize its use efficiency in agroecosystems and minimize its transport into natural waters. Understanding the magnitude and the dynamics of P in the soils of the world is critical to developing techniques to replenish it in ways that the inputs and the outputs are delicately balanced (Figure 13.1). With the focus on soil as a source of plant uptake

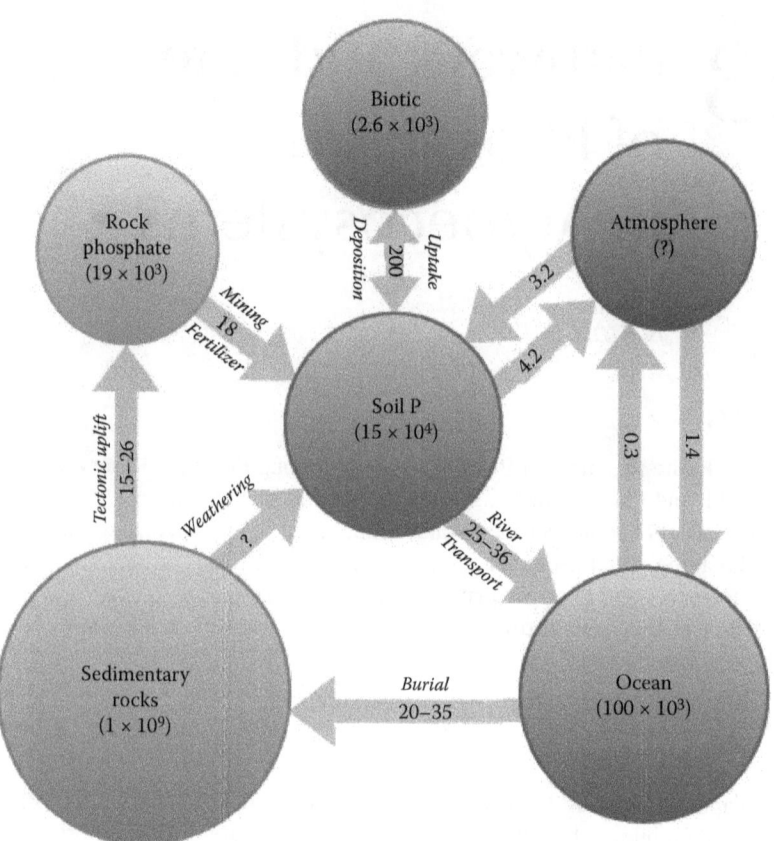

**FIGURE 13.1** The global carbon cycle: P pool (teragram) and P flux (teragram/year). (From Schlesinger, W. H., *Biogeochemistry: An Analysis of Global Change.* Academic Press, San Diego, CA, 1997; Filippelli, G. M., *Elements*, 4, 89–95, 2008; Smil, V., *Annual Review of Energy and Environment*, 25, 53–88, 2000. With permission.)

and for transport into aquatic ecosystems, P reservoir ($10^3$ Tg = $10^{12}$ g = million metric ton) in soil is 15 compared with 2.6 in all biota (plants and animals), and 19 in rocks as a minable ore (Smil 2000). The annual flux (teragram per year) between soil and biotic pools is 200, and it is a two-way transport. Soil receives input from the rock phosphate as a fertilizer (18 Tg/year) and is a source of eutrophication through P-laden sediments from the agroecosystems (Filippeli 2008). It is this transport from the soil to the surface waters that is the cause of algal blooms. Major reservoirs of P are oceans and sedimentary rocks (Figure 13.1).

Indiscriminate use of P in soils of agroecosystems is an important cause of environmental pollution and water eutrophication. Increased input of fertilizers (N and P) (Table 13.1) has aggravated the transport of these nutrients into natural waters by runoff and sediment transport. The transport of P from soils of the agroecosystems is primarily via sediments (Figure 13.2). Therefore, experiments designed to study the P budget in conjunction with that of the sediment flux from a landscape or a

**TABLE 13.1**
**Estimates of Global N and P Use**

| | Global Fertilizer Use (Mt/year) | | | |
|---|---|---|---|---|
| Year | N | P | K | Total |
| 1950 | <10 | – | – | 10 |
| 1960 | 11.6 | 10.9 | 8.7 | 31.2 |
| 1970 | 31.8 | 21.1 | 16.6 | 73.3 |
| 1980 | 60.8 | 31.7 | 24.2 | 116.7 |
| 1990 | 77.2 | 36.3 | 26.5 | 138.0 |
| 2000 | 80.9 | 32.5 | 21.8 | 135.2 |
| 2003 | 84.7 | 33.6 | 23.2 | 141.6 |
| 2010 | 113.0 | 44.0 | 27.0 | 184.0 |
| 2012 | 123.0 | 46.0 | 28.0 | 194.0 |
| 2020 | 135.0 | 47.6 | 30.0[a] | 212.6 |
| 2050 | 236.0 | 83.7 | 35.0[a] | 354.7 |

*Source:* International Fertilizer Development Center, Muscle Shoals, Alabama, 2004; FAO, Food and Agriculture Organization, Rome, Italy, 2015. With permission.
[a] Estimated. The values of P are in terms of $P_2O_5$.

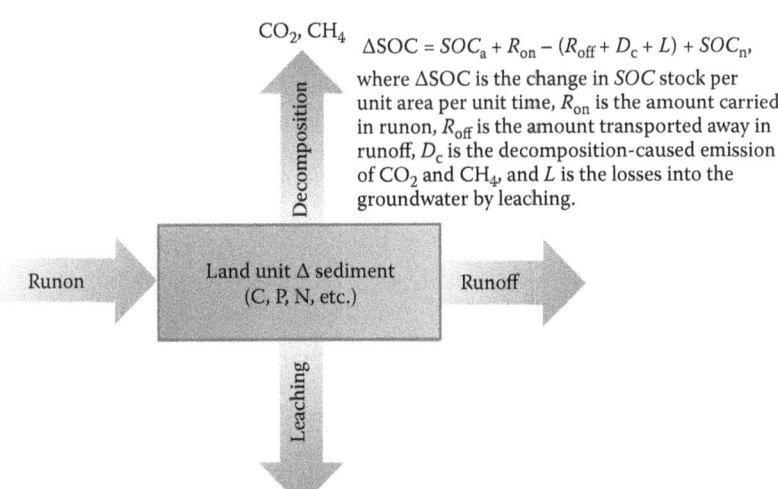

$$\Delta SOC = SOC_a + R_{on} - (R_{off} + D_c + L) + SOC_n,$$

where $\Delta SOC$ is the change in $SOC$ stock per unit area per unit time, $R_{on}$ is the amount carried in runon, $R_{off}$ is the amount transported away in runoff, $D_c$ is the decomposition-caused emission of $CO_2$ and $CH_4$, and $L$ is the losses into the groundwater by leaching.

**FIGURE 13.2** Mass balance approach is needed to assess the change in sediment and sediment-laden and water transported dissolved and suspended constituents (e.g., C, N, P, Ca, Mg, K). SOC, soil organic carbon.

watershed is an important researchable priority. Indeed, it is an excess of P in the surface layer, over and above the demand for crop uptake, that is, the source of eutrophication of natural waters. On the contrary, deficiency of P in agricultural soils can adversely affect crop growth and agronomic production. Furthermore, there exists a strong interaction between N and P availabilities in terms of crop growth and yield and risks of environmental pollution. Optimal N:P ratios depend on species, growth rate, plant age, and plant parts. In vegetation, N:P ratios <10 and >20 correspond with N- and P-limited biomass production (Güsewell 2004). Relative changes in biomass C:P can be indicative of nutrient limitation (Griffiths et al. 2012).

The Haber–Bosch process has greatly enhanced the availability of reactive nitrogen (N). However, the availability of P is limited because of its finite reserves in geological materials. The problem of P scarcity is further aggravated by the changes in the stoichiometry of C and N relative to P since the middle of the twentieth century because of the anthropogenic disturbance of $H_2O$, C, N, and P cycles. Long-term availability of rock phosphate must be critically assessed (Smil 2000).

## 13.2   SOIL P AND PLANT GROWTH

The global P use has increased from 10 Mt (1 Mt = $10^6$ t) in 1960 to 46 Mt in 2012 and is projected to be 84 Mt by 2050 (Table 13.1). The enhanced use of N, P, and K on agricultural soils has increased food production but also drastically altered the stoichiometry of terrestrial ecosystems and aggravated the risks of environmental pollution. Yet P deficiency is a major problem in a large proportion of soils managed by the resource-poor farmers of sub-Saharan Africa (SSA), South Asia, the Caribbean, and Central America. As much as 80% of the world's population lived in developing countries until the early 2000s. Yet the rate of fertilizer use is low especially in Africa. The data on the regional distribution of fertilizer use (kilogram per hectare of cropland) for 2009 were five for SSA minus South Africa, 10 for SSA, 17 for Africa, 96 for Latin America, 85 for North America, and 196 for East and Southeast Asia (Syers et al. 2011). The P deficiency in soils under arable lands in South Asia and elsewhere is exacerbated by subsidies provided for nitrogenous fertilizers but not for phosphorus fertilizers. Thus, relatively higher procurement price has been a constraint to the application of P fertilizers at the recommended rates by the small landholders of the tropics and subtropics. The deficiency of P, along with that of some critical micronutrients (Zn), is an important constraint affecting the crop yields and the nutritional quality of the food produced by small landholders on depleted and degraded soils.

## 13.3   TRANSPORT OF P INTO SURFACE WATERS

The application of N and P fertilizers to agricultural soils leads to the transport of these nutrients into aquatic ecosystems and adversely affects the water quality through eutrophication. The transport of P from agricultural lands is primarily bound to sediments, and it is a principal factor causing nonpoint source pollution. Thus, studies at landscape or watershed levels are needed for a mass balance of P along with that of sediment (Figure 13.2). Soil erosion and sediment transport must be controlled on agricultural lands. The surface flux of P from overfertilized soils may be an important factor responsible for the eutrophication of lakes (Carpenter 2005). Thus, there is a strong need to quantify the P transport at the watershed scale and identify the critical areas or the major sources of P-laden sediments (Anderson et al. 2002; Smith 2003; Schindler 2006; Leone et al. 2008). It is pertinent to conduct periodic surveys of topsoil P levels in croplands on a regional basis, which can be a useful tool to monitor future changes of P levels in soil (Tóth et al. 2014).

Algal blooms are a primary source of deteriorating water quality by hypoxia, toxicity, fish kill, and food web alterations (Paerl et al. 2001). The algal bloom in the Ohio River stretched >100 km during the summer of 2015 (Arenschield 2015). Dodds et al. (2009) reported that over 90% of the rivers in 12 of 16 ecoregions of the United States exceed the reference mean value of P. The economic losses from eutrophication (e.g., recreational water usage, waterfront real estate, spending on recovery of threatened species, drinking water) are estimated about $2.2 billion per annum (Dodds et al. 2009).

Climate change is another factor affecting the proliferation and the expansion of harmful algal blooms (O'Neil et al. 2012). Wagner and Adrian (2009) also reported that the incidence of cyanobacteria blooms will increase in many lakes under future climate scenarios. Rivers, lakes, and reservoirs may experience increased temperatures, more intense and larger duration of thermal stratification, and altered nutrient loading that increases the growth of cyanobacteria (Carey et al. 2012). However, the interactive effects of eutrophication and climate change on harmful cyanobacteria blooms are complex, and many exacerbate the blooming hazard. Indeed, temperature and nutrient loadings affect the growth and the dynamics of toxic and nontoxic strains (Davis et al. 2009).

## 13.4 MANAGING ALGAL BLOOMS

Agricultural lands are an important source of P in the natural waters, and in some regions, there has been a transfer of P from land under grain production to animal production, thereby creating a regional surplus in P inputs over outputs and buildup of soil P and transport to aquatic bodies. Thus, an important strategy is to balance the P inputs and outputs at both farm and watershed levels (Sharpley et al. 2001). Furthermore, reducing the losses of water runoff and sediment through the adoption of conservation-effective measures is critical to reducing the risks of eutrophication. Paerl et al. (2004) and Chen et al. (2009) suggested the importance of dual nutrient (N and P) reductions at watershed scale as a long-term strategy of reducing estuarine eutrophication.

Based on their research in the rivers of southeastern Australia, Davis and Koop (2006) emphasized the importance of stratification and light penetration, which trigger the algal bloom. Thus, in addition to sediment nutrient release, turbidity is another factor to be considered in managing soil, water, and nutrient resources at a watershed scale.

## 13.5 P RESERVES AND FOOD SECURITY

Being a finite resource, the global need of P as a fertilizer is a pertinent issue that deserves an objective consideration. The *peak P* concept emerged in 2007 (Ulrich and Schnug 2013). There is a need for informed policy decisions based on the better understanding of the nature and the amount of P flux through diverse land use systems at different spatial and temporal scales (Chowdhury et al. 2014). Sattari et al. (2012) estimated that between 2008 and 2050, a global cumulative P application of

700–790 kg/ha of cropland or a total of 1070–1200 Tg of P (1 Tg = $10^{12}$ g = 1 million metric ton) is required to achieve the needed crop production. Sattari et al. (2012) estimated that the global P fertilizer use may change from 17.8–16.8 to 20.8 Tg/year in 2050, about 50% less than the estimates commonly reported in the literature. Sustainable management, including recycling of P from urban ecosystem (human waste), is an important issue that needs to be addressed (Lal 2014).

Matsubae et al. (2011) opined that the data on virtual P ore requirement indicate the direct and indirect demands for P as transformed into agricultural produce. Matsubae et al. observed that P contained in eaten agricultural product was only 12% of the virtual P ore requirement for Japan. Urban permaculture based on regenerative agriculture (Rhodes 2012), and recycling of all nutrients brought into the cities, is a very pertinent strategy. Future planning and priority setting for sustainable strategies in critical to sustainable agriculture (Ulrich and Schnug 2013).

## 13.6 EFFICIENT USE OF P FERTILIZER

Being a chemical element, P cannot be manufactured and must be obtained from P-bearing rocks. Because of its relatively low solubility, P is often a limiting factor affecting NPP. The weathering of P-bearing rocks can also release a small amount of P. The countries with significant deposits of P are Morocco, Western Sahara, Russia, South Africa, Jordan, the United States, and China. Most of the P deposits are from marine sediments (Lamprecht et al. 2011).

The input and output of P into agroecosystems must be critically balanced. The residual soil P, the amount remaining in soil after crop harvest (Sattari et al. 2012), and the contained in waste (i.e., crop, animal, and human waste) must be effectively utilized. Because P losses from agroecosystems have severe environmental impact, it is important to close the P cycle through recycling. However, recycling (animal bones and human waste) may have health risks, which must be addressed (Lamprecht et al. 2012). Thus, the identification of more efficient uses of fertilization techniques can reduce nonpoint P losses (Smil 2000). The environmental sustainability of agricultural systems should be related to the amount of P (along with N and C) lost in runoff and sediments (Sharpley et al. 1995).

## 13.7 P AND NUTRITIONAL SECURITY

As an essential nutrient, P makes up about 1% of human body weight. It is an essential constituent of ATP. P interacts with calcium in the formation of bones and teeth and is also essential for the repair of cells and tissues. The human body uses it in several metabolic functions including metabolism of carbohydrates and fats and for synthesis of protein. Thus, its deficiency in human food can lead to several health issues. Because of the strong interconnectivity between the "health of soil, plants, animals, and people" (Albrecht 1947, 1966), judicious management of P availability in arable and pastoral lands is also critical to human health and well-being. With the projected increase of 2.4 billion people between 2015 and 2050 and changes in dietary preferences toward animal-based diet, achieving food and nutritional securities in developing countries remains a major global concern.

## 13.8 CONCLUSIONS

Phosphorus is a finite but an essential resource. Natural reserves of P as rock phosphate are found only in China, Morocco, Western Sahara, Jordan, and the United States. Despite the revised estimates of the available P reserves, it is a finite resource because it occurs as an element and cannot be manufactured.

Yet the indiscriminate use of P on soils of agroecosystems in some regions of the world has caused the eutrophication of natural waters leading to algal blooms, anoxia, and loss of aquatic life and creating problems with drinking water supply. Contrarily, the deficiency of P in some soils, especially those managed by small landholders in the tropics and subtropics and in soils with a high P-fixation capacity, limits crop yields and agronomic productivity. In addition to soil and fertilizer management, it is also pertinent to develop crop plants (varieties) and select species that have high attributes for PUE such as root architecture (Zhang et al. 2014).

Food (grains and livestock) grown on P-deficient soils can exacerbate the problems of hidden hunger and jeopardize human health and well-being because P is an essential element for several metabolic functions in the human body.

Judicious management of P in soils of agroecosystems is essential to minimize environmental hazards, enhance agronomic productivity, and improve human health and well-being. The projected global use of P fertilizer by 2050 (Table 13.1) can be reduced through better management and reduction of the losses. Soil application of animal waste (manure) should be done in a manner that does not pollute the environment. Similarly, the flux of sediments and P transport through sediments must be minimized to reduce the risks of nonpoint source pollution.

Being a finite resource, the recycling of P in human, agricultural, and industrial waste (by-products) is a prudent strategy. The world population is being rapidly urbanized. By 2050, 70% of the world population may live in urban centers. Indeed, the number of megacities (with more than 10 million people) is projected to increase from 28 in 2015 to 41 by 2030. A city of 10 million people requires 6000 Mg (metric ton) of food per day. Thus, vast amounts of nutrients and especially P are being brought into the urban centers and not being returned to the land where they came from. Thus, the recycling of P in human waste is a critical issue and a high research and development priority. Rather than being a liability, P in human waste (solid and liquid) can be a resource for growing food within the urban centers. Gray water (urban water) or black water (that contaminated by human feces), after sanitization and appropriate precautionary measures, can be used to produce food by sky farming techniques using aquaculture, aquaponics, hydroponics, and aeroponics. As much as 10–20% of the fresh produce consumed in the cities can be locally grown within the urban centers.

## REFERENCES

Albrecht, A. 1947. Our teeth and our soil. *Annuals of Dentistry* 8 (4):199–213.
Albrecht, A. 1966, February. Plant, animal and human health vary with soil nutrition. *Modern Nutrition* 19:18–20.
Anderson, D. M., P. M. Glibert, and J. M. Burkholder. 2002. Harmful algal blooms and eutrophication: Nutrient sources, composition, and consequences. *Estuaries* 25 (4B):704–726.

Arenschield, L. 2015, October 3. Algae along the Ohio. *Columbus Dispatch*. Columbus, OH.

Carey, C. C., B. W. Ibelings, E. P. Hoffmann, D. P. Hamilton, and J. D. Brookes. 2012. Ecophysiological adaptations that favour freshwater cyanobacteria in a changing climate. *Water Research* 46 (5):1394–1407.

Carpenter, S. R. 2005. Eutrophication of aquatic ecosystems: Bistability and soil phosphorus. *Proceedings of the National Academy of Sciences of the United States of America* 102 (29):10002–10005.

Chen, S. N., X. L. Chen, Y. Peng, and K. D. Peng. 2009. A mathematical model of the effect of nitrogen and phosphorus on the growth of blue-green algae population. *Applied Mathematical Modelling* 33 (2):1097–1106.

Chowdhury, R. B., G. A. Moore, A. J. Weatherley, and M. Arora. 2014. A review of recent substance flow analyses of phosphorus to identify priority management areas at different geographical scales. *Resources Conservation and Recycling* 83:213–228.

Cordell, D., J. Darget, and S. White. 2009. The story of phosphorus: Global food security and food for thought. *Global Environment Change* 19:292–305.

Davis, J. R., and K. Koop. 2006. Eutrophication in Australian rivers, reservoirs and estuaries—A southern hemisphere perspective on the science and its implications. *Hydrobiologia* 559:23–76.

Davis, T. W., D. L. Berry, G. L. Boyer, and C. J. Gobler. 2009. The effects of temperature and nutrients on the growth and dynamics of toxic and non-toxic strains of Microcystis during cyanobacteria blooms. *Harmful Algae* 8 (5):715–725.

Dodds, W. K., W. W. Bouska, J. L. Eitzmann, T. J. Pilger, K. L. Pitts, A. J. Riley, J. T. Schloesser, and D. J. Thornbrugh. 2009. Eutrophication of US freshwaters: Analysis of potential economic damages. *Environmental Science & Technology* 43 (1):12–19.

Filippelli, G. M. 2008. The global phosphorus cycle: Past, present and future. *Elements* 4 (2):89–95.

Food and Agriculture Organization (FAO). 2015. World fertilizer trends and outlook to 2018. FAO, Rome, Italy, 66 pp.

Griffiths, B. S., A. Spilles, M. Bonkowski. 2012. C:N:P stoichiometry and nutrient limitation of the soil microbial biomass in grazed grassland site under experimental P limitation or excess. *Ecological Processes* 1:6.

Güsewell, S. 2004. N:P ratios in terrestrial plants: Variation and functional significance. *New Physiologist* 164:243–266.

International Fertilizer Development Center (IFDC). 2004. World Fertilizer Consumption. IFDC, Muscle Shoals, AL, 90 pp.

International Fertilizer Industry Association. 2009. *Annual Phosphate Rock Statistics*. International Fertilizer Industry Association, Paris.

Lal, R. 2014. Book Review: Scholz, R. W., A. H. Roy, F. S. Brand, D. T. Hellums, A. E. Ulrich (Eds.): Sustainable Phosphorus Management: A Global Transdisciplinary Roadmap. *Journal of Plant Nutrition and Soil Science* 177 (6):934–935.

Lamprecht, H., D. J. Lang, C. R. Binder, and R. W. Scholz. 2011. The trade-off between phosphorus recycling and health protection during the BSE crisis in Switzerland: A "disposal dilemma." *Gaia-Ecological Perspectives For Science and Society* 20 (2):112–121.

Leone, A., M. N. Ripa, L. Boccia, and A. Lo Porto. 2008. Phosphorus export from agricultural land: A simple approach. *Biosystems Engineering* 101 (2):270–280.

Matsubae, K., J. Kajiyama, T. Hiraki, and T. Nagasaka. 2011. Virtual phosphorus ore requirement of Japanese economy. *Chemosphere* 84 (6):767–772.

O'Neil, J. M., T. W. Davis, M. A. Burford, and C. J. Gobler. 2012. The rise of harmful cyanobacteria blooms: The potential roles of eutrophication and climate change. *Harmful Algae* 14:313–334.

Paerl, H. W., R. S. Fulton, P. H. Moisander, and J. Dyble. 2001. Harmful freshwater algal blooms, with an emphasis on cyanobacteria. *The Scientific World Journal* 1:76–113.

Paerl, H. W., L. M. Valdes, A. R. Joyner, and M. F. Piehler. 2004. Solving problems resulting from solutions: Evolution of a dual nutrient management strategy for the eutrophying Neuse river estuary, North Carolina. *Environmental Science & Technology* 38 (11):3068–3073.

Rhodes, C. J. 2012. Feeding and healing the world: Through regenerative agriculture and permaculture. *Science Progress* 95 (4):345–446.

Sattari, S. Z., A. F. Bouwman, K. E. Giller, and M. K. van Ittersum. 2012. Residual soil phosphorus as the missing piece in the global phosphorus crisis puzzle. *Proceedings of the National Academy of Sciences of the United States of America* 109 (16):6348–6353.

Schindler, D. W. 2006. Recent advances in the understanding and management of eutrophication. *Limnology and Oceanography* 51 (1):356–363.

Schlesinger, W. H. 1997. *Biogeochemistry: An Analysis of Global Change.* Academic Press, San Diego.

Scholz, R. W., A. H. Roy, F. S. Brand, D. T. Hellums, and A. E. Ulrich (Eds). 2014. *Sustainable Phosphorus Management: A Global Transdisciplinary Roadmap.* Springer, Dordrecht.

Sharpley, A. N., R. W. McDowell, and P. J. A. Kleinman. 2001. Phosphorus loss from land to water: Integrating agricultural and environmental management. *Plant and Soil* 237 (2):287–307.

Sharpley, A., J. S. Robinson, and S. J. Smith. 1995. Assessing environmental sustainability systems by simulation of nitrogen and phosphorus loss in runoff. *European Journal of Agronomy* 4:453–464.

Smil, V. 2000. Phosphorus in the environment: Natural flows and human interferences. *Annual Review of Energy and Environment* 25:53–88.

Smith, V. H. 2003. Eutrophication of freshwater and coastal marine ecosystems—A global problem. *Environmental Science and Pollution Research* 10 (2):126–139.

Syers, K., M. Bekunda, D. Cordell, J. Carman, J. Johnson, A. Rosemarin, and I. Salcedo. 2011. Phosphorus and food production. *UNEP Yearbook 2011.* United Nations Environment Programme (UNEP), Nairobi, pp. 34–45.

Tóth, G., R. A. Guicharnaud, B. Tóth, and T. Hermann. 2014. Phosphorus level in cropland of the European Union with implication for P fertilizer use. *European Journal of Agronomy* 55:42–52.

Ulrich, A. E., and E. Schnug. 2013. The modern phosphorus sustainability movement: A profiling experiment. *Sustainability* 5 (11):4523–4545.

Wagner, C., and R. Adrian. 2009. Cyanobacteria dominance: Quantifying the effects of climate change. *Limnology and Oceanography* 54 (6):2460–2468.

Zhang, Z., H. Kiao, and W. J. Lucas. 2014. Molecular mechanisms underlying phsopate sensing, signaling and adaptation in plants. *Journal of Integrated Plant Biology* 56:192–220.

# Index